Association for Women in Mathematics Series

Volume 11

Series Editor
Kristin Lauter
Microsoft Research
Redmond, Washington, USA

Association for Women in Mathematics Series

Focusing on the groundbreaking work of women in mathematics past, present, and future, Springer's Association for Women in Mathematics Series presents the latest research and proceedings of conferences worldwide organized by the Association for Women in Mathematics (AWM). All works are peer-reviewed to meet the highest standards of scientific literature, while presenting topics at the cutting edge of pure and applied mathematics. Since its inception in 1971, The Association for Women in Mathematics has been a non-profit organization designed to help encourage women and girls to study and pursue active careers in mathematics and the mathematical sciences and to promote equal opportunity and equal treatment of women and girls in the mathematical sciences. Currently, the organization represents more than 3000 members and 200 institutions constituting a broad spectrum of the mathematical community, in the United States and around the world.

More information about this series at http://www.springer.com/series/13764

Irene I. Bouw • Ekin Ozman
Jennifer Johnson-Leung • Rachel Newton
Editors

Women in Numbers Europe II

Contributions to Number Theory and Arithmetic Geometry

AWM
ASSOCIATION FOR WOMEN IN MATHEMATICS

Springer

Editors
Irene I. Bouw
Institut für Reine Mathematik
Universität Ulm
Ulm, Baden-Württemberg, Germany

Ekin Ozman
Department of Mathematics
Bogazici University
Istanbul, Istanbul, Turkey

Jennifer Johnson-Leung
Department of Mathematics
University of Idaho
Moscow, ID, USA

Rachel Newton
Department of Mathematics and Statistics
University of Reading
Reading, UK

ISSN 2364-5733 ISSN 2364-5741 (electronic)
Association for Women in Mathematics Series
ISBN 978-3-030-09124-8 ISBN 978-3-319-74998-3 (eBook)
https://doi.org/10.1007/978-3-319-74998-3

Printed on acid-free paper

This Springer imprint is published by the registered company Springer International Publishing AG part of Springer Nature.
The registered company address is: Gewerbestrasse 11, 6330 Cham, Switzerland

Women in Numbers Europe II, Lorentz Center, September 2016

Preface

This volume grew out of the second *Women in Numbers Europe* conference held at the Lorentz Center, Leiden, from 26th to 30th September 2016. This conference was one of an ongoing series of collaboration conferences for women in mathematics, which began with the *Women in Numbers* conference held at the Banff International Research Station in 2008. This volume includes research reports from projects that were started at the conference, expository papers describing ongoing research begun at the conference, and contributed papers from other women number theorists. All of the papers included were reviewed by anonymous referees who verified them as worthy and valid contributions to these proceedings. We are proud that this volume continues the tradition of supporting and highlighting the contributions of women number theorists at a variety of career stages from graduate students to leading experts in the field.

The articles collected here span a wide range of topics in contemporary number theory and arithmetic geometry such as arithmetic dynamics, failure of local-global principles, geometry in positive characteristic, and heights of algebraic integers. The use of tools from algebra, analysis, and geometry, as well as computational methods, exemplifies the wealth of techniques available to modern researchers in number theory.

Several papers in this volume stem from collaborations between authors with different mathematical backgrounds and expertise. These papers naturally explore connections between different branches of mathematics and combine different points of view. Another common theme is the provision of explicit examples to complement new theorems and to provide evidence for conjectures. These concrete examples provide an easily accessible introduction to the field and have the potential to inspire future work.

Ulm, Germany — Irene I. Bouw
Moscow, ID, USA — Jennifer Johnson-Leung
Reading, UK — Rachel Newton
Istanbul, Turkey — Ekin Ozman

Acknowledgement

We would like to thank the following organizations for their generous support of the conference and this resulting volume: Clay Mathematics Institute, Foundation Compositio Mathematica, Heilbronn Institute for Mathematical Research, National Science Foundation as part of the Advance grant for the Association for Women in Mathematics, Microsoft Research, and the Number Theory Foundation.

We are particularly thankful for the additional financial support we received when we were forced to relocate our workshop, originally planned to take place in the Nesin Mathematics Village, due to instability in Turkey. The Lorentz Center was very generous in accommodating our conference at short notice and provided us with extremely valuable organizational support. We also would like to extend our sincere gratitude to the Nesin Mathematics Village for their understanding when we needed to relocate at the very last minute.

We are deeply grateful to the referees for their careful and timely review of the papers included in this volume.

Contents

Lower Bounds for Heights in Relative Galois Extensions 1
Shabnam Akhtari, Kevser Aktaş, Kirsti D. Biggs, Alia Hamieh, Kathleen Petersen, and Lola Thompson

Reductions of Algebraic Integers II 19
Antonella Perucca

Reductions of One-Dimensional Tori II 35
Antonella Perucca

On the Carlitz Rank of Permutation Polynomials Over Finite Fields: Recent Developments 39
Nurdagül Anbar, Almasa Odžak, Vandita Patel, Luciane Quoos, Anna Somoza, and Alev Topuzoğlu

Dynamical Belyi Maps 57
Jacqueline Anderson, Irene I. Bouw, Ozlem Ejder, Neslihan Girgin, Valentijn Karemaker, and Michelle Manes

Discriminant Twins 83
Alyson Deines

The a-Number of Hyperelliptic Curves 107
Sarah Frei

Non-ordinary Curves with a Prym Variety of Low p-Rank 117
Turku Ozlum Celik, Yara Elias, Burçin Güneş, Rachel Newton, Ekin Ozman, Rachel Pries, and Lara Thomas

Elliptic Fibrations on Covers of the Elliptic Modular Surface of Level 5 159
Francesca Balestrieri, Julie Desjardins, Alice Garbagnati, Céline Maistret, Cecília Salgado, and Isabel Vogt

On Birch and Swinnerton-Dyer's Cubic Surfaces 199
Mckenzie West

Lower Bounds for Heights in Relative Galois Extensions

Shabnam Akhtari, Kevser Aktaş, Kirsti D. Biggs, Alia Hamieh, Kathleen Petersen, and Lola Thompson

Abstract The goal of this paper is to obtain lower bounds on the height of an algebraic number in a relative setting, extending previous work of Amoroso and Masser. Specifically, in our first theorem, we obtain an effective bound for the height of an algebraic number α when the base field $\mathbb{K}$ is a number field and $\mathbb{K}(\alpha)/\mathbb{K}$ is Galois. Our second result establishes an explicit height bound for any nonzero element α which is not a root of unity in a Galois extension $\mathbb{F}/\mathbb{K}$, depending on the degree of $\mathbb{K}/\mathbb{Q}$ and the number of conjugates of α which are multiplicatively independent over $\mathbb{K}$. As a consequence, we obtain a height bound for such α that is independent of the multiplicative independence condition.

S. Akhtari (✉)
Department of Mathematics, University of Oregon, Eugene, OR, USA
e-mail: akhtari@uoregon.edu

K. Aktaş
Department of Mathematics Education, Gazi University, Ankara, Turkey
e-mail: kevseraktas@gazi.edu.tr

K. D. Biggs
School of Mathematics, University of Bristol, Clifton, Bristol, UK
e-mail: kirsti.biggs@bristol.ac.uk

A. Hamieh
Department of Mathematics and Statistics, The University of Northern British Columbia, Prince George, BC, Canada
e-mail: alia.hamieh@unbc.ca

K. Petersen
Department of Mathematics, Florida State University, Tallahassee, FL, USA
e-mail: petersen@math.fsu.edu

L. Thompson
Department of Mathematics, Oberlin College, Oberlin, OH, USA
e-mail: lola.thompson@oberlin.edu

I. I. Bouw et al. (eds.), *Women in Numbers Europe II*, Association for Women in Mathematics Series 11, https://doi.org/10.1007/978-3-319-74998-3_1

Keywords Height of algebraic numbers · Lehmer's problem

2010 Mathematics Subject Classification. 11G50

1 Introduction

Consider the nonconstant polynomial

$$P(x) = c_d x^d + c_{d-1}x^{d-1} + \cdots + c_1 x + c_0 = c_d \prod_{i=1}^{d}(x - r_i).$$

The Mahler measure of $P(x)$ is defined as

$$M(P) = \exp\Big(\int_0^1 \log|P(e^{2\pi it})|\,dt\Big),$$

the geometric mean of $|P(z)|$ for z on the unit circle. By Jensen's formula, this is equivalent to

$$M(P) = |c_d| \prod_{|r_i|\geq 1} |r_i|.$$

If $P(x)$ has integer coefficients, then $M(P) \geq 1$; by a result of Kronecker, $M(P) = 1$ exactly when $P(x)$ is a power of x times a product of cyclotomic polynomials.

Given an algebraic number α, we let $d = [\mathbb{Q}(\alpha) : \mathbb{Q}]$ be its degree over $\mathbb{Q}$. We will use $M(\alpha)$ to denote the Mahler measure of the minimal polynomial of α over $\mathbb{Z}$. We will formulate our results in terms of the Weil height of α, defined to be

$$h(\alpha) = \tfrac{1}{d} \log M(\alpha).$$

In [16] Lehmer asked whether there are monic integer polynomials whose Mahler measure is arbitrarily close to 1. For the polynomial $L(x) = x^{10} + x^9 - x^7 - x^6 - x^5 - x^4 - x^3 + x + 1$ (now called Lehmer's polynomial), he calculated $M(L) = 1.176280818\ldots$, which is still the smallest value of $M(P) > 1$ known for $P \in \mathbb{Z}[x]$. Although he did not make a conjecture, the statement that there exists a constant $\delta > 0$ such that the Mahler measure of any polynomial in $\mathbb{Z}[x]$ is either 1 or is greater than $1 + \delta$ has become known as Lehmer's conjecture. In terms of height, Lehmer's conjecture states that there is a universal constant $c_0 > 0$ such that if α is a nonzero algebraic number of degree d which is not a root of unity, then

$$h(\alpha) \geq \frac{c_0}{d}.$$

In 1971 Blanksby and Montgomery [8] and later Stewart [23] produced bounds for the Mahler measure of such algebraic numbers. These bounds inspired the work of Dobrowolski [12] who, in 1979, proved for $d \geq 2$ that

$$M(\alpha) > 1 + \frac{1}{1200}\left(\frac{\log\log d}{\log d}\right)^3.$$

Many of the best bounds are modifications of Dobrowolski's bound. The constants in these bounds have been improved over the years, but the dependence on the degree (for general polynomials) has remained. Of note, in 1996 Voutier [24] used elementary techniques to show that for $d \geq 2$, we have

$$h(\alpha) > \frac{1}{4d}\left(\frac{\log\log d}{\log d}\right)^3. \tag{1}$$

(Dobrowolski's bound, when translated into a statement about Weil height, has a similar form.) Voutier also showed that for $d \geq 2$, we have

$$h(\alpha) > \frac{2}{d\,(\log 3d)^3}, \tag{2}$$

which gives a better lower bound than (1) for small values of d. For more details on the history of Lehmer's conjecture and related problems, see the excellent survey paper of Smyth [21].

Lehmer's conjecture has been proven in certain settings. Notably, Breusch [10] and Smyth [20] independently proved it for nonreciprocal polynomials. More recently, Borwein, Dobrowolski, and Mossinghoff [9] proved it for many infinite families of polynomials, including polynomials with no cyclotomic factors and all odd coefficients. (Their result therefore proves Lehmer's conjecture for the Littlewood polynomials, namely, those polynomials whose coefficients are ± 1.)

Results also exist concerning height bounds for α with certain properties. For example, Amoroso and David [1] have proven that there is an absolute constant c such that if $\mathbb{Q}(\alpha)/\mathbb{Q}$ is Galois, and α is not a root of unity, then $h(\alpha) \geq cd^{-1}$. This proves Lehmer's conjecture for such α. Moreover, if α is any nonzero algebraic number that lies in an abelian extension of $\mathbb{Q}$, then Amoroso and Dvornicich [3] have shown that the height of α is greater than the constant $(\log 5)/12$.

Amoroso and Masser [4] improved upon the bounds in [1] for the case where $\mathbb{Q}(\alpha)/\mathbb{Q}$ is Galois. They showed that, for any ε, the height of α is bounded below by $c(\varepsilon)d^{-\varepsilon}$. Our first theorem is a generalization of this result to the case when α generates a Galois extension of an arbitrary number field.

Theorem 1 *Let $\varepsilon > 0$ be given. Let α be a nonzero algebraic number, not a root of unity, such that $[\mathbb{Q}(\alpha) : \mathbb{Q}] \geq 2$ and $\mathbb{K}(\alpha)/\mathbb{K}$ is Galois for some number field $\mathbb{K}$. Let δ be the degree of α over $\mathbb{K}$. Then there is an effectively computable constant $c(\varepsilon, \mathbb{K}) > 0$ such that*

$$h(\alpha) \geq c(\varepsilon, \mathbb{K})\delta^{-2\varepsilon}.$$

Relative height bounds for α in a number field $\mathbb{K}$ which is abelian over $\mathbb{L}$ are given in [6] and [2]. These bounds are similar in shape to Dobrowolski's bound.

Theorem 1 determines bounds for $h(\alpha)$ when $\mathbb{K}(\alpha)/\mathbb{K}$ is Galois and therefore when $\mathbb{Q}(\alpha)/\mathbb{K}$ is Galois. Our next theorem determines height bounds for any element α in a Galois extension $\mathbb{F}$ of $\mathbb{K}$ which is nonzero and not a root of unity. This is a generalization of Theorem 3.1 in [4].

Theorem 2 *Let $\mathbb{K}$ be a number field with degree τ over $\mathbb{Q}$. For any positive integer $r \geq 1$ and any $\varepsilon > 0$, there is a positive effective constant $c(\varepsilon, r, \tau)$ with the following property. Let $\mathbb{F}/\mathbb{K}$ be a Galois extension of relative degree η, and suppose $\alpha \in \mathbb{F}^*$ is not a root of unity. Assume that r conjugates of α over $\mathbb{K}$ are multiplicatively independent. Then,*

$$h(\alpha) \geq c(\varepsilon, r, \tau)\eta^{-\frac{1}{r+1}-\varepsilon}.$$

Theorem 2 is proven in Sect. 5, where the explicit constants are presented. Taking $r = 1$, we have the following corollary as an immediate consequence.

Corollary 1 For any $\varepsilon > 0$ there is a positive effective constant $c(\varepsilon, \tau)$ with the following property. Let $\mathbb{F}/\mathbb{K}$ be a Galois extension, with $[\mathbb{F} : \mathbb{K}] = \eta$, and suppose $\alpha \in \mathbb{F}^*$ is not a root of unity. Then,

$$h(\alpha) \geq c(\varepsilon, \tau)\eta^{-\frac{1}{2}-\varepsilon}.$$

The present paper closely follows and builds on the work of Amoroso and Masser in [4].

2 Preliminaries

In this section, we collect results that will be used in the proofs of Theorems 1 and 2.

2.1 *Finite Linear Groups*

We will require a bound on the size of finite subgroups of $\mathrm{GL}_n(\mathbb{Z})$ in the proof of Lemma 2. We now establish this bound, following the work of Serre [19].

Proposition 1 (Serre) *Let A be an abelian variety, and let u be an automorphism of A of finite order. Let $n \geq 2$ be a positive integer such that $u \equiv 1 \mod n$. If $n = 2$, then $u^2 = 1$. Otherwise, we have $u = 1$.*

The proof of Lemma 2 will use the following well-known corollary to Proposition 1 (see also [4, Remark 2.3]).

Corollary 2 *Let H be a finite subgroup of* $\mathrm{GL}_\rho(\mathbb{Z})$. *The reduction modulo 3 homomorphism* $\phi_3 : H \to \mathrm{GL}_\rho(\mathbb{Z}/3\mathbb{Z})$ *is injective. As a result, the order of a finite subgroup of* $\mathrm{GL}_\rho(\mathbb{Z})$ *is less than* 3^{ρ^2}.

Proof Let u be an element in $\ker(\phi_3) \subset H$. Then u has finite order and $u \equiv I_\rho \mod 3$, where I_ρ is the $\rho \times \rho$ identity matrix. By Proposition 1, we have $u = I_\rho$. This establishes that ϕ_3 is injective. We conclude that the order of H is at most $|\mathrm{GL}_\rho(\mathbb{Z}/3\mathbb{Z})|$, which is less than 3^{ρ^2}.

Remark 1 In an unpublished paper from 1995, Feit [13] shows that the maximal order of a finite subgroup of $\mathrm{GL}_\rho(\mathbb{Q})$ is $2^\rho \rho!$, except when $\rho = 2, 4, 6, 7, 8, 9, 10$. He further shows that for these exceptional cases, the maximal order is

$$12, 1152, 103680, 2903040, 696729600, 1393459200, 8360755200,$$

respectively. Therefore, the maximal order of a finite subgroup is at most $\frac{135}{2} 2^\rho \rho!$ for all ρ. See [7] for more information about these subgroups. Additionally, in 1997, Friedland showed in [14] that the orthogonal groups are the maximal subgroups for ρ large enough.

2.2 Height of Algebraic Numbers

We will use the following auxiliary height bounds in our proofs of Theorems 1 and 2.

The first is Corollary 1.6 of [5].

Proposition 2 (Amoroso-Viada) *Let $\alpha_1, \ldots, \alpha_n$ be multiplicatively independent algebraic numbers in a number field $\mathbb{A}$ of degree $D = [\mathbb{A} : \mathbb{Q}]$. Then,*

$$h(\alpha_1) \ldots h(\alpha_n) \geq D^{-1}(1050 n^5 \log 3D)^{-n^2(n+1)^2}.$$

The following result is Théorème 1.3 from [2].

Proposition 3 (Amoroso-Delsinne) *Let α be a nonzero algebraic number which is not a root of unity. For every abelian extension $\mathbb{A}$ of $\mathbb{B}$, we have*

$$h(\alpha) \geq \frac{(g(\tau)\Delta)^{-c}}{D} \frac{(\log\log 5D)^3}{(\log 2D)^4},$$

where c is an absolute, strictly positive constant, Δ is the absolute value of the discriminant of $\mathbb{B}$ over $\mathbb{Q}$, $\tau = [\mathbb{B} : \mathbb{Q}]$, $D = [\mathbb{A}(\alpha) : \mathbb{A}]$, and $g(\tau) = 1$ if there exists a tower of extensions

$$\mathbb{Q} = \mathbb{B}_0 \subset \mathbb{B}_1 \subset \cdots \subset \mathbb{B}_m = \mathbb{B}$$

with $\mathbb{B}_i/\mathbb{B}_{i-1}$ Galois for $i = 1, \ldots m$, and $g(\tau) = \tau!$ otherwise.

Remark 2 The constant c in Proposition 3 depends on a number of constants defined in [2], as well as on constants from papers of Friedlander [15] and Stark [22].

Finally, we will use Théorème 1.6 of [11], in which $\mathbb{Q}^{ab}$ denotes the maximal abelian extension of $\mathbb{Q}$, and $\mathbb{G}_m(\overline{\mathbb{Q}})$ denotes the multiplicative group of $\overline{\mathbb{Q}}$.

Proposition 4 (Delsinne) *For any positive integer n, there exists an effectively computable constant $c(n) > 0$ depending only on n for which the following property holds. Let $\boldsymbol{\alpha} = (\alpha_1, \dots, \alpha_n) \in \mathbb{G}_m^n(\overline{\mathbb{Q}})$. If*

$$\prod_{i=1}^{n} h(\alpha_i) \leq \left(c(n)[\mathbb{Q}^{ab}(\boldsymbol{\alpha}) : \mathbb{Q}^{ab}](\log(3[\mathbb{Q}^{ab}(\boldsymbol{\alpha}) : \mathbb{Q}^{ab}]))^{\kappa(n)}\right)^{-1},$$

where $\kappa(n) = 3n(2(n+1)^2(n+1)!)^n$, then $\boldsymbol{\alpha}$ is contained in a torsion subvariety B for which

$$(\deg B)^{1/\operatorname{codim}(B)} \leq c(n)[\mathbb{Q}^{ab}(\boldsymbol{\alpha}) : \mathbb{Q}^{ab}]^{\eta(n)}(\log(3[\mathbb{Q}^{ab}(\boldsymbol{\alpha}) : \mathbb{Q}^{ab}]))^{\mu(n)},$$

where

$$\eta(n) = (n-1)!\left(\sum_{i=0}^{n-3} \frac{1}{i!} + 1\right) + n - 1$$

and $\mu(n) = 8m!(2(n+1)^2(n+1)!)^n$.

In fact, we may take $c(n) = (2n^2)^n \exp\left(64n^2 n! \left(2(n+1)^2(n+1)!\right)^{2n}\right)$.

Notice that if $\alpha_1, \dots, \alpha_n$ are multiplicatively independent, then $\boldsymbol{\alpha} = (\alpha_1, \dots, \alpha_n)$ cannot be contained in a torsion subvariety. This simple observation yields the following corollary to Proposition 4.

Corollary 3 *Let n be a positive integer, and let $\alpha_1, \dots, \alpha_n$ be multiplicatively independent algebraic numbers. Then there exists an effectively computable constant $c(n) > 0$ depending only on n for which*

$$\prod_{i=1}^{n} h(\alpha_i) > \left(c(n)[\mathbb{Q}^{ab}(\boldsymbol{\alpha}) : \mathbb{Q}^{ab}](\log(3[\mathbb{Q}^{ab}(\boldsymbol{\alpha}) : \mathbb{Q}^{ab}]))^{\kappa(n)}\right)^{-1},$$

where $\kappa(n) = 3n(2(n+1)^2(n+1)!)^n$.

2.3 Estimates for $\phi(n)/n$

We will make use of the following lower bound for Euler's totient function, which is a slightly weaker version of [18, Theorem 15].

Proposition 5 *For all natural numbers $n \geq 3$, we have*

$$\frac{\phi(n)}{n} > \frac{1}{\exp(\gamma)\log\log n + \frac{3}{\log\log n}},$$

where γ is Euler's constant.

The following lower bound for $\phi(n)^{1+\varepsilon}/n$ will be useful in making the lower bound constants explicit in the proofs of both of our main theorems.

Lemma 1 *For any $\varepsilon > 0$, there is an effective constant $C(\varepsilon)$ such that*

$$\frac{\phi(n)^{1+\varepsilon}}{n} \geq C(\varepsilon)$$

for all $n \geq 3$. Specifically, one can take

$$C(\varepsilon) = \left(\frac{(\log\log 3)\left(\exp(2)\frac{\varepsilon}{2+2\varepsilon}\right)^{\sqrt{\frac{\varepsilon}{2+2\varepsilon}}+1}}{\exp(\gamma) + 3^{\frac{1}{1+\varepsilon}}\left(\exp(2)\frac{\varepsilon}{2+2\varepsilon}\right)^{\sqrt{\frac{\varepsilon}{2+2\varepsilon}}+1}} \right)^{1+\varepsilon}.$$

Proof By Proposition 5, for all $n \geq 3$ we have

$$\frac{\phi(n)}{n} > \frac{\log\log n}{\exp(\gamma)(\log\log n)^2 + 3}.$$

We use the fact that $\log x \leq \frac{x^\theta}{\exp(1)\theta}$ for any $\theta > 0$ to replace the power of $\log\log n$ in the denominator and conclude that

$$\frac{\phi(n)}{n} > \frac{\log\log 3}{\exp(\gamma)\frac{n^{2\theta^2}}{(\exp(1)\theta)^{2\theta+2}} + 3}.$$

Hence,

$$\frac{\phi(n)}{n^{1-2\theta^2}} \geq \frac{\log\log 3}{\frac{\exp(\gamma)}{(\exp(1)\theta)^{2\theta+2}} + \frac{3}{n^{2\theta^2}}} \geq \frac{\log\log 3}{\frac{\exp(\gamma)}{(\exp(1)\theta)^{2\theta+2}} + \frac{3}{3^{2\theta^2}}},$$

which implies that

$$\frac{\phi(n)^{1+\frac{2\theta^2}{1-2\theta^2}}}{n} \geq \left(\frac{\log\log 3}{\frac{\exp(\gamma)}{(\exp(1)\theta)^{2\theta+2}} + 3^{1-2\theta^2}} \right)^{1+\frac{2\theta^2}{1-2\theta^2}}.$$

Choosing θ such that $2\theta^2 = \frac{\varepsilon}{1+\varepsilon}$ completes the proof.

Remark 3 Our lemma holds for all $n \geq 3$. By Mertens' theorem (see, e.g., [17, Theorem 3.15]), $\phi(n)/n \sim 1/(\exp(\gamma)\log\log n)$ as $n \to \infty$. Using this, one can obtain sharper lower bounds for n "sufficiently large."

3 Some Useful Lemmas

In this section we prove two lemmas that will be useful in the proof of Theorem 1.

Lemma 2 *Let $\mathbb{F}/\mathbb{K}$ be a Galois extension. Assume that $\alpha \in \mathbb{F}^*$ is not a root of unity, let $\alpha_1, \ldots, \alpha_\delta$ be the conjugates of α over $\mathbb{K}$, and let ρ be the multiplicative rank of this set of conjugates. Let e be the order of the group of roots of unity in $\mathbb{F}$, so that $\mathbb{Q}(\zeta_e) \subset \mathbb{F}$. Then there exists a subfield $\mathbb{L}$ of $\mathbb{F}$ which is Galois over $\mathbb{K}$ of relative degree $[\mathbb{L} : \mathbb{K}] = n \leq n(\rho) < 3^{\rho^2}$, and $\alpha^e \in \mathbb{L}$.*

Proof Let $\beta_i = \alpha_i^e$ and $\mathbb{L} = \mathbb{K}(\beta_1, \ldots, \beta_\delta) \subseteq \mathbb{F}$. Then, by construction, $\mathbb{L}$ is Galois over $\mathbb{K}$ and $\alpha^e \in \mathbb{L}$. Consider the multiplicative group

$$\mathscr{M} = \{\beta_1^{a_1} \cdot \cdots \cdot \beta_\delta^{a_\delta} : a_i \in \mathbb{Z}\},$$

which is a $\mathbb{Z}$-module that is multiplicatively spanned by $\{\beta_1, \beta_2, \ldots, \beta_\delta\}$. First, we will show that $\mathscr{M}$ is a free $\mathbb{Z}$-module of rank ρ. It is enough to show that $\mathscr{M}$ is torsion-free as the fact that ρ is the multiplicative rank of $\{\alpha_1, \ldots, \alpha_\delta\}$ implies that it is also the multiplicative rank of $\{\beta_1, \ldots, \beta_\delta\}$. Assume for the sake of contradiction that there exists an $x \in \mathscr{M}$ such that $x \neq 1$ and $x^n = 1$ for some positive integer $n > 1$. Then $x = \beta_1^{a_1} \cdot \cdots \cdot \beta_\delta^{a_\delta}$ for some $a_1, \ldots, a_\delta \in \mathbb{Z}$. Since $x^n = 1$, we get

$$\left(\beta_1^{a_1} \cdot \cdots \cdot \beta_\delta^{a_\delta}\right)^n = \left(\alpha_1^{a_1} \cdot \cdots \cdot \alpha_\delta^{a_\delta}\right)^{ne} = 1.$$

Hence, $y = \alpha_1^{a_1} \cdot \cdots \cdot \alpha_\delta^{a_\delta}$ is a root of unity. Since $y \in \mathbb{F}$ and e is the order of the group of roots of unity in $\mathbb{F}$, it follows that $x = y^e = 1$, contrary to our assumption.

Since $\mathrm{Gal}(\mathbb{L}/\mathbb{K})$ acts on $\mathscr{M}$ by permuting the α_i, this action defines an injective homomorphism from $\mathrm{Gal}(\mathbb{L}/\mathbb{K})$ to $\mathrm{GL}_\rho(\mathbb{Z})$. This implies that the finite group $\mathrm{Gal}(\mathbb{L}/\mathbb{K})$ is isomorphic to a finite subgroup of $\mathrm{GL}_\rho(\mathbb{Z})$. By Corollary 2 the order of a finite subgroup of $\mathrm{GL}_\rho(\mathbb{Z})$ is bounded by $n(\rho)$ which is at most 3^{ρ^2}. We conclude that $[\mathbb{L} : \mathbb{K}] \leq n(\rho) < 3^{\rho^2}$.

Lemma 3 *Let $\varepsilon > 0$ be given. Let $\mathbb{K}$ be a number field. Assume that α is a nonzero algebraic number, not a root of unity, such that $\mathbb{K}(\alpha)/\mathbb{K}$ is Galois. Let δ be the degree of α over $\mathbb{K}$. Further, let e be the order of the group of roots of unity in $\mathbb{K}(\alpha)$, f be the order of the group of roots of unity in $\mathbb{K}$, $\tau = [\mathbb{K} : \mathbb{Q}]$, and ρ be the multiplicative rank of the conjugates of α over $\mathbb{K}$. Then,*

$$[\mathbb{K}(\alpha) : \mathbb{K}(\zeta_e)] \leq \delta^\varepsilon C_4(\mathbb{K}, \varepsilon),$$

with $C_4(\mathbb{K}, \varepsilon) = \frac{1}{C(\varepsilon)} n(\rho) f \tau^{1+\varepsilon}$. *We have* $C(\varepsilon)$ *as in Lemma 1 unless* $e/f \in \{1, 2\}$ *in which case we take* $C(\varepsilon) = 1/2$.

Proof We begin by obtaining a few inequalities, proving (3) and (4) below. By the second isomorphism theorem, we have

$$[\mathbb{K}(\zeta_e) : \mathbb{K}] = [\mathbb{Q}(\zeta_e) : \mathbb{K} \cap \mathbb{Q}(\zeta_e)] = \frac{[\mathbb{Q}(\zeta_e) : \mathbb{Q}(\zeta_f)]}{[\mathbb{K} \cap \mathbb{Q}(\zeta_e) : \mathbb{Q}(\zeta_f)]}$$
$$= \frac{\phi(e)/\phi(f)}{[\mathbb{K} \cap \mathbb{Q}(\zeta_e) : \mathbb{Q}(\zeta_f)]}.$$

Since $[\mathbb{K} \cap \mathbb{Q}(\zeta_e) : \mathbb{Q}(\zeta_f)] \leq [\mathbb{K} : \mathbb{Q}(\zeta_f)]$, we conclude that

$$[\mathbb{K}(\zeta_e) : \mathbb{K}] \geq \frac{\phi(e)/\phi(f)}{[\mathbb{K} : \mathbb{Q}(\zeta_f)]}.$$

It follows from the fact that ϕ is multiplicative that

$$[\mathbb{K}(\zeta_e) : \mathbb{K}] \geq \frac{\phi(e)/\phi(f)}{[\mathbb{K} : \mathbb{Q}(\zeta_f)]} \geq \frac{\phi(\frac{e}{f})}{[\mathbb{K} : \mathbb{Q}(\zeta_f)]}. \tag{3}$$

Next, we will show

$$\frac{\frac{e}{f}}{\phi(\frac{e}{f})} \leq \frac{[\mathbb{K} : \mathbb{Q}(\zeta_f)]^\varepsilon}{C(\varepsilon)} \delta^\varepsilon. \tag{4}$$

By Lemma 1, if $e/f \geq 3$ we have the upper bound

$$\frac{\frac{e}{f}}{\phi(\frac{e}{f})} \leq \frac{1}{C(\varepsilon)} \left(\phi(\tfrac{e}{f})\right)^\varepsilon.$$

If $e/f \in \{1, 2\}$, we can take $C(\varepsilon) = 1/2$, and this is still satisfied. Again appealing to the multiplicativity of ϕ, we have

$$\phi(\tfrac{e}{f}) \leq \frac{\phi(e)}{\phi(f)} = [\mathbb{Q}(\zeta_e) : \mathbb{Q}(\zeta_f)] \leq [\mathbb{K}(\zeta_e) : \mathbb{K}][\mathbb{K} : \mathbb{Q}(\zeta_f)] \leq \delta[\mathbb{K} : \mathbb{Q}(\zeta_f)].$$

Hence, (4) follows by combining these two inequalities.

Now we will proceed to prove the bound for $[\mathbb{K}(\alpha) : \mathbb{K}(\zeta_e)]$ stated in the lemma. By Lemma 2 there is a subfield $\mathbb{L}$ of $\mathbb{K}(\alpha)$ which is Galois over $\mathbb{K}$, which contains α^e and

$$[\mathbb{L} : \mathbb{K}] = n \leq n(\rho). \tag{5}$$

Thus, we have $\mathbb{K} \subseteq \mathbb{L} \subseteq \mathbb{K}(\alpha)$, so $\mathbb{K}(\alpha) = \mathbb{L}(\alpha)$.

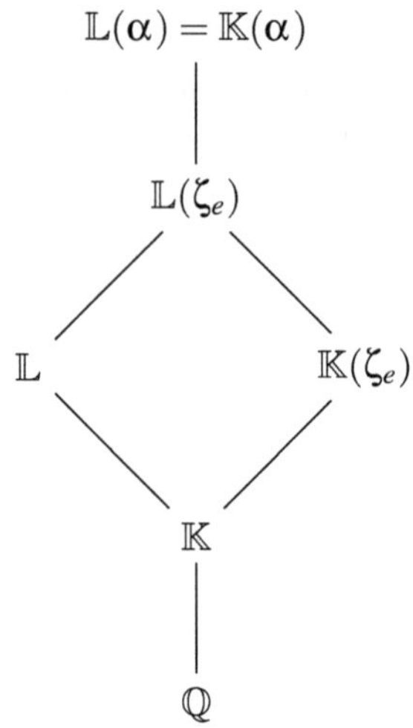

Let $e' = [\mathbb{L}(\alpha) : \mathbb{L}]$. Since the minimal polynomial for α over $\mathbb{L}$ divides $x^e - \alpha^e$, we conclude that $e' \leq e$. Using multiple applications of the tower law, we have

$$\begin{aligned}
[\mathbb{K}(\alpha) : \mathbb{K}(\zeta_e)] &= [\mathbb{L}(\alpha) : \mathbb{L}(\zeta_e)][\mathbb{L}(\zeta_e) : \mathbb{K}(\zeta_e)] \\
&= [\mathbb{L}(\zeta_e) : \mathbb{K}(\zeta_e)] \frac{[\mathbb{L}(\alpha) : \mathbb{L}]}{[\mathbb{L}(\zeta_e) : \mathbb{L}]} \\
&= e' \frac{[\mathbb{L}(\zeta_e) : \mathbb{K}(\zeta_e)]}{[\mathbb{L}(\zeta_e) : \mathbb{L}]} = e' \frac{[\mathbb{L} : \mathbb{K}]}{[\mathbb{K}(\zeta_e) : \mathbb{K}]}.
\end{aligned}$$

By (5), we see that

$$[\mathbb{K}(\alpha) : \mathbb{K}(\zeta_e)] \leq e' \frac{n(\rho)}{[\mathbb{K}(\zeta_e) : \mathbb{K}]}.$$

Using (3) we have

$$[\mathbb{K}(\alpha) : \mathbb{K}(\zeta_e)] \leq e' \frac{n(\rho)}{[\mathbb{K}(\zeta_e) : \mathbb{K}]} \leq \frac{e'}{\phi(\frac{e}{f})} n(\rho)[\mathbb{K} : \mathbb{Q}(\zeta_f)].$$

Since $e' \leq e$, we conclude that $e' \leq \frac{e}{f} f$, and hence

$$[\mathbb{K}(\alpha) : \mathbb{K}(\zeta_e)] \leq \frac{\frac{e}{f}}{\phi(\frac{e}{f})} n(\rho) f [\mathbb{K} : \mathbb{Q}(\zeta_f)].$$

Combining this bound with (4) shows that

$$[\mathbb{K}(\alpha):\mathbb{K}(\zeta_e)] \leq \frac{1}{C(\varepsilon)}\delta^{\varepsilon} n(\rho) f[\mathbb{K}:\mathbb{Q}(\zeta_f)]^{1+\varepsilon} \leq \delta^{\varepsilon} C_4(\mathbb{K},\varepsilon)$$

with $C_4(\mathbb{K},\varepsilon) = \frac{1}{C(\varepsilon)} n(\rho) f \tau^{1+\varepsilon}$, as needed.

4 Proof of Theorem 1

In this section, we present the proof of Theorem 1, which generalizes Theorem 3.3 of [4].

Proof Let $\varepsilon > 0$ be given, let r be the smallest integer greater than $1/\varepsilon$, and let $\tau = [\mathbb{K}:\mathbb{Q}]$. First consider the case when $r > \delta$, so that

$$\delta \leq r-1 \leq \tfrac{1}{\varepsilon} < r.$$

We will show that $h(\alpha) \gg_{\tau,\varepsilon} 1$. For $d \geq 2$, using Eq. (2), we obtain

$$h(\alpha) \geq \frac{2}{d(\log 3d)^3}.$$

The function $f_1(x) = \frac{2}{x(\log 3x)^3}$ is decreasing for $x \geq 1$. Since $\mathbb{Q}(\alpha) \subset \mathbb{K}(\alpha)$ we have

$$d = [\mathbb{Q}(\alpha):\mathbb{Q}] \leq [\mathbb{K}(\alpha):\mathbb{Q}] = [\mathbb{K}(\alpha):\mathbb{K}][\mathbb{K}:\mathbb{Q}] = \delta\tau \leq \tau/\varepsilon.$$

Therefore,

$$h(\alpha) \geq f_1(\tau/\varepsilon) = \frac{2}{\tau/\varepsilon(\log 3\tau/\varepsilon)^3}.$$

We can often improve upon this lower bound. Using Eq. (1) for $d \geq 2$ yields

$$h(\alpha) \geq \frac{1}{4d}\left(\frac{\log\log d}{\log d}\right)^3.$$

Let $g_1(x) = \frac{1}{4x}\left(\frac{\log\log x}{\log x}\right)^3$. The function $g_1(x)$ is positive for all $x \geq 3$ and decreasing for $x \geq 7$. For $x = 3, 4, 5, 6$, we see that $g_1(x)$ achieves its minimum at $g_1(3) = 0.00005227953369\ldots$. For $d \geq 7$, since $d \leq \tau/\varepsilon$, we have

$$h(\alpha) \geq g_1(\tau/\varepsilon).$$

There exists $a \in (184, 185)$ such that for $x \le a$, we have $f_1(x) > g_1(x)$, but for $x \ge a$, $g_1(x) > f_1(x)$. (In fact, $a = 184.615\ldots$.) We also note that $f_1(6) > g_1(3)$. We conclude that when $r > \delta$, we have $h(\alpha) \ge C_1(\varepsilon, \tau)$, where

$$C_1(\varepsilon, \tau) = \begin{cases} f_1(6), & \text{if } 3 \le d \le 6 \text{ and } \tau/\varepsilon < 6, \\ f_1(\tau/\varepsilon), & \text{if } 3 \le d \le 6 \text{ and } \tau/\varepsilon \ge 6, \text{ or if } d \ge 7 \text{ and } \tau/\varepsilon \le a, \\ g_1(\tau/\varepsilon), & \text{if } d \ge 7 \text{ and } \tau/\varepsilon \ge a. \end{cases}$$

For $d = 2$, we can use $C_1 = f_1(2)$, and for $d = 1$, we can use $C_1 = \log 2$.

We may now assume that $r \le \delta$. Let ρ be the multiplicative rank of the conjugates of α over $\mathbb{K}$.

First, consider the case when $\rho \ge r$. (That is, r of the conjugates of α over $\mathbb{K}$ are multiplicatively independent.) By Proposition 2, with $D = [\mathbb{K}(\alpha) : \mathbb{Q}] = \delta\tau$, we have

$$h(\alpha)^r \ge (\delta\tau)^{-1}(1050r^5 \log(3\delta\tau))^{-r^2(r+1)^2}$$

using the fact that the Weil heights of conjugate algebraic numbers are equal. Since $r > \frac{1}{\varepsilon}$, it follows that $-\frac{1}{r} > -\varepsilon$, so $(\delta\tau)^{-\frac{1}{r}} > (\delta\tau)^{-\varepsilon}$. Therefore, upon taking the rth roots, with $f_2(\delta, \tau, r) = (1050r^5 \log(3\delta\tau))^{-r(r+1)^2}$, we have

$$h(\alpha) > \delta^{-\varepsilon}\tau^{-\varepsilon} f_2(\delta, \tau, r). \tag{6}$$

Since $r - 1 \le \frac{1}{\varepsilon}$, it follows that $r \le 1 + \frac{1}{\varepsilon}$ so

$$f_2(\delta, \tau, r) \ge \Big(1050(1 + \tfrac{1}{\varepsilon})^5 \log(3\delta\tau)\Big)^{-(1+\frac{1}{\varepsilon})(2+\frac{1}{\varepsilon})^2}.$$

Using the inequality $\log(x) \le \frac{1}{\varepsilon_1 \exp(1)} x^{\varepsilon_1}$, which holds for any $\varepsilon_1 > 0$, we see that

$$f_2(\delta, \tau, r) \ge \Big(1050(1 + \tfrac{1}{\varepsilon})^5 \frac{1}{\varepsilon_1 \exp(1)} (3\delta\tau)^{\varepsilon_1}\Big)^{-(1+\frac{1}{\varepsilon})(2+\frac{1}{\varepsilon})^2}.$$

Taking $\varepsilon_1 = \varepsilon/(1 + \frac{1}{\varepsilon})(2 + \frac{1}{\varepsilon})^2$, we have

$$f_2(\delta, \tau, r) \ge \Big(\frac{1050}{\varepsilon \exp(1)}(1 + \tfrac{1}{\varepsilon})^6(2 + \tfrac{1}{\varepsilon})^2\Big)^{-(1+\frac{1}{\varepsilon})(2+\frac{1}{\varepsilon})^2} (3\delta\tau)^{-\varepsilon}.$$

We conclude, from (6), that $h(\alpha) \ge C_2(\varepsilon, \tau)\delta^{-2\varepsilon}$, where

$$C_2(\varepsilon, \tau) = \tau^{-2\varepsilon} 3^{-\varepsilon}\Big(\frac{1050}{\varepsilon \exp(1)}(1 + \tfrac{1}{\varepsilon})^6(2 + \tfrac{1}{\varepsilon})^2\Big)^{-(1+\frac{1}{\varepsilon})(2+\frac{1}{\varepsilon})^2}.$$

Now we may assume that $r \leq \delta$ and $\rho \leq r - 1$. First, let us establish some notation. Let e be the order of the group of roots of unity in $\mathbb{K}(\alpha)$, let f be the order of the group of roots of unity in $\mathbb{K}$, and let $D = [\mathbb{K}(\alpha) : \mathbb{K}(\zeta_e)]$. By Proposition 3, taking $\mathbb{A} = \mathbb{K}(\zeta_e)$ and $\mathbb{B} = \mathbb{K}$, we conclude that there is an absolute positive constant c such that

$$h(\alpha) \geq \frac{(g(\tau)\Delta)^{-c}}{D} \frac{(\log\log 5D)^3}{(\log 2D)^4},$$

where Δ is the absolute value of the discriminant of $\mathbb{K}$ over $\mathbb{Q}$ and $g(\tau) = 1$ if there exists a tower of successive Galois extensions $\mathbb{Q} = \mathbb{K}_0 \subset \mathbb{K}_1 \subset \cdots \subset \mathbb{K}_m = \mathbb{K}$, and $g(\tau) = \tau!$ otherwise.

Notice that the function $f(x) = \frac{1}{x}\frac{(\log\log 5x)^3}{(\log 2x)^4}$ is decreasing for all $x \geq 1$. By Lemma 3 we have

$$D = [\mathbb{K}(\alpha) : \mathbb{K}(\zeta_e)] \leq \delta^{\varepsilon} C_4(\mathbb{K}, \varepsilon)$$

with $C_4(\mathbb{K}, \varepsilon) = \frac{1}{C(\varepsilon)} n(\rho) f \tau^{1+\varepsilon}$. Therefore,

$$h(\alpha) \geq \frac{(g(\tau)\Delta)^{-c}}{C_4(\mathbb{K}, \varepsilon)\delta^{\varepsilon}} \frac{\left(\log\log(5C_4(\mathbb{K}, \varepsilon)\delta^{\varepsilon})\right)^3}{\left(\log(2C_4(\mathbb{K}, \varepsilon)\delta^{\varepsilon})\right)^4}.$$

It remains to show that this is $\gg_{\rho,\mathbb{K}} \delta^{-2\varepsilon}$. The constant $C(\varepsilon)$ from Lemma 1 is easily seen to be positive and less than 1, which implies that $C_4(\mathbb{K}, \varepsilon) \geq 1$. Moreover, for all $y \geq 1$, we have

$$\frac{y\,(\log\log 5y)^3}{(\log 2y)^4} \geq \frac{1}{4}.$$

Since $C_4(\mathbb{K}, \varepsilon)\delta^{\varepsilon} \geq 1$, we conclude that

$$\frac{(\log\log(5C_4(\mathbb{K}, \varepsilon)\delta^{\varepsilon}))^3}{(\log(2C_4(\mathbb{K}, \varepsilon)\delta^{\varepsilon}))^4} \geq \frac{1}{4C_4(\mathbb{K}, \varepsilon)\delta^{\varepsilon}}.$$

We have shown that

$$h(\alpha) \geq \delta^{-2\varepsilon} C_5(\mathbb{K}, \varepsilon),$$

where $C_5(\mathbb{K}, \varepsilon) = \frac{(g(\tau)\Delta)^{-c} C(\varepsilon)^2}{4(n(\rho) f \tau^{1+\varepsilon})^2}$ and $C(\varepsilon)$ is the constant from Lemma 1. Since we are assuming that $\rho \leq r - 1 < 1/\varepsilon$, then $n(\rho) < 3^{\rho^2} < 3^{(1/\varepsilon)^2}$. Thus, we have

$$h(\alpha) \geq \delta^{-2\varepsilon} C_3(\mathbb{K}, \varepsilon),$$

where $C_3(\mathbb{K}, \varepsilon) = \frac{(g(\tau)\Delta)^{-c} C(\varepsilon)^2}{4(3^{(1/\varepsilon)^2} f \tau^{1+\varepsilon})^2}$.

Remark 4 Recall that, if $\mathbb{Q}(\alpha)/\mathbb{K}$ is Galois, then since $\mathbb{K} \subset \mathbb{Q}(\alpha)$, it follows that $\mathbb{K}(\alpha) = \mathbb{Q}(\alpha)$. Therefore, this theorem also applies to the case where $\mathbb{Q}(\alpha)/\mathbb{K}$ is Galois.

5 Proof of Theorem 2

We prove Theorem 2 below.

Proof Let $\alpha_1, \ldots, \alpha_\delta$ be the conjugates of α over $\mathbb{K}$, and let ρ be their multiplicative rank. As a result, $\delta \leq \eta$. Since we assume that r conjugates of α over $\mathbb{K}$ are multiplicatively independent, we know that $\rho \geq r$.

Case 1 ($\rho > r$)**:** If the multiplicative rank of the conjugates of α over $\mathbb{K}$ is strictly larger than r, we know that there exists a subset

$$\{\alpha_{i_1}, \alpha_{i_2}, \ldots, \alpha_{i_{r+1}}\} \subset \{\alpha_1, \alpha_2, \ldots, \alpha_\delta\}$$

such that $\alpha_{i_1}, \alpha_{i_2}, \ldots, \alpha_{i_{r+1}}$ are distinct and multiplicatively independent. By Proposition 2,

$$h(\alpha_{i_1})h(\alpha_{i_2})\cdots h(\alpha_{i_{r+1}}) \geq D^{-1}\left(1050(r+1)^5 \log(3D)\right)^{-(r+1)^2(r+2)^2},$$

where $D = [\mathbb{Q}(\alpha_{i_1}, \ldots, \alpha_{i_{r+1}}) : \mathbb{Q}]$. Since the α_i are all conjugates, they all have the same height, so the left hand side of this inequality is $h(\alpha)^{r+1}$. In addition,

$$D = [\mathbb{Q}(\alpha_{i_1}, \ldots, \alpha_{i_{r+1}}) : \mathbb{Q}] \leq [\mathbb{F} : \mathbb{Q}] = [\mathbb{F} : \mathbb{K}][\mathbb{K} : \mathbb{Q}] = \eta\tau.$$

Upon taking $(r+1)^{\text{st}}$ roots, it follows that

$$h(\alpha) \geq \tau^{-\frac{1}{r+1}}\eta^{-\frac{1}{r+1}}\left(1050(r+1)^5 \log(3\tau\eta)\right)^{-(r+1)(r+2)^2}.$$

Recall that $\log x \leq \frac{1}{\varepsilon_1 \exp(1)} x^{\varepsilon_1}$ for any $\varepsilon_1 > 0$. By applying this inequality with $\varepsilon_1 = \frac{\varepsilon}{(r+1)(r+2)^2}$, we get an explicit lower bound for $h(\alpha)$ in the desired form,

$$h(\alpha) \geq C_1(\varepsilon, r, \tau)\eta^{-\frac{1}{r+1}-\varepsilon},$$

where

$$C_1(\varepsilon, r, \tau) = 3^{-\varepsilon}\left(\frac{1050(r+1)^6(r+2)^2}{\varepsilon \exp(1)}\right)^{-(r+1)(r+2)^2} \tau^{-\frac{1}{r+1}-\varepsilon}.$$

Case 2 ($\rho = r$): Let $\alpha_{i_1}, \alpha_{i_2}, \ldots, \alpha_{i_r}$ be multiplicatively independent conjugates of α over $\mathbb{K}$. We denote by e the order of the group of roots of unity in $\mathbb{F}$ so that $\mathbb{Q}(\zeta_e) \subset \mathbb{F}$. By Lemma 2 we know that there exists a subfield $\mathbb{L}$ of $\mathbb{F}$ which is Galois over $\mathbb{K}$ such that $\alpha^e \in \mathbb{L}$ and $[\mathbb{L} : \mathbb{K}] = n \le n(r) < 3^{r^2}$. By (1), we have

$$h(\alpha^e) \ge \frac{1}{4[\mathbb{Q}(\alpha^e) : \mathbb{Q}]} \left(\frac{\log\log([\mathbb{Q}(\alpha^e) : \mathbb{Q}])}{\log([\mathbb{Q}(\alpha^e) : \mathbb{Q}])} \right)^3,$$

provided $\alpha^e \notin \mathbb{Q}$. Now,

$$[\mathbb{Q}(\alpha^e) : \mathbb{Q}] \le [\mathbb{L} : \mathbb{Q}] = [\mathbb{L} : \mathbb{K}][\mathbb{K} : \mathbb{Q}] \le n(r)\tau.$$

As in the proof of Theorem 1, we can use the properties of the function $g_1(x)$ previously defined to obtain

$$h(\alpha) = \frac{1}{e} h(\alpha^e) \ge \frac{1}{e} \frac{1}{4n(r)\tau} \left(\frac{\log\log(n(r)\tau)}{\log(n(r)\tau)} \right)^3 \tag{7}$$

whenever $[\mathbb{Q}(\alpha^e) : \mathbb{Q}] \ge 7$. When $n(r)\tau \le 184$, this can be improved by using $f_1(x)$ in place of $g_1(x)$, as before, and similarly, we use $f_1(x)$ when $3 \le [\mathbb{Q}(\alpha^e) : \mathbb{Q}] \le 6$. For $[\mathbb{Q}(\alpha^e) : \mathbb{Q}] = 2$, we have $h(\alpha) \ge \frac{1}{e} f_1(2)$, and for $[\mathbb{Q}(\alpha^e) : \mathbb{Q}] = 1$ we have $h(\alpha) \ge \frac{1}{e} \log 2$. For the remainder of the proof, we focus on the case given in Eq. (7) and trust the reader to make the appropriate substitutions.

On the other hand, Corollary 3 implies that with $\boldsymbol{\alpha} = (\alpha_{i_1}, \ldots, \alpha_{i_r})$, we have

$$h(\alpha)^r > \left(c_2(r)[\mathbb{Q}^{\mathrm{ab}}(\boldsymbol{\alpha}) : \mathbb{Q}^{\mathrm{ab}}]\Big(\log\big(3[\mathbb{Q}^{\mathrm{ab}}(\boldsymbol{\alpha}) : \mathbb{Q}^{\mathrm{ab}}]\big)\Big)^{\kappa_2(r)} \right)^{-1},$$

where $\kappa_2(r) = 3r\left(2(r+1)^2(r+1)!\right)^r$ and

$$c_2(r) = (2r^2)^r \exp\left(64r^2 r! \left(2(r+1)^2(r+1)!\right)^{2r} \right).$$

Using the bound

$$[\mathbb{Q}^{\mathrm{ab}}(\boldsymbol{\alpha}) : \mathbb{Q}^{\mathrm{ab}}] \le [\mathbb{Q}(\zeta_e)(\boldsymbol{\alpha}) : \mathbb{Q}(\zeta_e)] \le [\mathbb{F} : \mathbb{Q}(\zeta_e)] = \frac{[\mathbb{F} : \mathbb{Q}]}{[\mathbb{Q}(\zeta_e) : \mathbb{Q}]} = \frac{\tau\eta}{\phi(e)},$$

we conclude that

$$h(\alpha)^r > \left(c_2(r)\frac{\tau\eta}{\phi(e)}\Big(\log\big(3\frac{\tau\eta}{\phi(e)}\big)\Big)^{\kappa_2(r)} \right)^{-1}. \tag{8}$$

Combining Eqs. (7) and (8) yields

$$h(\alpha)^{r+1} > C_1(r,\tau,e)\eta^{-1}\left(\log(3\tfrac{\tau\eta}{\phi(e)})\right)^{-\kappa_2(r)},$$

where $C_1(r,\tau,e) = c_2(r)^{-1}\left(\frac{\log\log(n(r)\tau)}{\log(n(r)\tau)}\right)^3 \frac{1}{4n(r)\tau^2}\frac{\phi(e)}{e}$. We now apply the inequality $\log x \leq \frac{1}{\varepsilon_1 \exp(1)} x^{\varepsilon_1}$ with $\varepsilon_1 = \varepsilon/\kappa_2(r)$ and $x = 3\frac{\tau\eta}{\phi(e)}$ and conclude that

$$h(\alpha)^{r+1} > C_1(r,\tau,e)\eta^{-1}\left(\frac{\kappa_2(r)}{\varepsilon\exp(1)}\right)^{-\kappa_2(r)}\left(3\frac{\tau\eta}{\phi(e)}\right)^{-\varepsilon}.$$

This simplifies to

$$h(\alpha)^{r+1} > C_2(\varepsilon,r,\tau,e)\eta^{-1-\varepsilon}$$

with $C_2(\varepsilon,r,\tau,e) = \left(\frac{\log\log(n(r)\tau)}{\log(n(r)\tau)}\right)^3\left(\frac{\kappa_2(r)}{\varepsilon\exp(1)}\right)^{-\kappa_2(r)}\frac{\phi(e)^{1+\varepsilon}/e}{4n(r)c_2(r)3^\varepsilon\tau^{2+\varepsilon}}$. By Lemma 1

$$\phi(e)^{1+\varepsilon}/e \geq C(\varepsilon),$$

so that we can replace $C_2(\varepsilon,r,\tau,e)$ in the inequality by

$$C_3(\varepsilon,r,\tau) = \left(\frac{\log\log(n(r)\tau)}{\log(n(r)\tau)}\right)^3\left(\frac{\kappa_2(r)}{\varepsilon\exp(1)}\right)^{-\kappa_2(r)}\frac{C(\varepsilon)}{4n(r)c_2(r)3^\varepsilon\tau^{2+\varepsilon}},$$

and upon taking $(r+1)^{\mathrm{st}}$ roots we have

$$h(\alpha) > C_3(\varepsilon,r,\tau)^{\frac{1}{r+1}}\eta^{-\frac{1}{r+1}-\frac{\varepsilon}{r+1}}.$$

Using the fact that $\eta^{-\varepsilon/(r+1)} \geq \eta^{-\varepsilon}$, we get the bound

$$h(\alpha) > C_3(\varepsilon,r,\tau)^{\frac{1}{r+1}}\eta^{-\frac{1}{r+1}-\varepsilon}.$$

Acknowledgements This work began as a research project for the working group *Heights of Algebraic Integers* at the Women in Numbers Europe 2 workshop held at the Lorentz Center at the University of Leiden. The authors would like to thank the organizers of the workshop and the Lorentz Center for their hospitality.

Research of Shabnam Akhtari is supported by the NSF grant DMS-1601837. Kirsti Biggs is supported by an EPSRC Doctoral Training Partnership. Research of Alia Hamieh is partially supported by a PIMS postdoctoral fellowship. Research of Kathleen Petersen is supported by Simons Foundation Collaboration grant number 209226 and 430077; she would like to thank the Tata Institute of Fundamental Research for their hospitality while preparing this manuscript. Lola Thompson is supported by an AMS Simons Travel Grant, by a Max Planck Institute fellowship during the Fall 2016 semester, and by the NSF grant DMS-1440140 while in residence at the Mathematical Sciences Research Institute during the Spring 2017 semester.

References

1. F. Amoroso, S. David, Le problème de Lehmer en dimension supérieure. J. Reine Angew. Math. **513**, 145–179 (1999)
2. F. Amoroso, E. Delsinne, Une minoration relative explicite pour la hauteur dans une extension d'une extension abélienne, in *Diophantine Geometry*. CRM Series, vol. 4 (Edizioni della Normale, Pisa, 2007), pp. 1–24
3. F. Amoroso, R. Dvornicich, A lower bound for the height in abelian extensions. J. Number Theory **80**(2), 260–272 (2000)
4. F. Amoroso, D. Masser, Lower bounds for the height in Galois extensions. Bull. Lond. Math. Soc. **48**(6), 1008–1012 (2016)
5. F. Amoroso, E. Viada, Small points on rational subvarieties of tori. Comment. Math. Helv. **87**(2), 355–383 (2012)
6. F. Amoroso, U. Zannier, A relative Dobrowolski lower bound over abelian extensions. Ann. Sc. Norm. Super. Pisa Cl. Sci. (4) **29**(3), 711–727 (2000)
7. N. Berry, A. Dubickas, N.D. Elkies, B. Poonen, C. Smyth, The conjugate dimension of algebraic numbers. Q. J. Math. **55**(3), 237–252 (2004)
8. P.E. Blanksby, H.L. Montgomery, Algebraic integers near the unit circle. Acta Arith. **18**, 355–369 (1971)
9. P. Borwein, E. Dobrowolski, M.J. Mossinghoff, Lehmer's problem for polynomials with odd coefficients. Ann. Math. (2) **166**(2), 347–366 (2007)
10. R. Breusch, On the distribution of the roots of a polynomial with integral coefficients. Proc. Am. Math. Soc. **2**, 939–941 (1951)
11. E. Delsinne, Le problème de Lehmer relatif en dimension supérieure. Ann. Sci. Éc. Norm. Supér. (4) **42**(6), 981–1028 (2009)
12. E. Dobrowolski, On a question of Lehmer and the number of irreducible factors of a polynomial. Acta Arith. **34**(4), 391–401 (1979)
13. W. Feit, The orders of finite linear groups (1995, Preprint)
14. S. Friedland, The maximal orders of finite subgroups in $GL_n(\mathbf{Q})$. Proc. Am. Math. Soc. **125**(12), 3519–3526 (1997)
15. J.B. Friedlander, Estimates for prime ideals. J. Number Theory **12**(1), 101–105 (1980)
16. D.H. Lehmer, Factorization of certain cyclotomic functions. Ann. Math. (2) **34**, 461–479 (1933)
17. P. Pollack, *Not Always Buried Deep: A Second Course in Elementary Number Theory* (American Mathematical Society, Providence, 2009)
18. J.B. Rosser, L. Schoenfeld, Approximate formulas for some functions of prime numbers. Ill. J. Math. **6**, 64–94 (1962)
19. J.P. Serre, Rigidité du foncteur de Jacobi d'échelon $n \geq 3$, Appendix to A. Grothendieck, Techniques de construction en géométrie analytique, X. Construction de l'espace de Teichmüller. *Appendice à l'exposé* 17 *du séminaire Cartan*, (17), pp. 1960–61
20. C.J. Smyth, On the product of the conjugates outside the unit circle of an algebraic integer. Bull. Lond. Math. Soc. **3**, 169–175 (1971)
21. C. Smyth, The Mahler measure of algebraic numbers: a survey, in *Number Theory and Polynomials*. London Mathematical Society Lecture Note Series, vol. 352 (Cambridge University Press, Cambridge, 2008), pp. 322–349
22. H.M. Stark, Some effective cases of the Brauer-Siegel theorem. Invent. Math. **23**, 135–152 (1974)
23. C.L. Stewart, Algebraic integers whose conjugates lie near the unit circle. Bull. Soc. Math. France **106**(2), 169–176 (1978)
24. P. Voutier, An effective lower bound for the height of algebraic numbers. Acta Arith. **74**(1), 81–95 (1996)

Reductions of Algebraic Integers II

Antonella Perucca

Abstract Let K be a number field, and let G be a finitely generated subgroup of $K^\times$. Fix some positive integer m, and consider the set of primes $\mathfrak{p}$ of K satisfying the following condition: the reduction of G modulo $\mathfrak{p}$ is well-defined and has size coprime to m. We show that the natural density of this set is a computable rational number by reducing to the case where m is prime, case which has been treated in the previous work *Reductions of algebraic integers* (joint with Christophe Debry, J. Number Theory, 2016).

Keywords Number field · Algebraic integer · Reduction · Kummer theory · Density

2010 Mathematics Subject Classification. Primary 11R44; Secondary 11R18, 11R20, 11Y40

1 Introduction

This paper is the continuation of [1] by Christophe Debry and the author; therefore, we refer to this other work for the history of the problem and for further references. Let K be a number field, and let G be a finitely generated subgroup of $K^\times$. Up to excluding finitely many primes $\mathfrak{p}$ of K, we always assume that the reduction of G modulo $\mathfrak{p}$ is well-defined. Fix some prime number ℓ, and consider the set of primes $\mathfrak{p}$ of K satisfying the following condition: the reduction of G modulo $\mathfrak{p}$ has size coprime to ℓ. In [1] it is proven that this set admits a natural density, which is a computable rational number.

We now deal with the generalization that consists of replacing ℓ by some positive integer m, which we may as well suppose to be square-free. So our aim is to show

A. Perucca (✉)
University of Luxembourg, Mathematics Research Unit, Esch-sur-Alzette, Luxembourg
e-mail: antonella.perucca@uni.lu

I. I. Bouw et al. (eds.), *Women in Numbers Europe II*, Association for Women in Mathematics Series 11, https://doi.org/10.1007/978-3-319-74998-3_2

that the following natural density is a well-defined rational number and how one can compute it:

$$D_{K,G,m} := \operatorname{dens}\{\mathfrak{p} : \operatorname{ord}(G \bmod \mathfrak{p}) \text{ is coprime to } m\}\,.$$

The condition on $\mathfrak{p}$ means that for every prime factor ℓ of m, the group $(G \bmod \mathfrak{p})$ has order coprime to ℓ. We are thus requiring simultaneous conditions related to different prime numbers. The main question is whether those conditions are independent, which would heuristically give

$$D_{K,G,m} = \prod_{\ell} D_{K,G,\ell}\,. \tag{1}$$

Note that we may suppose that G is torsion-free because roots of unity of order coprime to m do not matter for the density, while if G contains a root of unity of order not coprime to m, then the order of $(G \bmod \mathfrak{p})$ is also not coprime to m for almost all $\mathfrak{p}$.

Write K_x for the cyclotomic extension of K obtained by adding the x-th roots of unity. The method of [1] relies on the fact that the Kummer extension $K_{\ell^n}(\sqrt[\ell^n]{G})$ over K_{ℓ^n} has maximal degree ℓ^{rn} (where r is the rank of G), unless the elements of G have some divisibility property in $K^\times$. There is one small exception for the case $\ell = 2$, which is related to the fact that the cyclotomic extension K_8/K need not to be cyclic. In short [1] relied on the fact that the Kummer extensions related to ℓ have nothing to do with the cyclotomic extensions related to ℓ.

However, Kummer extensions may in general be contained in cyclotomic extensions because if ℓ and ℓ' are distinct prime numbers, then the field $K_{\ell'}$ could contain a cyclic extension of K of degree a power of ℓ (see Sect. 3). It is exactly this interplay between cyclotomic and Kummer extensions that we must treat delicately in the present paper.

We prove in Theorem 9 that the density $D_{K,G,m}$ can be expressed as an infinite sum involving splitting conditions in cyclotomic-Kummer extensions of K. Then we show that the density is always a computable rational number. In fact by Theorem 15, we know that $D_{K,G,m}$ can be written in terms of densities related to one single prime number, and those are known from [4] (for G of rank 1) and from [1] (for G of arbitrary rank).

In Theorem 11 we show that the product formula (1) is true if the following condition holds for every $n \geqslant 1$ and for every prime divisor ℓ of m: *the extensions* $K_{\ell^n \cdot \frac{m}{\ell}}$ *and* $K_{\ell^n}(\sqrt[\ell^n]{G})$ *are linearly disjoint over* K_{ℓ^n}.

The product formula (1) is then true under the assumption $K_m = K$ (see Corollary 12) or under the assumption that m is odd and that $K_\ell \neq K$ holds for every prime divisor ℓ of m (see Proposition 13). The last condition holds in particular (if m is odd) for $\mathbb{Q}$ and for every quadratic field, unless $K = \mathbb{Q}(\zeta_3)$ and 3 divides m. We also answer in the negative the question whether (1) holds for m odd; see Example 18.

We have tested our results in several explicit examples, for which an approximated density (by considering the primes of small norm) has been computed with Sage [6].

2 Preliminaries on the Chebotarev Density Theorem

Let K be a number field, and call P_K the set of primes of K. For $\mathfrak{p} \in P_K$ we denote by $N(\mathfrak{p})$ the cardinality of the residue field at $\mathfrak{p}$. If $\Gamma \subseteq P_K$ and the following limit exists, we call it the Dirichlet density of Γ:

$$\mathrm{dens}_{Dir}(\Gamma) = \lim_{s\to 1^+} \frac{\sum_{\mathfrak{p}\in\Gamma} N(\mathfrak{p})^{-s}}{\sum_{\mathfrak{p}\in P_K} N(\mathfrak{p})^{-s}}\,.$$

If the following limit exists, we call it the natural density of Γ:

$$\mathrm{dens}(\Gamma) = \lim_{n\to+\infty} \frac{\#\{\mathfrak{p}\in\Gamma : N(\mathfrak{p}) \leqslant n\}}{\#\{\mathfrak{p}\in P_K : N(\mathfrak{p}) \leqslant n\}}\,.$$

By the upper and lower density, we, respectively, mean the limit inferior and superior: these exist, and if they coincide, then the density exists. Note that if the natural density exists, then the Dirichlet density also exists, and they coincide (however, there are sets having a Dirichlet density and for which the natural density does not exist).

The following general result will allow us (in certain cases) to extend the base field:

Proposition 1 *Let K be a number field and let L be a finite Galois extension of K. Let Γ be a set of primes of K that split completely in L. Call Γ_L the set of primes of L which lie over the primes in Γ. If Γ_L has a Dirichlet density, then the same holds for Γ, and we have*

$$\mathrm{dens}_{Dir}(\Gamma) = [L:K]^{-1} \cdot \mathrm{dens}_{Dir}(\Gamma_L)\,. \tag{2}$$

Proof Call S the set of primes of K which split completely in L, and call S_L the set of primes of L which lie over the primes of S. If $\mathfrak{p} \in S$ and $\mathfrak{q}$ is one of the $[L:K]$ primes of S_L lying over it, then $\mathfrak{p}$ and $\mathfrak{q}$ have the same norm. On one hand, the Chebotarev Density Theorem gives

$$[L:K]^{-1} = \mathrm{dens}_{Dir}(S) = \lim_{s\to 1^+} \frac{\sum_{\mathfrak{p}\in S} N(\mathfrak{p})^{-s}}{\sum_{\mathfrak{p}\in P_K} N(\mathfrak{p})^{-s}}$$

and on the other hand, we know

$$1 = \mathrm{dens}_{Dir}(S_L) = \lim_{s\to 1^+} \frac{\sum_{\mathfrak{q}\in S_L} N(\mathfrak{q})^{-s}}{\sum_{\mathfrak{q}\in P_L} N(\mathfrak{q})^{-s}} = [L:K] \cdot \lim_{s\to 1^+} \frac{\sum_{\mathfrak{p}\in S} N(\mathfrak{p})^{-s}}{\sum_{\mathfrak{q}\in P_L} N(\mathfrak{q})^{-s}}$$

so we deduce

$$\lim_{s\to 1^+} \sum_{\mathfrak{p}\in P_K} N(\mathfrak{p})^{-s} = \lim_{s\to 1^+} \sum_{\mathfrak{q}\in P_L} N(\mathfrak{q})^{-s}.$$

We conclude because we have

$$\frac{\sum_{\mathfrak{p}\in\Gamma} N(\mathfrak{p})^{-s}}{\sum_{\mathfrak{p}\in P_K} N(\mathfrak{p})^{-s}} = [L:K]^{-1}\cdot\frac{\sum_{\mathfrak{q}\in\Gamma_L} N(\mathfrak{q})^{-s}}{\sum_{\mathfrak{p}\in P_K} N(\mathfrak{p})^{-s}} \xrightarrow{s\to 1^+} [L:K]^{-1}\cdot\mathrm{dens}_{Dir}(\Gamma_L).$$

□

The following result and its corollary are variants of the Chebotarev Density Theorem, where several field extensions are considered:

Theorem 2 *Let K be a number field. If $F_1, \ldots, F_n$ are linearly disjoint finite Galois extensions of K, then the set consisting of the primes of K that do not split completely in any of those extensions has a natural density, and this equals*

$$\prod_{i=1}^{n}\Big(1-\frac{1}{[F_i:K]}\Big). \tag{3}$$

Proof Call Γ the set of the primes of K that do not split completely in any of the extensions $F_1, \ldots, F_n$. By working with the compositum $F := F_1 \cdots F_n$ (and excluding the finitely many primes that ramify in F), we can interpret Γ as the primes of K whose F/K-Frobenius conjugacy class is contained in a certain conjugacy-invariant subset of $\mathrm{Gal}(F/K)$. The existence of the natural density for Γ then follows from the Chebotarev Density Theorem.

We prove the formula in the statement by induction on n, the case $n = 1$ being clear by the Chebotarev Density Theorem. For the induction step, consider linearly disjoint extensions $F_1, \ldots, F_{n+1}$ of K, and write for convenience $L = F_{n+1}$.

By the inductive hypothesis, we know

$$\mathrm{dens}(\Gamma) = \prod_{i=1}^{n}(1-[F_i:K]^{-1}).$$

Write $\Gamma' \subseteq \Gamma$ for the subset of the primes that split completely in L, and let Γ'_L be the set of primes of L lying over the primes in Γ'. One can argue as above and show that Γ' has a natural density.

Any set consisting of primes of L that lie over primes of K which do not split completely in L has Dirichlet density 0. The fields $LF_1, \ldots, LF_n$ are linearly disjoint over L, and Γ'_L consists of the primes of L that do not split completely in any of those fields (up to a set of primes of L of Dirichlet density 0). So by the inductive hypothesis, we get

$$\text{dens}_{Dir}(\Gamma'_L) = \prod_{i=1}^{n} \Big(1 - \frac{1}{[LF_i : L]}\Big) = \prod_{i=1}^{n} \Big(1 - \frac{1}{[F_i : K]}\Big).$$

By Proposition 1 we then have

$$\text{dens}_{Dir}(\Gamma') = [L : K]^{-1} \cdot \text{dens}_{Dir}(\Gamma'_L) = \frac{1}{[F_{n+1} : K]} \cdot \prod_{i=1}^{n} \Big(1 - \frac{1}{[F_i : K]}\Big).$$

Since $\Gamma' \subseteq \Gamma$ and these two sets have a natural density, then the same holds for their difference, and we conclude the induction step because we have

$$\text{dens}(\Gamma \setminus \Gamma') = \text{dens}(\Gamma) - \text{dens}(\Gamma') = \prod_{i=1}^{n+1} \Big(1 - \frac{1}{[F_i : K]}\Big).$$

□

Corollary 3 *Let K be a number field, and let L be a finite Galois extension of K. If $F_1, \dots, F_n$ are linearly disjoint finite Galois extensions of L, then the set of primes of K that split completely in L and do not split completely in any of the extensions $F_1, \dots, F_n$ has a natural density, and this equals*

$$[L : K]^{-1} \cdot \prod_{i=1}^{n} \Big(1 - \frac{1}{[F_i : L]}\Big). \tag{4}$$

Proof For the existence of the natural density, we may apply the Chebotarev Density Theorem. To prove the formula, it suffices to combine Proposition 1 and Theorem 2 (applied to L). □

3 Cyclotomic and Kummer Extensions

Let K be a number field, and fix some algebraic closure $\bar{K}$. We write K_x for the cyclotomic extension of K obtained by adding the x-th roots of unity. If ℓ is a prime number, we use the notation K_{ℓ^∞} to denote the union of the fields K_{ℓ^n} for $n \geqslant 1$.

If G is a finitely generated subgroup of $K^\times$, we also write $K_x(\sqrt[y]{G})$ for the extension of K_x obtained by adding all elements of $\bar{K}$ whose y-th power belongs to G.

We make use of the following result of Schinzel:

Theorem 4 (Schinzel [5, Thm. 2], with an Alternative Proof in [3, 7]) *Let K be a number field, and let $a \in K^\times$. For $n \geqslant 1$ the extension $K_n(\sqrt[n]{a})/K$ is abelian if and only if $a^t = b^n$ holds for some $b \in K^\times$ and for some divisor t of n satisfying $K = K_t$.*

We now recall the definition of *strongly ℓ-indivisible* element. In the remaining of the section, we use such elements to investigate the Kummer extensions.

Definition 5 Let K be a number field and ℓ a prime number. We say that $a \in K^\times$ is *strongly ℓ-indivisible* if, for every root of unity $\xi \in K$, $a\xi$ has no ℓ-th roots in K.

Theorem 6 *Consider integers $n, d \geqslant 1$ such that ℓ^d divides n. If the condition*

$$K_n(\sqrt[\ell^d]{a}) = K_n \tag{5}$$

holds for some strongly ℓ-indivisible $a \in K^\times$, then we have $K_{\ell^d} = K$, and there is some odd prime factor q of n such that ℓ^d divides $[K_q : K]$, unless $\ell = 2$ and $K \neq K_4$ and $d = 1$ and $K(\sqrt{a}) \subseteq K_{2^\infty}$.

Proof We know that the field $K_{\ell^d}(\sqrt[\ell^d]{a})$, which is contained in K_n by assumption, is an abelian extension of K. Thus by Theorem 4 we have $a^{\ell^e} = b^{\ell^d}$ for some $b \in K^\times$ and for some $e \geqslant 0$ satisfying $K_{\ell^e} = K$. Since a is strongly ℓ-indivisible, we must have $d \leqslant e$ and hence $K_{\ell^d} = K$ holds.

Again since a is strongly ℓ-indivisible, we know by Perucca [4, Theorems 11 and 13] that unless $\ell = 2$ and $K \neq K_4$ and $d = 1$ and $K(\sqrt{a}) \subseteq K_{2^\infty}$, we must have

$$[K_{\ell^\infty}(\sqrt[\ell^d]{a}) : K_{\ell^\infty}] = \ell^d\,.$$

Let this be the case, and denote by n' the product of all odd prime factors q of n distinct from ℓ. We deduce that the extension $K_{n'}/K$ contains a cyclic subextension of degree ℓ^d. Since $\mathrm{Gal}(K_{n'}/K)$ is the product of the groups $\mathrm{Gal}(K_q/K)$, at least one of these has exponent divisible by ℓ^d, and hence ℓ^d divides $[K_q : K]$. □

Theorem 7 *Consider integers $n, d \geqslant 1$ such that ℓ^d divides n. If $K_{\ell^d} = K$ holds, and if ℓ^d divides $[K_q : K]$ for some odd prime factor q of n, then there is some strongly ℓ-indivisible $a \in K^\times$ satisfying (for all choices of the ℓ^{d+1}-th root)*

$$K_n(\sqrt[\ell^d]{a}) = K_n \qquad \text{and} \qquad \sqrt[\ell^{d+1}]{a} \notin K_n\,. \tag{6}$$

Proof By assumption there is a cyclic extension C of K of degree ℓ^d contained in K_q. Since $K_{\ell^d} = K$ holds, there is some $c \in K^\times$ satisfying $C = K(\sqrt[\ell^d]{c})$. The element c is strongly ℓ-indivisible because the field $K(\sqrt[\ell]{c})$ is contained in K_q but not in K, and hence it is not contained in K_{ℓ^∞}. The field C is contained in K_n, so we are done if $\sqrt[\ell^{d+1}]{c} \notin K_n$ holds for all choices of the ℓ^{d+1}-th root. If not, take $b \in K^\times$ such that $K(\sqrt[\ell]{b})$ is not contained in $K_{n\ell}$: such an element exists because $K_{n\ell}$ contains only finitely many subextensions of degree ℓ while $K^\times/K^{\times\ell}$ is infinite. Then, cb^{ℓ^d} is strongly ℓ-indivisible, and we have $\sqrt[\ell^d]{cb^{\ell^d}} \in K_n$. By construction $\sqrt[\ell^{d+1}]{cb^{\ell^d}}$ is not contained in K_n for any choice of the ℓ^{d+1}-th root. □

4 Prescribed Torsion in the Reductions

The aim of this section is to compute the density of reductions that have some prescribed valuations for the size of the multiplicative group of the residue field.

Let $m \geqslant 2$ be a square-free integer, and write $m = \ell_1 \cdots \ell_f$ as a product of prime numbers. We define the m-adic valuation as the f-tuple of the ℓ_i-adic valuations. We then consider f-tuples of nonnegative integers

$$A = (a_1, \ldots, a_f)\,.$$

We write $A + 1$ if we increase all entries by 1 and $S_i A$ if we increase only the i-th entry by 1, i.e., $(S_i A)_j = a_j$ for $j \neq i$ and $(S_i A)_i = a_i + 1$. In particular we have $A + 1 = S_1 \cdots S_f A$. We also define

$$m^A := \prod_{i=1}^{f} \ell_i^{a_i} \tag{7}$$

Let K be a number field, and let $\mathfrak{p}$ be a prime of K. If we have for the m-adic valuation $v_m(\#k_\mathfrak{p}^\times) = A$, then this means that $v_{\ell_i}(\#k_\mathfrak{p}^\times) = a_i$ holds for every $i = 1, \ldots, f$. In other words, we can write $\#k_\mathfrak{p}^\times = m^A \cdot m'$ with m' coprime to m.

We first write down a formula for the natural density of the set of primes $\mathfrak{p}$ of K such that the m-adic valuation of $\#k_\mathfrak{p}^\times$ equals A. We may neglect the finitely many primes of K that ramify in $K_{m^{A+1}}$, so we are looking for the primes that split completely in K_{m^A} and that for every i do not split completely in $K_{m^{S_i A}}$.

Proposition 8 *The set of primes $\mathfrak{p}$ of K such that the m-adic valuation of $k_\mathfrak{p}^\times$ equals A has a natural density, which we call δ_{K,m^A}. Define $\delta_{K,\ell_i^{a_i}}$ similarly by requiring the ℓ_i-adic valuation of $k_\mathfrak{p}^\times$ to be a_i. We then have*

$$\delta_{K,m^A} = \prod_{i=1}^{f} \left([K_{\ell^{a_i}} : K]^{-1} - [K_{\ell^{a_i+1}} : K]^{-1} \right) = \prod_{i=1}^{f} \delta_{K,\ell_i^{a_i}}\,. \tag{8}$$

Proof The existence of the natural density follows from Corollary 3. The second equality is clear by the Chebotarev Density Theorem because for $\delta_{K,\ell_i^{a_i}}$ we count the primes of K that split completely in $K_{\ell_i^{a_i}}$ and that do not split completely in $K_{\ell_i^{a_i+1}}$.

For δ_{K,m^A} we count the primes of K that split completely in $L := K_{m^A}$ and that for every $i = 1, \ldots, f$ do not split completely in $K_{m^{S_i A}}$. By Corollary 3 we get

$$\begin{aligned}\delta_{K,m^A} &= [L : K]^{-1} \cdot \prod_{i=1}^{f} \left(1 - [K_{m^{S_i A}} : L]^{-1}\right) \\ &= [L : K]^{-1} \cdot \prod_{i=1}^{f} \left(1 - [K_{\ell^{a_i+1}} : K_{\ell^{a_i}}]^{-1}\right).\end{aligned} \tag{9}$$

We conclude because we have $[L : K] = \prod_{i=1}^{f} [K_{\ell^{a_i}} : K]$. □

5 General Formulas for the Density

We first investigate the existence of the density under consideration and write it as an infinite sum (according to the size of the multiplicative group of the residue field).

Let K be a number field, let G be a finitely generated and torsion-free subgroup of $K^\times$, and let $m \geqslant 2$ be a square-free integer. We make use of the notation introduced in the beginning of Sect. 4. We always tacitly exclude the finitely many primes that ramify in the cyclotomic-Kummer extensions that we consider. Indeed there are only finitely many primes of K that ramify in $K_{m^n}(\sqrt[m^n]{G})$ for some $n \geqslant 1$; see [2, Lemma C.1.7].

Theorem 9 *Let $\Gamma_{K,G,m}$ be the set of primes of K for which the reduction of G has order coprime to m. Let Γ_{K,G,m^A} be the set of primes of K that split completely in $K_{m^A}(\sqrt[m^A]{G})$ and that for every i do not split completely in $K_{m^{S_i A}}$. Then $\Gamma_{K,G,m}$ and Γ_{K,G,m^A} have a natural density, which we call $D_{K,G,m}$ and Δ_{K,G,m^A}, respectively, and we have*

$$D_{K,G,m} = \sum_A \Delta_{K,G,m^A} . \tag{10}$$

Proof To ease notation, we remove the subindex (K, G, m), and we write A for the subindex (K, G, m^A). We first prove that Δ_A is well-defined. Write $L := K_{m^A}(\sqrt[m^A]{G})$. For Γ_A we may equivalently consider the primes that split completely in L and that for every i do not split completely in $F_i = K_{m^{S_i A}}(\sqrt[m^A]{G})$. We conclude by applying Corollary 3.

Consider a prime $\mathfrak{p}$ of K such that the m-adic valuation of $k_\mathfrak{p}^\times$ equals A. Then $\mathfrak{p} \in \Gamma_{K,G,m}$ if and only if $\mathfrak{p} \in \Gamma_A$ because an element of (G mod $\mathfrak{p}$) has order coprime to m if and only if it has m^A-th roots in $k_\mathfrak{p}^\times$.

We have proven that $\Gamma = \cup_A \Gamma_A$ holds. Write $A \leqslant n$ if $a_i \leqslant n$ holds for all $i = 1, \ldots, f$. Since the sets Γ_A are pairwise disjoint and each of them has natural density Δ_A, we get that $\cup_{A \leqslant n} \Gamma_A$ has a natural density, given by

$$\mathrm{dens}\Big(\bigcup_{A\leqslant n} \Gamma_A\Big) = \sum_{A\leqslant n} \Delta_A .$$

Since this holds for every n, then the lower natural density satisfies

$$\mathrm{dens}_-(\Gamma) \geqslant \sum_A \Delta_A .$$

To conclude we show that for the upper natural density, we have

$$\mathrm{dens}_+(\Gamma) \leqslant \varepsilon(n) + \sum_{A\leqslant n} \Delta_A$$

for some function $\varepsilon(n)$ that goes to zero for n going to infinity. It then suffices to prove that the difference

$$\Gamma' := \Gamma \setminus \bigcup_{A\leqslant n} \Gamma_A$$

is contained in a set whose upper density goes to zero with n. This is true because the primes in Γ' split completely in $K_{\ell_i^n}$ for some $i = 1, \dots, f$, and hence Γ' is contained in a finite union of sets that have a natural density that goes to zero with n. □

In the remaining of the section, we investigate cases in which the density $D_{K,G,m}$ is the product of the densities related to the prime divisors of m.

Lemma 10 *Let ℓ vary over the prime divisor of m. We have*

$$D_{K,G,m} = \prod_{\ell} D_{K,G,\ell} \tag{11}$$

if for every $n \geqslant 1$ the following two conditions hold, where $w = \frac{m}{\ell}$:

(i) the extensions $K_{\ell^n\cdot w}$ and $K_{\ell^n}(\sqrt[\ell^n]{G})$ are linearly disjoint over K_{ℓ^n};
(ii) the extensions $K_{\ell\cdot w^n}$ and $K_{w^n}(\sqrt[w^n]{G})$ are linearly disjoint over K_{w^n}.

Proof Recall the notation $m = \ell_1 \cdots \ell_f$. We claim that we have

$$\Delta_{K,G,m^A} = \prod_{i=1}^{f} \Delta_{K,G,\ell_i^{a_i}} \,.$$

Then we can write by Theorem 9:

$$D_{K,G,m} = \sum_A \Delta_{K,G,m^A} = \sum_A \prod_{i=1}^{f} \Delta_{K,G,\ell_i^{a_i}} = \prod_{i=1}^{f} \sum_{a_i\geqslant 0} \Delta_{K,G,\ell_i^{a_i}} = \prod_{i=1}^{f} D_{K,G,\ell_i} \,.$$

So we are left to prove the claim. By Corollary 3 we can write

$$\Delta_{K,G,\ell_i^{a_i}} = [K_{\ell_i^{a_i}}(\sqrt[\ell_i^{a_i}]{G}) : K]^{-1} \cdot \left(1 - \frac{1}{[K_{\ell_i^{a_i+1}}(\sqrt[\ell_i^{a_i}]{G}) : K_{\ell_i^{a_i}}(\sqrt[\ell_i^{a_i}]{G})]}\right)$$

and also

$$\Delta_{K,G,m^A} = [K_{m^A}(\sqrt[m^A]{G}) : K]^{-1} \cdot \prod_{i=1}^{f} \left(1 - \frac{1}{[K_{m^{S_i A}}(\sqrt[m^A]{G}) : K_{m^A}(\sqrt[m^A]{G})]}\right).$$

We are then just left to prove

$$[K_{m^A}(\sqrt[m^A]{G}) : K] = \prod_{i=1}^{f}[K_{\ell_i^{a_i}}(\sqrt[\ell_i^{a_i}]{G}) : K] \tag{12}$$

and

$$[K_{m s_i A}(\sqrt[m^A]{G}) : K_{m^A}(\sqrt[m^A]{G})] = [K_{\ell_i^{a_i+1}}(\sqrt[\ell_i^{a_i}]{G}) : K_{\ell_i^{a_i}}(\sqrt[\ell_i^{a_i}]{G})]\,. \tag{13}$$

We always have

$$[K_{m^A}(\sqrt[m^A]{G}) : K_{m^A}] = \prod_{i=1}^{f}[K_{m^A}(\sqrt[\ell_i^{a_i}]{G}) : K_{m^A}]$$

and $[K_{m^A} : K] = \prod_{i=1}^{f}[K_{\ell_i^{a_i}} : K]$, so (12) reduces to the equality

$$[K_{m^A}(\sqrt[\ell^{a_i}]{G}) : K_{m^A}] = [K_{\ell_i^{a_i}}(\sqrt[\ell_i^{a_i}]{G}) : K_{\ell_i^{a_i}}]\,.$$

Since the extension $K_{m^A}/K_{\ell_i^{a_i}\cdot\frac{m}{\ell_i}}$ has degree coprime to ℓ_i, we are done by condition (i).

For better readability, we write $L := K_{\ell_i^{a_i}}(\sqrt[\ell_i^{a_i}]{G})$, and we call B the tuple obtained from A by removing a_i.

If $a_i = 0$ then $L = K$ and (13) is equivalent to knowing for every B:

$$[K_{\ell_i w_i^B}(\sqrt[w_i^B]{G}) : K_{w_i^B}(\sqrt[w_i^B]{G})] = [K_{\ell_i} : K]\,.$$

This exactly means that the extensions K_{ℓ_i} and $K_{w_i^B}(\sqrt[w_i^B]{G})$ are linearly disjoint over K. We may suppose that all entries of B are equal and write $w_i^B = w_i^b$ for some integer b. Since ℓ_i and w_i^b are coprime, it suffices that the extensions $K_{\ell_i w_i^b}$ and $K_{w_i^b}(\sqrt[w_i^b]{G})$ are linearly disjoint over $K_{w_i^b}$. This is ensured by condition (ii).

Now suppose $a_i > 0$. The right-hand side of (13) is a power of ℓ and hence we are left to show

$$[L_{\ell_i^{a_i+1} w^B} : L_{w^B}] = [L_{\ell_i^{a_i+1}} : L]\,,$$

which is true because w^B is coprime to ℓ_i. □

Theorem 11 *Let ℓ vary over the prime divisor of m. We have*

$$D_{K,G,m} = \prod_{\ell} D_{K,G,\ell} \tag{14}$$

if for every $n \geqslant 1$ the extensions $K_{\ell^n \cdot \frac{m}{\ell}}$ and $K_{\ell^n}(\sqrt[\ell^n]{G})$ are linearly disjoint over K_{ℓ^n}.

Proof It suffices to prove that in Lemma 10 Condition (ii) is implied by Condition (i). If Condition (ii) does not hold, then there is some prime divisor ℓ of m such that the extensions $K_{\ell \cdot w^n}$ and $K_{w^n}(\sqrt[w^n]{G})$ are not linearly disjoint over K_{w^n}. Then there must be a prime divisor q of w such that the same holds for $K_{\ell \cdot w^n}$ and $K_{w^n}(\sqrt[q^n]{G})$. Consequently, the extensions $K_{\ell \cdot q^n}$ and $K_{q^n}(\sqrt[q^n]{G})$ are not linearly disjoint over K_{q^n}. In particular $K_{\frac{m}{q} \cdot q^n}$ and $K_{q^n}(\sqrt[q^n]{G})$ are not linearly disjoint over K_{q^n}, which contradicts Condition (i) for q. □

Corollary 12 *If $K_m = K$ then the product formula of Theorem 11 holds.*

Proof The assumption of Theorem 11 holds because we have $K_{\frac{m}{\ell}} = K$. □

Proposition 13 *Suppose that m is odd and that $K_\ell \neq K$ holds for all prime factors ℓ of m. Then the product formula of Theorem 11 holds.*

Proof We prove that the condition of Theorem 11 holds. Write $F := K_{\ell^n}$ and $w := \frac{m}{\ell}$. Suppose that there is some prime divisor ℓ of m such that the extensions F_w and $F(\sqrt[\ell^n]{G})$ are not linearly disjoint over F. Clearly we have $n > 0$. The first extension is cyclic, so there is some $g \in G$ such that F_w and $F(\sqrt[\ell^n]{g})$ are not linearly disjoint over F. There is some maximal $d < n$ such that there is some $a \in K^\times$ satisfying $a^{\ell^d} = g$. For any choice of $\sqrt[\ell]{a}$, the field $K(\sqrt[\ell]{a})$ is different from K, but it is contained in F_w. Then, a is strongly ℓ-indivisible because $K_\ell \neq K$. By Theorem 4 the identity $K_{\ell^n w}(\sqrt[\ell]{a}) = K_{\ell^n w}$ implies $K_\ell = K$, contradicting the assumption in the statement. □

6 Formulas to Reduce to Known Cases

We develop a strategy to reduce the computation of a general density $D_{K,G,m}$ to the computation of finitely many densities that concern only one prime divisor of m.

Let K be a number field, let G be a finitely generated and torsion-free subgroup of $K^\times$, and let $m \geqslant 2$ be a square-free integer. We also use the notation introduced in the beginning of Sect. 4. We want to reduce the calculation of

$$D_{K,G,m} := \text{dens}\,\{\mathfrak{p} : \text{ord}(G \bmod \mathfrak{p}) \text{ is coprime to } m\}$$

to the case where m is a prime number. We can accomplish this in finitely many steps up to increasing the base field, as the following results show. Since all densities are related to G, we have removed G from the notation for better readability.

Theorem 14 *If m is composite and ℓ is a prime factor of m, we have*

$$D_{K,m} = D_{K,\frac{m}{\ell}} + [K_\ell : K]^{-1} \cdot (D_{K_\ell,m} - D_{K_\ell,\frac{m}{\ell}})\,. \tag{15}$$

Proof We use the notation of Theorem 9. We suppose w.l.o.g. $\ell = \ell_f$ and we write for convenience $L := K_\ell$ and $n := \frac{m}{\ell}$. By Theorem 9 we have

$$D_{K,m} = \sum_A \Delta_{K,m^A} = \sum_{A:a_f=0} \Delta_{K,m^A} + \sum_{A:a_f>0} \Delta_{K,m^A}\,.$$

If $a_f > 0$, we are considering primes that split completely in L, and we can apply Proposition 1. Moreover, we have $\Delta_{L,m^A} = 0$ if $a_f = 0$. So we get

$$\sum_{A:a_f>0} \Delta_{K,m^A} = [L : K]^{-1} \cdot \sum_{A:a_f>0} \Delta_{L,m^A} = \frac{D_{L,m}}{[L : K]}\,.$$

Write $A = (B, a_f)$ where $B = (a_1, \ldots, a_{f-1})$, and call Γ_B the set of primes $\mathfrak{p}$ of K that split completely in $K_{n^B}(\sqrt[n^B]{G})$ and for every $i = 1, \ldots, f-1$ do not split completely in $K_{n^{S_i B}}$. The set Γ_B has natural density Δ_{K,n^B}. The primes in Γ_B which split completely in L have density $[L : K]^{-1} \cdot \Delta_{L,n^B}$ by Proposition 1. The primes in Γ_B which do not split completely in L have density $\Delta_{K,m^{(B,0)}}$, because having ℓ^0-th roots is an empty condition. So we get

$$D_{K,n} = \sum_B \Delta_{K,n^B} = \sum_B \left(\frac{\Delta_{L,n^B}}{[L : K]} + \Delta_{K,m^{(B,0)}}\right) = \frac{D_{L,n}}{[L : K]} + \sum_{a_f=0} \Delta_{K,m^A}$$

and we may easily recover the formula in the statement. □

Theorem 15 *Let $m = \ell_1 \cdots \ell_f$ be a product of distinct prime numbers. If $f \geqslant 2$ we have*

$$D_{K,m} = \sum_{i=1}^{f} \left(\varepsilon_{i-1} \cdot D_{L_{i-1},\frac{m}{\ell_i}} - \varepsilon_i \cdot D_{L_i,\frac{m}{\ell_i}}\right) + \varepsilon_f \cdot \prod_{i=1}^{f} D_{L_f,\ell_i} \tag{16}$$

where the notations are as follows: we write $\varepsilon_0 := 1$ and $L_0 := K$, and for $i = 1, \ldots, f$, we set

$$\varepsilon_i := \prod_{1 \leqslant j \leqslant i} [K_{\ell_j} : K]^{-1} \qquad \text{and} \qquad L_i := \prod_{1 \leqslant j \leqslant i} K_{\ell_j}\,. \tag{17}$$

Proof By Corollary 12 we have $D_{L_f,m} = \prod_{i=1}^{f} D_{L_f,\ell_i}$. We then prove that for $1 \leqslant n \leqslant f$ we have

$$D_{K,m} = \sum_{i=1}^{n} \left(\varepsilon_{i-1} \cdot D_{L_{i-1},\frac{m}{\ell_i}} - \varepsilon_i \cdot D_{L_i,\frac{m}{\ell_i}} \right) + \varepsilon_n \cdot D_{L_n,m} \,.$$

We prove this formula by induction on n. The case $n = 1$ can be obtained by applying Theorem 14 to ℓ_1:

$$D_{K,m} = D_{K,\frac{m}{\ell_1}} + \varepsilon_1 \cdot D_{L_1,m} - \varepsilon_1 \cdot D_{L_1,\frac{m}{\ell_1}} = \left(\varepsilon_0 \cdot D_{L_0,\frac{m}{\ell_1}} - \varepsilon_1 \cdot D_{L_1,\frac{m}{\ell_1}} \right) + \varepsilon_1 \cdot D_{L_1,m} \,.$$

Now suppose that we know the inductive assumption

$$D_{K,m} = \sum_{i=1}^{n-1} \left(\varepsilon_{i-1} \cdot D_{L_{i-1},\frac{m}{\ell_i}} - \varepsilon_i \cdot D_{L_i,\frac{m}{\ell_i}} \right) + \varepsilon_{n-1} \cdot D_{L_{n-1},m} \,.$$

We can achieve the induction step by applying Theorem 14 to ℓ_n, which gives

$$\varepsilon_{n-1} \cdot D_{L_{n-1},m} = \varepsilon_{n-1} \cdot D_{L_{n-1},\frac{m}{\ell_n}} + \varepsilon_n \cdot D_{L_n,m} - \varepsilon_n \cdot D_{L_n,\frac{m}{\ell_n}} \,.$$

□

By the above result, we can reduce to computing densities which involve one prime factor less. With finitely many applications of this result, we have reduced to the case of exactly one prime number, and we can make use of the formulas of [1, 4].

7 Examples

Example 16 *Let* $K = \mathbb{Q}$ *and* $m = 6$. *By Corollary 12, we know* $D_{\mathbb{Q}(\zeta_3),6} = D_{\mathbb{Q}(\zeta_3),2} \cdot D_{\mathbb{Q}(\zeta_3),3}$, *and by Theorem 14 applied to* $\ell = 3$, *we have*

$$D_{\mathbb{Q},6} = D_{\mathbb{Q},2} + [\mathbb{Q}(\zeta_3) : \mathbb{Q}]^{-1} \cdot (D_{\mathbb{Q}(\zeta_3),6} - D_{\mathbb{Q}(\zeta_3),2}) \,. \tag{18}$$

We evaluate the right-hand side of this expression (with [4, Theorems 16 and 17] for rank 1 *and [1, Theorems 3 and 4] otherwise) in some explicit examples, which are listed in the Table 1. We also compute* $D_{\mathbb{Q},3}$ *so that one can easily see that the formula* $D_{\mathbb{Q},6} = D_{\mathbb{Q},2} D_{\mathbb{Q},3}$ *holds for the examples in the upper part of the table (and in general it does not hold). Notice that the field* $\mathbb{Q}(\zeta_3)$ *contains* $\sqrt{-3}$.

Example 17 *Let* $K = \mathbb{Q}$ *and* $m = 15$. *If we apply Theorem 14 for* $\ell = 3$, *we have*

$$D_{\mathbb{Q},15} = D_{\mathbb{Q},5} + [\mathbb{Q}(\zeta_3) : \mathbb{Q}]^{-1} \cdot (D_{\mathbb{Q}(\zeta_3),15} - D_{\mathbb{Q}(\zeta_3),5})$$

Table 1 All examples have been tested with Sage [6]

G	$D_{\mathbb{Q},6}$	$D_{\mathbb{Q},2}$	$D_{\mathbb{Q},3}$	$D_{\mathbb{Q}(\zeta_3),2}$	$D_{\mathbb{Q}(\zeta_3),3}$
$\langle 2\rangle$	35/192	7/24	5/8	7/24	1/4
$\langle 5\rangle$	5/24	1/3	5/8	1/3	1/4
$\langle 2, 5\rangle$	29/416	29/224	7/13	29/224	1/13
$\langle -3\rangle$	1/12	1/3	5/8	2/3	1/4
$\langle 3\rangle$	13/48	1/3	5/8	1/6	1/4
$\langle 9\rangle$	17/48	2/3	5/8	5/6	1/4
$\langle -9\rangle$	13/96	1/6	5/8	1/12	1/4
$\langle -27\rangle$	1/4	1/3	7/8	2/3	3/4
$\langle 27\rangle$	5/16	1/3	7/8	1/6	3/4
$\langle 2, 3\rangle$	365/2912	29/224	7/13	1/112	1/13
$\langle 2, -3\rangle$	29/2912	29/224	7/13	29/112	1/13

and by Theorem 14 for $\ell = 5$, we can write

$$D_{\mathbb{Q}(\zeta_3),15} = D_{\mathbb{Q}(\zeta_3),3} + [\mathbb{Q}(\zeta_{15}) : \mathbb{Q}(\zeta_3)]^{-1} \cdot (D_{\mathbb{Q}(\zeta_{15}),15} - D_{\mathbb{Q}(\zeta_{15}),3})$$

so by recalling $D_{\mathbb{Q}(\zeta_{15}),15} = D_{\mathbb{Q}(\zeta_{15}),3} \cdot D_{\mathbb{Q}(\zeta_{15}),5}$, we get

$$D_{\mathbb{Q},15} = D_{\mathbb{Q},5} + \frac{1}{2} \cdot (D_{\mathbb{Q}(\zeta_3),3} - D_{\mathbb{Q}(\zeta_3),5}) - \frac{1}{8} D_{\mathbb{Q}(\zeta_{15}),3} + \frac{1}{8} \cdot D_{\mathbb{Q}(\zeta_{15}),3} \cdot D_{\mathbb{Q}(\zeta_{15}),5}\,.$$

If we apply Theorem 14 for $\ell = 5$ and then expand $D_{\mathbb{Q}(\zeta_5),15}$ (by Theorem 14 for $\ell = 3$), we similarly get

$$D_{\mathbb{Q},15} = D_{\mathbb{Q},3} + \frac{1}{4} (D_{\mathbb{Q}(\zeta_5),5} - D_{\mathbb{Q}(\zeta_5),3}) - \frac{1}{8} D_{\mathbb{Q}(\zeta_{15}),5} + \frac{1}{8} \cdot D_{\mathbb{Q}(\zeta_{15}),3} \cdot D_{\mathbb{Q}(\zeta_{15}),5}\,.$$

Both methods give of course the same value for $D_{\mathbb{Q},15}$ (tested with Sage [6] for the following examples):

G	$D_{\mathbb{Q},15}$	$D_{\mathbb{Q},3}$	$D_{\mathbb{Q},5}$	$D_{\mathbb{Q}_3,3}$	$D_{\mathbb{Q}_3,5}$	$D_{\mathbb{Q}_5,3}$	$D_{\mathbb{Q}_5,5}$	$D_{\mathbb{Q}_{15},3}$	$D_{\mathbb{Q}_{15},5}$
$\langle 2\rangle$	95/192	5/8	19/24	1/4	19/24	5/8	1/6	1/4	1/6
$\langle 2^9\rangle$	437/576	23/24	19/24	11/12	19/24	23/24	1/6	11/12	1/6
$\langle 2^{15}\rangle$	161/192	7/8	23/24	3/4	23/24	7/8	5/6	3/4	5/6

Example 18 *This example is in particular a counterexample to* (1) *with m odd. Let $K = \mathbb{Q}(\zeta_3)$ and $m = 21$. If C denotes the cyclic subextension of $\mathbb{Q}(\zeta_7)$ of degree 3, then $C(\zeta_3)$ is a Kummer extension of $\mathbb{Q}(\zeta_3)$, and we can write it as $\mathbb{Q}(\zeta_3, \sqrt[3]{g})$ for some $g \in \mathbb{Q}(\zeta_3)^{\times}$. We claim that for the group $G = \langle g\rangle$, we have*

$$D_{\mathbb{Q}(\zeta_3),21} \neq D_{\mathbb{Q}(\zeta_3),3} \cdot D_{\mathbb{Q}(\zeta_3),7}\,.$$

We may suppose that g is strongly 7-indivisible in $\mathbb{Q}(\zeta_3)$ and hence also in $\mathbb{Q}(\zeta_{21})$. We know that it is strongly 3-indivisible in $\mathbb{Q}(\zeta_3)$. We deduce that $\sqrt[3]{g}$ is strongly 3-indivisible in $\mathbb{Q}(\zeta_{21})$.

We have $D_{\mathbb{Q}(\zeta_3),3} = 1/4$ and $D_{\mathbb{Q}(\zeta_3),7} = 41/48$. By Theorem 14 applied to $\ell = 7$, we get

$$D_{\mathbb{Q}(\zeta_3),21} = D_{\mathbb{Q}(\zeta_3),3} + [\mathbb{Q}(\zeta_{21}) : \mathbb{Q}(\zeta_3)]^{-1} \cdot (D_{\mathbb{Q}(\zeta_{21}),21} - D_{\mathbb{Q}(\zeta_{21}),3}) .$$

We have $D_{\mathbb{Q}(\zeta_{21}),3} = 3/4$ and $D_{\mathbb{Q}(\zeta_{21}),7} = 1/8$ and hence $D_{\mathbb{Q}(\zeta_{21}),21} = 3/32$. We deduce $D_{\mathbb{Q}(\zeta_3),21} = 9/64$ and the claim follows.

Acknowledgements The author would like to thank Christophe Debry, Franziska Schneider, and Franziska Wutz for their support and the referee for their careful reading of the paper. The project originated as a mentor/mentee collaboration in the WIN style (the mentees left academia).

References

1. C. Debry, A. Perucca, Reductions of algebraic integers. J. Number Theory **167**, 259–283 (2016)
2. M. Hindry, J. Silverman, *Diophantine Geometry. An Introduction*. Graduate Texts in Mathematics, vol. 201 (Springer, New York, 2000), xiv+558 pp.
3. H.W. Lenstra Jr., *Commentary on H: Divisibility and Congruences*. Andrzej Schinzel Selecta, vol. II (European Mathematical Society, Zürich, 2007), pp. 901–902
4. A. Perucca, The order of the reductions of an algebraic integer. J. Number Theory **148**, 121–136 (2015)
5. A. Schinzel, Abelian binomials, power residues and exponential congruences. Acta Arith. **32**(3), 245–274 (1977). Addendum, ibid. **36**, 101–104 (1980). See also Andrzej Schinzel Selecta Vol. II (European Mathematical Society, Zürich, 2007), pp. 939–970
6. W.A. Stein et al., Sage Mathematics Software (Version 5.7). The Sage Development Team, 2013. http://www.sagemath.org
7. J. Wójcik, Criterion for a field to be abelian. Colloq. Math. **68**(2), 187–191 (1995)

Reductions of One-Dimensional Tori II

Antonella Perucca

Abstract Consider a one-dimensional torus defined over a number field, and fix a finitely generated group of rational points. How often is the size of the reduction of this group coprime to some given (square-free) integer? In this short note, we prove a formula that allows us to reduce to the case of a prime number.

Keywords Number field · Reduction · Order · Density · One-dimensional torus

2010 Mathematics Subject Classification. Primary 11R44, 11R20, 11Y40

Consider a one-dimensional torus defined over a number field K, and fix a finitely generated subgroup G of K-rational points (which we assume w.l.o.g. to be torsion-free and non-trivial). Up to excluding finitely many primes $\mathfrak{p}$ of K, we suppose that the reduction of G modulo $\mathfrak{p}$ is well-defined. Since the group ($G \bmod \mathfrak{p}$) is finite, we can ask whether its size is coprime to some given square-free integer m. We thus aim at understanding the natural density

$$D_m(K) := \mathrm{dens}\{\mathfrak{p} : \#(G \bmod \mathfrak{p}) \text{ is coprime to } m\}.$$

The case in which m is a prime number was treated in [3], and the same problem for split tori had been solved in [2] with a different approach. The result contained in this short note presents a very general closed formula. The existence of such an elegant formula, which combines the densities with respect to the different prime divisors of m, was quite surprising because the corresponding number fields are intertwined. Moreover, the splitting field of the torus can be contained in the involved torsion and Kummer extensions, but apparently no case distinction was needed.

A. Perucca (✉)
University of Luxembourg, Mathematics Research Unit, Esch-sur-Alzette, Luxembourg
e-mail: antonella.perucca@uni.lu

I. I. Bouw et al. (eds.), *Women in Numbers Europe II*, Association for Women in Mathematics Series 11, https://doi.org/10.1007/978-3-319-74998-3_3

Theorem 1 *If $m = \ell_1 \cdots \ell_f$ is the product of distinct prime numbers, we have*

$$D_m(K) = \sum_{\substack{A,B \subseteq \{1,\cdots,f\} \\ A \cap B = \emptyset}} (-1)^{\#B} \cdot [K_{A\cup B} : K]^{-1} \cdot \prod_{a \in A} D_{\ell_a}(K_{A\cup B})$$

where $K_{A\cup B}$ is the $(\prod_{i\in A\cup B} \ell_i)$-th torsion field of the torus.

By the results of [3], we may then compute $D_m(K)$ for all one-dimensional tori.

If $n \geqslant 1$, we write K_n for the n-th torsion field of the torus, which is the smallest extension of K over which all points of the torus having order n are defined. We also write $K(n^{-1}G)$ for the n-th division field of G, which is the smallest extension of K over which all n-th division points of G (namely, all points of the torus x such that $nx \in G$) are defined. We write $N = (n_1, \ldots, n_f)$ for an f-tuple of nonnegative integers, and we set $m^N := \ell_1^{n_1} \cdots \ell_f^{n_f}$. If $A \subseteq \{1, \ldots, f\}$, we write $m_A := \prod_{a\in A} \ell_a$ and $K_A := K_{m_A}$.

Lemma 2 *Consider the set $\Gamma_{K,m} := \{\mathfrak{p} : \#(G \bmod \mathfrak{p})$ is coprime to $m\}$.*

1. *The natural density $D_m(K)$ of $\Gamma_{K,m}$ is well-defined.*
2. *The set $S_{m,N}$ consisting of the primes of K that split completely in $K(m^{-N}G)$ and that for each prime divisor ℓ of m do not split completely in $K_{\ell m^N}$ has a natural density. We have $\Gamma_{K,m} = \cup_N S_{m,N}$ and $D_m(K) = \sum_N \mathrm{dens}(S_{m,N})$.*
3. *Let $P_{K,A}$ be the set of primes of K that split completely in K_A and that for each prime divisor ℓ of $\frac{m}{m_A}$ do not split completely in K_ℓ. Then $\Gamma_{K,m} \cap P_{K,A}$ has a natural density, and we have $D_m(K) = \sum_A \mathrm{dens}(\Gamma_{K,m} \cap P_{K,A})$.*
4. *If $K = K_m$, then considering all prime divisors ℓ of m, we have*

$$D_m(K) = \prod_\ell D_\ell(K).$$

5. *If ℓ is a prime factor of m, we have*

$$D_{\frac{m}{\ell}}(K) - D_m(K) = [K_\ell : K]^{-1} \cdot \left(D_{\frac{m}{\ell}}(K_\ell) - D_m(K_\ell)\right).$$

Proof The first three assertions can be proven as for split tori; see [2, Theorem 9]. We obtain (4) by applying (2) to m and to each of its prime divisors:

$$\begin{aligned} D_m(K) &= \sum_N \mathrm{dens}(S_{m,N}) = \sum_N \prod_i \mathrm{dens}(S_{\ell_i,n_i}) = \prod_i \sum_{n_i \geqslant 0} \mathrm{dens}(S_{\ell_i,n_i}) \\ &= \prod_i D_{\ell_i}(K). \end{aligned}$$

Note, the second equality holds because $S_{m,N}$ consists of the primes that for every i split completely in $K(\ell_i^{-n_i}G)$ but not in $K_{\ell_i^{n_i+1}}$ (by Perucca [3, Section 2] the degree of these fields is a power of ℓ_i).

In (5) we count the primes of K for which the reduction of G has order coprime to $\frac{m}{\ell}$ and divisible by ℓ: these primes split completely in K_ℓ, and we may apply [2, Proposition 1]. □

Proof (Proof of the Theorem) By (4) we have $\prod_{a\in A} D_{\ell_a}(K_{A\cup B}) = D_{m_A}(K_{A\cup B})$, so by (3), it suffices to prove

$$\mathrm{dens}(\Gamma_{K,m} \cap P_{K,A}) = \sum_{B\subseteq\{1,\cdots,f\}\setminus A} (-1)^{\#B} \cdot [K_{A\cup B} : K]^{-1} \cdot D_{m_A}(K_{A\cup B})\,.$$

We clearly have $\Gamma_{K,m} \cap P_{K,A} = \Gamma_{K,m_A} \cap P_{K,A}$. These primes split completely in $F := K_A$, so by Perucca [2, Proposition 1], we are left to prove

$$\mathrm{dens}(\Gamma_{F,m_A} \cap P_{F,A}) = \sum_{B\subseteq\{1,\cdots,f\}\setminus A} (-1)^{\#B} \cdot [F_B : F]^{-1} \cdot D_{m_A}(F_B)\,.$$

The primes of F that split completely in F_B contribute to $D_{m_A}(F)$ with density

$$[F_B : F]^{-1} D_{m_A}(F_B)$$

by Perucca [2, Proposition 1]. The above formula can then be obtained with the inclusion-exclusion principle with respect to the finitely many conditions defining $P_{F,A}$: we are restricting to the primes of F which for every prime divisor ℓ of $\frac{m}{m_A}$ do not split completely in F_ℓ.

The formula of the theorem can also be obtained by repeatedly applying (5).

Corollary 3 *If F is a Galois extension of K which is linearly disjoint from the division field $K(m^{-\infty}G)$, then we have $D_m(F) = D_m(K)$.*

Proof We may reduce to the case where m is prime and hence apply [1, Proposition 14]. □

References

1. D. Lombardo, A. Perucca, Reductions of points on algebraic groups, arXiv:1612.02847
2. A. Perucca, Reductions of algebraic integers II, in *Women in Numbers Europe II: Contributions to Number Theory and Arithmetic Geometry*, ed. by I. I. Bouw, E. Ozman, J. Johnson-Leung, R. Newton. Association for Women in Mathematics Series, vol. 11, 1st edn. (Springer, New York, 2018). https://doi.org/10.1007/978-3-319-74998-3_2
3. A. Perucca, Reductions of one-dimensional tori. Int. J. Number Theory **13**(6), 1473–1489 (2017)

On the Carlitz Rank of Permutation Polynomials Over Finite Fields: Recent Developments

Nurdagül Anbar, Almasa Odžak, Vandita Patel, Luciane Quoos, Anna Somoza, and Alev Topuzoğlu

Abstract The Carlitz rank of a permutation polynomial over a finite field $\mathbb{F}_q$ is a simple concept that was introduced in the last decade. In this survey article, we present various interesting results obtained by the use of this notion in the last few years. We emphasize the recent work of the authors on the permutation behavior of polynomials $f + g$, where f is a permutation over $\mathbb{F}_q$ of a given Carlitz rank, and $g \in \mathbb{F}_q[x]$ is of prescribed degree. The relation of this problem to the well-known Chowla–Zassenhaus conjecture is described. We also present some initial observations on the iterations of a permutation polynomial $f \in \mathbb{F}_q[x]$ and hence on the order of f as an element of the symmetric group S_q.

N. Anbar
Johann Radon Institute for Computational and Applied Mathematics, Austrian Academy of Sciences, Linz, Austria

A. Odžak
University of Sarajevo, Sarajevo, Bosnia and Herzegovina
e-mail: almasa.odzak@pmf.unsa.ba

V. Patel
University of Warwick, Coventry, UK
e-mail: vandita.patel@warwick.ac.uk

L. Quoos
Universidade Federal do Rio de Janeiro, Cidade Universitária, Rio de Janeiro, Brazil
e-mail: luciane@im.ufrj.br

A. Somoza
Universitat Politècnica de Catalunya, Barcelona, Spain
Leiden University, Leiden, RA, Netherlands
e-mail: anna.somoza@upc.edu

A. Topuzoğlu (✉)
Sabancı University, MDBF, Orhanlı, Tuzla, Istanbul, Turkey
e-mail: alev@sabanciuniv.edu

I. I. Bouw et al. (eds.), *Women in Numbers Europe II*, Association for Women in Mathematics Series 11, https://doi.org/10.1007/978-3-319-74998-3_4

Keywords Carlitz rank · Permutation polynomials · Finite fields

2010 Mathematics Subject Classification. 11T06, 14H05

1 Introduction

Let $\mathbb{F}_q$ be the finite field with $q = p^r$ elements, where p is a prime and $r \geq 1$. For simplicity, we assume p is odd, although most results below hold for even q also. We recall that any map from $\mathbb{F}_q$ to itself can be represented uniquely by a polynomial $f \in \mathbb{F}_q[x]$ of degree less than q. A polynomial f is called a *permutation* polynomial if it induces a bijection from $\mathbb{F}_q$ to $\mathbb{F}_q$.

Permutation polynomials over finite fields have been studied widely in the last decades, especially due to their applications in combinatorics, coding theory, and symmetric cryptography. In order to meet the specific requirements of individual applications, methods to construct various types of permutations and/or alternative ways of classifying them are needed. Although the work on permutation polynomials goes back to the nineteenth century, they are still of theoretical interest today, with the research area posing many open problems. We refer to [27, 29, 31, 38] for a detailed exposition of permutation polynomials over finite fields.

We recall that S_q, the symmetric group on q letters, is isomorphic to the group of permutation polynomials over $\mathbb{F}_q$ of degree less than q, under the operation of composition and subsequent reduction modulo $x^q - x$, hence we identify them. A well-known result of Carlitz [7] states that S_q is generated by linear polynomials $ax + b$, $a, b \in \mathbb{F}_q$, $a \neq 0$, and x^{q-2}. Hence any permutation f over $\mathbb{F}_q$ can be represented by a polynomial of the form

$$P_n(x) = \left(\dots \left((a_0 x + a_1)^{q-2} + a_2 \right)^{q-2} \dots + a_n \right)^{q-2} + a_{n+1}, \tag{1}$$

for some $n \geq 0$, where $a_i \neq 0$ for $i = 0, 2, \dots, n$. Note that $f(c) = P_n(c)$ holds for all $c \in \mathbb{F}_q$; however, this representation is not unique, and n is not necessarily minimal. Accordingly, the authors of [3] define the *Carlitz rank* of a permutation polynomial f over $\mathbb{F}_q$ to be the smallest integer $n \geq 0$ satisfying $f = P_n$ for a permutation P_n of the form (1) and denote it by $\mathrm{Crk}(f)$.

Some of the natural questions concerning the concept of Carlitz rank were addressed in [3, 8, 9] and [15]. In Sect. 2, we introduce the notation and some of the basic tools used in these studies. We also describe the relation of $\mathrm{Crk}(f)$ to other polynomial invariants of f and to the cycle structure of the permutation induced on $\mathbb{F}_q$ by f. In Sect. 3, we focus on the permutation behavior of polynomials $f + g$, where f is a permutation over $\mathbb{F}_q$ of a given Carlitz rank and $g \in \mathbb{F}_q[x]$ is of prescribed degree. The case $g(x) = x$ is of particular interest. A polynomial f is called a *complete mapping polynomial* whenever f and $f + x$

are both permutation polynomials. Complete mappings have been widely studied due to their various applications, for example, in [26, 30, 32, 36, 37] and [40]. In Theorems 1 and 2, we assume f and $f+g$ to be permutations over $\mathbb{F}_q$ and give lower bounds for the Carlitz rank of f in terms of q and the degree of g. The bounds established in Theorems 1 and 2 are analogous to a well-known result of Cohen, Mullen, and Shiue, where they obtain a lower bound for $\deg(g)$ in terms of $\deg(f) = \deg(f+g) = d$ when the cardinality of the field is sufficiently large with respect to d [13, Theorem 1]. The bound in Theorem 1 also generalizes the main result of [24] on the nonexistence of complete mappings. Finally, Sect. 4 focuses on iterations of permutation polynomials over finite fields of odd characteristic, which yield results on the order of a permutation polynomial over $\mathbb{F}_q$ as an element of S_q.

2 The Carlitz Rank: Basic Properties

We start this section by describing some of the key questions concerning permutation polynomials over finite fields. One line of research is on the construction of new classes of permutation polynomials, possibly with additional properties. We refer the interested reader to [21] for a survey on various recent methods. A challenging problem is to develop criteria for checking the permutation behavior of a given polynomial over $\mathbb{F}_q$; see [1, 2, 21, 25] and the references therein for some new criteria. One may be interested, for example, in having sparse permutation polynomials. It is very easy to check if a monomial is a permutation: x^k is a permutation polynomial of $\mathbb{F}_q$ if and only if $\gcd(k, q-1) = 1$. In contrast, supplying theorems (even in certain special cases) to decide whether a binomial or a trinomial is a permutation is not at all a trivial task; see, for instance, [18–20]. To determine the cycle structure of the induced permutation is a difficult open problem. Indeed, the cycle structure is known only for a few special classes of permutation polynomials.

Classifying permutations over $\mathbb{F}_q$ with respect to their Carlitz rank has various advantages which we illustrate below. An effective tool that is used often is the *approximation* of a permutation polynomial. The approximation is given by a rational linear transformation derived from the representation as stated in (1). Recall that $x^{q-2} = x^{-1}$ for $x \neq 0$ and $x^{q-2} = 0$ if $x = 0$. Hence the representation in (1) can be expressed as a continued fraction for suitable $x \in \mathbb{F}_q$. We consider the function

$$(\ldots((a_0x + a_1)^{-1} + a_2)^{-1} \ldots + a_n)^{-1} + a_{n+1}$$

and its continued fraction expansion,

$$a_{n+1} + 1/(a_n + 1/(\ldots/a_2 + 1/(a_0x + a_1)\ldots)).$$

Letting $R_0 = a_0x + a_1$ and $R_k(x) = a_{k+1} + \frac{1}{R_{k-1}(x)}$ for $k \geq 1$, we obtain the nth *convergent*

$$R_n(x) = \frac{\alpha_{n+1}x + \beta_{n+1}}{\alpha_n x + \beta_n}, \tag{2}$$

where

$$\alpha_k = a_k\alpha_{k-1} + \alpha_{k-2} \quad \text{and} \quad \beta_k = a_k\beta_{k-1} + \beta_{k-2}, \tag{3}$$

for $k \geq 2$ and $\alpha_0 = 0, \alpha_1 = a_0, \beta_0 = 1, \beta_1 = a_1$.

The set of *poles* is denoted by $\mathcal{O}_n$, i.e.,

$$\mathcal{O}_n = \left\{ x_i = \frac{-\beta_i}{\alpha_i} : i = 1, \ldots, n \right\} \subset \mathbb{P}^1(\mathbb{F}_q) = \mathbb{F}_q \cup \{\infty\}. \tag{4}$$

The elements of $\mathcal{O}_n$ may not be distinct. When $a_{n+1} = 0$ in (1), R_n takes the form

$$R_n(x) = \frac{\alpha_{n-1}x + \beta_{n-1}}{\alpha_n x + \beta_n}. \tag{5}$$

It can easily be verified that

$$f(c) = P_n(c) = R_n(c) \text{ for all } c \in \mathbb{F}_q \setminus \mathcal{O}_n. \tag{6}$$

To every rational transformation $R_n(x)$ of the form (2), we can associate a permutation $F_n(x)$ of $\mathbb{F}_q$ defined by $F_n(x) = R_n(x)$ for $x \neq x_n$ and $F_n(x_n) = \alpha_{n+1}/\alpha_n$ when $x_n \in \mathcal{O}_n$.

We note that one could take an alternative view and think of $R_n \in \mathrm{PGL}(2, q)$ as a permutation of the projective line, hence an element of S_{q+1}. One could then relate the Carlitz rank to the Hamming distance between the elements of S_{q+1} in the subgroups S_q and $\mathrm{PGL}(2, q)$, i.e., distance between permutation polynomials in S_q and rational linear transformations in $\mathrm{PGL}(2, q)$.

Given a permutation polynomial f, and a representation P_n of f as in (1), one can decompose $P_n(x)$ as

$$P_n(x) = \tau_1 \circ \ldots \circ \tau_m \circ F_n(x) = \tau_1 \ldots \tau_m F_n(x), \tag{7}$$

where $\tau_1, \ldots, \tau_m$ are disjoint cycles in S_q; see [3]. In particular, if $\mathcal{O}_n \subset \mathbb{F}_q$ and the elements of $\mathcal{O}_n$ are distinct, then

$$P_n(x) = (F_n(x_{n-1}) \ \cdots \ F_n(x_1) \ F_n(x_n))F_n(x)$$

(see Lemma 1 in [9]). When $\mathcal{O}_n$ is an arbitrary subset of $\mathbb{F}_q \cup \{\infty\}$, a similar relation can be obtained; see [9].

Note also that, given a representation, P_n, of a permutation f, one can trivially determine the *string* $(x_1, x_2, \ldots, x_n)$ of poles and the associated rational transformation R_n by using the recurrence relation (3). At this point, one may wonder if and how the rational transformations corresponding to different representations are related. We have the following property, taken from [3].

Lemma 1 *Let $P_n^{(1)}$ and $P_m^{(2)}$ be two representations of a permutation of $\mathbb{F}_q$, with associated rational functions $R_n^{(1)}(x)$ and $R_m^{(2)}(x)$, respectively. If $m + n < q - 2$, then $R_n^{(1)}(x) = R_m^{(2)}(x)$.*

Remark 1 Lemma 1 implies that if $\mathrm{Crk}(f) = n < (q-1)/2$ and $P_n^{(1)}$ and $P_n^{(2)}$ are two representations of f, then the corresponding rational fractions are the same; $R_n^{(1)}(x) = R_n^{(2)}(x)$. In particular, α_n and the last pole x_n are uniquely defined when $\mathrm{Crk}(f) = n < (q-1)/2$.

Obviously, the approximation property mentioned above is particularly useful when $\mathrm{Crk}(f)$ is small with respect to the field size. If $\alpha_n = 0$, i.e., the last pole $x_n = \infty$, then $R_n(x) \in \mathbb{F}_q[x]$ and is linear. Following the terminology of [24], we define the *linearity* of $f \in \mathbb{F}_q[x]$ as $\mathscr{L}(f) = \max_{a,b \in \mathbb{F}_q} |\{c \in \mathbb{F}_q : f(c) = ac + b\}|$. Intuitively, $\mathscr{L}(f)$ is large when f is a permutation polynomial over $\mathbb{F}_q$ of $\mathrm{Crk}(f) = n$, R_n is linear, and n is small with respect to q.

It is interesting to note that the polynomials of small $\mathrm{Crk}(f) = n \geq 1$ with linear R_n have large degree and large *weight*, i.e., they have many non-zero coefficients, although they differ from a linear polynomial at most at n elements. Indeed, it was shown in [3] that

$$\deg(f) \geq q - \mathrm{Crk}(f) - 1 \tag{8}$$

when $1 < \deg(f) \leq q - 2$. It also follows from a recent result in [15] that if $1 < \deg(f) \leq q - 2$, then

$$w(f) \geq \frac{q}{\mathrm{Crk}(f)} - 2, \tag{9}$$

where $w(f)$ denotes the number of non-zero coefficients of f.

The inequalities (8) and (9) illustrate that permutation polynomials of small Carlitz rank ≥ 1 have large polynomial complexity. A polynomial invariant that attracted considerable attention recently is the *index* of a polynomial, which we briefly recall here. Let $f \in \mathbb{F}_q[x]$ be a polynomial of degree $n \leq q - 1$ of the form $f(x) = a(x^n + a_{n-i_1}x^{n-i_1} + \cdots + a_{n-i_k}x^{n-i_k}) + b$ with $a, a_{n-i_j} \neq 0$ for $j = 1, \ldots, k$ and $n - i_k > 0$. Set $r := n - i_k$. Then, f is uniquely written as $f(x) = a\big(x^r g(x^{(q-1)/l})\big) + b$ for some monic polynomial $g \in \mathbb{F}_q[x]$ with $l = (q-1)/\gcd(n-r, n-r-i_1, \ldots, n-r-i_{k-1}, q-1)$. The uniquely defined number l is called the index of f; see [1]. The relation between the Carlitz rank and the index of a permutation polynomial was studied recently in [23] where it is shown, for example, that the discrete logarithm map has large index and large Carlitz rank.

The problem of determining the number of permutation polynomials of a given degree is open. However, the number $\mathscr{B}(n)$ of permutation polynomials of a given

Carlitz rank n was obtained in [3], under the condition that $n < (q-1)/2$. The (recursive) formula for $\mathscr{B}(n)$ involves the Stirling numbers of the first kind ([14] and [39, id:A008275]). In order to avoid technical details, we only give the result in the following form. The number $\mathscr{B}(n)$ satisfies $\mathscr{B}(0) = q(q-1)$, $\mathscr{B}(1) = q^2(q-1)$, $\mathscr{B}(2) = q^2(q-1)^2$, and for $3 \leq n < (q-1)/2$,

$$\mathscr{B}(n) = q^2(q-1)^2(q-2)\dots(q-(2n-3)/3)f_n(q), \text{ if } n \equiv 0 \mod 3,$$

$$\mathscr{B}(n) = q^2(q-1)^2(q-2)\dots(q-(2n-2)/3)g_n(q), \text{ if } n \equiv 1 \mod 3,$$

$$\mathscr{B}(n) = q^2(q-1)^2(q-2)\dots(q-(2n-1)/3)h_n(q), \text{ if } n \equiv 2 \mod 3,$$

where f_n, g_n, and h_n are monic polynomials of degree $\lfloor \frac{n}{3} \rfloor$.

In connection with the cycle structure of permutations, one may be interested, for instance, in finding the Carlitz rank of the permutation polynomial which induces a cycle of a given length. The answer to this question, as well as the enumeration result that is mentioned above, follows from the lemma below. Let σ be a permutation in S_q. We denote its support, i.e., the set of elements of $\mathbb{F}_q$ moved by σ, by $\text{supp}(\sigma)$. We consider a decomposition of f as in (7).

Lemma 2 *Suppose f has a representation P_s as in* (1) *and P_s is decomposed as*

$$P_s(x) = \tau_1 \dots \tau_m F_s(x),$$

where $\tau_1, \dots, \tau_m$ are disjoint cycles of length $l(\tau_j) = l_j \geq 2$, $1 \leq j \leq m$. Put $\sigma = \tau_1 \dots \tau_m$. Then there exists a representation $P_n(x)$ of f, with $P_n(c) = P_s(c)$ for all $c \in \mathbb{F}_q$, $n \leq s$, where

(i) $n = m + \sum_{j=1}^{m} l_j - 1$ if F_s is not linear and $F_s(x_s) \in \text{supp}(\sigma)$.
(ii) $n = m + \sum_{j=1}^{m} l_j + 1$ if F_s is not linear and $F_s(x_s) \notin \text{supp}(\sigma)$.
(iii) $n = m + \sum_{j=1}^{m} l_j$ if F_s is linear.

In all three cases, $\text{Crk}(f) = n$ *if $n < (q-1)/2$.*

Therefore, when the cycle structure of f is known and the number of its fixed points is large with respect to the field size, Lemma 2 gives the Carlitz rank of f with minimum effort. Observe that the permutation $\sigma = (123)(56)$ of $\mathbb{F}_{17}$ has Carlitz rank 7 (put $F(x) = x$). Similarly, the Carlitz rank of a cycle of length ℓ is $\ell + 1$ when $\ell + 1 < (q-1)/2$. The cycle structure of permutation polynomials of Carlitz rank ≤ 3 can be found in [9]; see also [41, Section 6].

The classification of permutations with respect to their Carlitz rank has already found interesting applications, see [8, 10, 15, 41] and references therein. An unexpected application concerning the distribution behavior of infinite sequences of real numbers in the interval $[0, 1)$ was obtained recently in [35].

3 On the Difference of Permutation Polynomials

Let f be a permutation polynomial over $\mathbb{F}_q$. We recall from Sect. 1 that f is called a complete mapping polynomial, or a complete mapping, if $f(x) + x$ is also a permutation. Complete mappings were introduced by Mann in 1942, [28], in connection with construction of mutually orthogonal Latin squares. We refer the reader to [33] for a detailed study and to [26, 32, 36, 37, 40] for various applications. Recent work on generalizations of complete mappings can be found in [43]. See [5, 6, 16, 42, 44, 45] for some new classes of complete mappings.

A problem that naturally arises is the existence of complete mappings of small degree. Theorem A below was conjectured by Chowla and Zassenhaus [11] in 1968, and proved by Cohen [12] in 1990.

Theorem A *If $d \geq 2$ and $p > (d^2 - 3d + 4)^2$, then there is no complete mapping polynomial of degree d over $\mathbb{F}_p$.*

Similarly, one may ask if $f + x^2$ or $f + x^3$ are permutations when $p > (d^2-3d+4)^2$. In connection with such questions, a significant generalization of Theorem A was obtained by Cohen, Mullen, and Shiue [13] in 1995, which provides a lower bound for the degree of the difference of two permutation polynomials in $\mathbb{F}_p[x]$ of the same degree d, when $p > (d^2 - 3d + 4)^2$.

Theorem B *Suppose f and $f + g$ are monic permutation polynomials over $\mathbb{F}_p$ of degree $d \geq 3$, where $p > (d^2-3d+4)^2$. Then either* $\deg(g) = 0$ *or* $\deg(g) \geq 3d/5$.

In light of Theorem A, one may wonder if complete mappings, $f+x$, exist when f has small Carlitz rank. A nonexistence result, similar to that in Theorem A, was given in [24] for permutations of small linearity.

Theorem C *If $f(x)$ is a complete mapping over $\mathbb{F}_q$ and $\mathscr{L}(f) < \lfloor (q+5)/2 \rfloor$, then* $\mathrm{Crk}(f) \geq \lfloor q/2 \rfloor$.

The authors of the present article obtained a generalization of Theorem C recently; see [4]. Theorems 1 and 2 below give lower bounds for the Carlitz rank in terms of q and the degree of the difference between two permutation polynomials, analogous to Theorem B.

We remark that Theorems A and B hold over prime fields only, whereas Theorems C, 1 and 2 are true for any finite field.

Let f be a permutation polynomial over $\mathbb{F}_q$, $q \geq 3$, with $\mathrm{Crk}(f) = n \geq 1$. Suppose that f has a representation as in (1), and the fractional linear transformation R_n which is associated to f as in (6) is not linear. In other words, α_n as defined in (3) is not zero. We denote the set of all such permutations by $\mathscr{C}_{1,n}$, i.e., *the set $\mathscr{C}_{1,n}$ consists of all permutation polynomials f over $\mathbb{F}_q$, with* $\mathrm{Crk}(f) = \mathrm{n} \geq 1$ *and have a representation that satisfies $\alpha_n \neq 0$.* Clearly $\mathscr{L}(f) \leq n+2$, if $f \in \mathscr{C}_{1,n}$.

We note that the linearity condition in Theorem C, which corresponds to the case $\alpha_n \neq 0$, is crucial. Indeed, one can find examples of complete mappings f with small Carlitz rank, when $\mathscr{L}(f)$ is large. For instance, the polynomial given by

$f(x) = \left(\left(((-x/(d+1))^{q-2}+1)^{q-2}+d\right)^{q-2} - 1/(d+1)\right)^{q-2}$ is a complete mapping for every $q \equiv 1 \mod 3$ where d is a primitive 3-rd root of unity, see [22]. Therefore, we only consider permutations in $\mathscr{C}_{1,n}$. Theorems 1 and 2 and Corollary 2 below are from [4].

Theorem 1 *Let f and $f+g$ be permutation polynomials over $\mathbb{F}_q$, where $f \in \mathscr{C}_{1,n}$ and the degree k of $g \in \mathbb{F}_q[x]$ satisfies $1 \le k < q-1$. Then*

$$n \ge \frac{1}{k+1}\left(q - \frac{k(k-1)}{2}\lfloor 2\sqrt{q}\rfloor - \gcd(k, q-1)\right), \tag{10}$$

where $\lfloor a \rfloor$ denotes the integer part of the real number a.

Despite Theorem 1 generalizing Theorem C, the methods used to prove Theorems A, B, and C are not amenable in proving Theorem 1. Therefore, an outline of the proof is presented for the curious reader.

Proof (Outline) We first consider the representation of $f+g$ by $R_n(x)+g(x)$ for $x \in \mathbb{F}_q \setminus \mathscr{O}_n$, where $R_n(x)$ is as in (6) with $R_n(x) = (ax+b)/(x+d)$, $a, b, d \in \mathbb{F}_q$, $ad - b \neq 0$. Since $f+g$ is a permutation, $R_n(x)+g(x)$ is injective on $\mathbb{F}_q \setminus \mathscr{O}_n$. Therefore, we consider the solutions of the equation

$$G_n(x) = R_n(x) + g(x) = u \text{ for } u \in \mathbb{F}_q.$$

We denote the value set of G_n by V_{G_n}, put $\mathbb{F}_q{}^\circ = \mathbb{F}_q \setminus \{-d\}$, and define $S = \left\{(x, y) \in \mathbb{F}_q{}^\circ \times \mathbb{F}_q{}^\circ : x \neq y \text{ and } G_n(x) = G_n(y)\right\}$. Moreover, for $u \in V_{G_n}$, we denote the cardinality of the inverse image $G_n^{-1}(u)$ of u by n_u. This leads to the inequalities $n_u \le k+1$, and we conclude that the cardinality μ of S satisfies

$$\mu \le (k+1) \sum_{u \in V_{G_n}} (n_u - 1). \tag{11}$$

Since G_n is injective on $\mathbb{F}_q \setminus \mathscr{O}_n$, if there exist n_u distinct elements with $G_n(z) = u$, then $n_u - 1$ distinct elements lie in $\mathscr{O}_n$. Together with the fact that $-d \in \mathscr{O}_n$, Eq. (11) implies

$$n \ge |\mathscr{O}_n| \ge 1 + \sum_{u \in V_{G_n}} (n_u - 1) \ge 1 + \frac{\mu}{k+1}. \tag{12}$$

Hence, to give a lower bound for n in terms of q and k, it suffices to determine μ in terms of q and k. By considering the algebraic curve $\mathscr{C}$ defined by $(G_n(x) - G_n(y))/(x-y)$ and applying Serre's bound (see [17, Theorem 9.57]) to an absolutely irreducible component of $\mathscr{C}$, we obtain that the number $N(\mathscr{C})$ of rational points in PG$(2, q)$ of $\mathscr{C}$ satisfies

$$N(\mathscr{C}) \ge q + 1 - \frac{k(k-1)}{2}\lfloor 2\sqrt{q}\rfloor.$$

Note that $(x : y : 0)$ is a point on $\mathscr{C}$ only if $xy(x^k - y^k)/(x - y) = 0$, i.e., $(x : y : 0) = (0 : 1 : 0)$, $(1 : 0 : 0)$ or $x^k = y^k$ for some $x, y \in \mathbb{F}_q^*$. Consequently, there are at most $\nu + 2$ such rational points, where $\nu = \gcd(k, q - 1)$. Bezout's theorem allows us to conclude that there exist at most $k+1$ rational points $(x : y : z)$ with $x = y$. Thus,

$$\mu \geq q + 1 - \frac{k(k-1)}{2} \lfloor 2\sqrt{q} \rfloor - (\nu + k + 2).$$

We remark that we subtract $\nu + k + 2$ instead of $\nu + k + 3$ because of the point $(1 : 1 : 0)$. Then, Eq. (12) gives the desired result. □

Remark 2 For $k = 1$ (and hence $\gcd(k, q - 1) = 1$), we obtain Theorem C, which is the main result in [24].

Corollary 1 *Let f be a permutation polynomial over $\mathbb{F}_q$ with $f \in \mathscr{C}_{1,n}$. If $n < (q - 1)/2$, then f is not a complete mapping.*

Remark 3 We note that the bound given in (10) is nontrivial only when $q \geq k(k - 1)\lfloor 2\sqrt{q} \rfloor /2 + \gcd(k, q - 1) + k + 1$. In particular, the bound is not applicable if $k > q^{1/4}$.

When $g(x) = cx^k \in \mathbb{F}_q[x]$, $\gcd(k+1, q-1) = 1$, and $f \in \mathscr{C}_{1,n}$ where $x_n \in \mathscr{O}_n$ in (4) satisfies $x_n = 0$, the lower bound in (10) can be simplified significantly.

Theorem 2 *Let $f(x)$ and $f(x)+cx^k$ be permutation polynomials over $\mathbb{F}_q$. Suppose $1 \leq k < q - 1$, $c \in \mathbb{F}_q{}^*$, and $f \in \mathscr{C}_{1,n}$ have a representation with $x_n = 0$. Let $m = \gcd(k + 1, q - 1)$. Then*

$$n \geq \frac{1}{k+1}\left(q - \frac{(k-1)(m-1)}{2}\lfloor 2\sqrt{q} \rfloor - 3k\right).$$

Once more, we present the sketch of the proof.

Proof (Outline) Proceeding as in the proof of Theorem 1, we obtain

$$n \geq 1 + \mu/(k+1),$$

where μ in this case is the cardinality of the set S. Changing variables via $(x, y) \to (xy, y)$, the equation $(G_n(x) - G_n(y))/(x - y)$ gives rise to the curve

$$\mathscr{C} : y^{k+1} = \frac{b(x-1)}{cx(x^k - 1)}. \tag{13}$$

Since the monomial $T^{(k+1)/m}$ is a permutation over $\mathbb{F}_q{}^*$ with $m = \gcd(k+1, q-1)$, there is one-to-one correspondence between the affine solutions $(x, y) \in \mathbb{F}_q{}^* \times \mathbb{F}_q{}^*$ of the curves

$$\mathscr{X} : y^m = \frac{b(x-1)}{cx(x^k - 1)}, \tag{14}$$

and $\mathscr{C}$ in (13). Note that Eq. (14) defines a Kummer extension. We observe that $\mathscr{X}$ is a curve of genus at most $(k-1)(m-1)/2$. Serre's bound then implies that $\mathscr{C}$ has at least $q-(k-1)(m-1)\lfloor 2\sqrt{q}\rfloor/2-k$ affine points $(x, y) \in \mathbb{F}_q{}^* \times \mathbb{F}_q{}^*$. In this case, we delete points (x, y) of $\mathscr{C}$ such that (x, y) is of the form (γ^2, γ) for some $\gamma \in \mathbb{F}_q$. It can be shown that there can be at most $3k+1$ such points, hence

$$\mu \geq q - \frac{(k-1)(m-1)}{2}\lfloor 2\sqrt{q}\rfloor - (4k+1),$$

which gives the desired result. □

We end this section with a corollary of Theorem 2, which indicates the analogy between Theorems 2 and B.

Corollary 2 *Let $f(x)$ and $f(x)+cx^k$ be as in Theorem 2. If $m = \gcd(k+1, q-1) = 1$, then $k \geq (q-n)/(n+3)$.*

4 On Iterations of Permutation Polynomials

Dynamical systems generated by polynomials in $\mathbb{F}_q[x]$ have been studied widely. We refer the reader to a recent survey [34] for algebraic and number theoretic properties of algebraic dynamical systems over finite fields and some of their applications. The distribution of elements in orbits of permutation polynomials in $\mathscr{C}_{1,n}$ is studied in [15]. Authors use the approximation property in $\mathscr{C}_{1,n}$ in the sense of (5) to analyze the distribution behavior of pseudorandom sequences generated by $f \in \mathscr{C}_{1,n}$ efficiently. This approach enables them to avoid encountering the usual problem of degree growth when iterations of polynomials are considered.

We denote the m-th iteration of $f \in \mathbb{F}_q[x]$ by

$$f^{(m)}(x) = f^{(m-1)}(f(x)) \text{ for } m \geq 1, \text{ where } f^{(0)}(x) = x.$$

In this section, we examine the order of the permutation polynomial f, viewed as an element of the symmetric group S_q, i.e., we seek methods to determine the minimum $m \geq 1$ such that $f^{(m)} \equiv \mathrm{Id}$.

Let $\mathbb{F}_q$ be a finite field of odd characteristic. We study iterations of permutation polynomials over $\mathbb{F}_q$ of a given Carlitz rank n. Iterations of permutations of Carlitz rank 1 are easy to determine since their cycle structures can be described in a simple manner; see [9, Theorem 2]. On the contrary, the cycle structure of permutations of higher rank is difficult to describe; see [9, Theorems 6, 7, 11, 13, and 15] for the cases $n = 2, 3$. Therefore, in what follows, we consider $n \geq 2$.

For simplicity, we only consider the set $\mathscr{C}'_{1,n}$ of polynomials f with a representation as in (1) such that $a_0 = 1$, $a_{n+1} = 0$. In this case, there is a representation P_{nm} of $f^{(m)}$ that satisfies $a_k = a_i$ for $k \equiv i \mod n$, $1 \leq k \leq nm$.

Then, for $f(x) \in \mathscr{C}'_{1,n}$ and the associated rational fraction $R_n(x)$ as in (5) and (6), we denote the m-th iteration of $R_n(x)$ by

$$R_n^{(m)}(x) = \frac{\alpha_{nm-1}x + \beta_{nm-1}}{\alpha_{nm}x + \beta_{nm}},$$

as the rational transformation, associated to the permutation P_{nm} above. Therefore, we have $f^{(m)}(c) = R_n^{(m)}(c)$ for all $c \in \mathbb{F}_q \setminus \mathscr{O}_{nm}$.

As mentioned earlier, the set of fractional transformations $(ax + b)/(cx + d)$, for $a, b, c, d \in \mathbb{F}_q$ with $ad - bc \neq 0$, forms the projective general linear group $\mathrm{PGL}(2, q)$. Hence, our aim is to relate the order of f in S_q to the order of R_n in $\mathrm{PGL}(2, q)$ and to determine the latter in a simple way. We denote the orders of f in S_q and of R_n in $\mathrm{PGL}(2, q)$ by $\mathrm{ord}_{S_q}(f)$ and $\mathrm{ord}(R_n)$, respectively.

Theorem 3 *Let $f \in \mathscr{C}'_{1,n}$ and put $\mathrm{ord}_{S_q}(f) = m_f$ and $\mathrm{ord}(R_n) = m_R$. If $q > nm_f + 2$, then $m_R | m_f$.*

Proof We note that $f^{(m)}(x)$ and $R_n^{(m)}(x)$ differ at most at nm elements of $\mathbb{F}_q$. Therefore, if $q > nm_f+2$, then $R_n^{(m_f)}(c) = c$ for at least three distinct elements c in $\mathbb{F}_q$. However, this implies that $R_n^{(m_f)}(x) \equiv x$. Thus, $f^{(m_f)}(x) \equiv x$ and $q > nm_f+2$ imply that $R_n^{(m_f)}(x) \equiv x$ and hence $m_R | m_f$. □

Remark 4 Up to conjugacy, a complete list of the subgroups of $\mathrm{PGL}(2, q)$ is known; see [17, Theorem A.8]. This classification implies that the order of $R_n(x)$ is the characteristic of $\mathbb{F}_q$ or is a divisor of $q \pm 1$.

In order to find $\mathrm{ord}(R_n)$, we must find the iterations $R_n^{(m)}$, and hence we need to determine the values of α_{nm}, α_{nm-1}, β_{nm} and β_{nm-1}. Lemmata 3 and 5 below enable us to express them in terms of eigenvalues of the matrix corresponding to R_n.

Lemma 3 *Let $R(x) = (ax + b)/(cx + d) \in \mathbb{F}_q(x)$ and $M \in \mathrm{PGL(2, q)}$ be the corresponding matrix representation. If M is diagonalizable over $\mathbb{F}_{q^2}$, with eigenvalues λ_1 and λ_2, then $R^{(m)}(x) \equiv x$ if and only if $\lambda_1^m = \lambda_2^m$. Otherwise, m is equal to p, where p is the characteristic of $\mathbb{F}_q$.*

Proof On the one hand, if M is diagonalizable over $\mathbb{F}_{q^2}$, then the characteristic polynomial has two roots, λ_1 and λ_2 in $\mathbb{F}_{q^2}$, and there exists $P \in \mathrm{GL}(2, q^2)$ such that $M = PDP^{-1}$ for the diagonal matrix $D = \begin{pmatrix} \lambda_1 & 0 \\ 0 & \lambda_2 \end{pmatrix}$. Then the m-th iteration $R^{(m)}(x)$ of $R(x)$ is obtained by

$$M^m = PD^mP^{-1} = P\begin{pmatrix} \lambda_1^m & 0 \\ 0 & \lambda_2^m \end{pmatrix}P^{-1}. \tag{15}$$

Equation (15) allows us to conclude that $R^{(m)}(x) \equiv x$ if and only if $\lambda_1^m = \lambda_2^m$.

On the other hand, if M is not diagonalizable over $\mathbb{F}_{q^2}$, then the characteristic polynomial has a double root λ in $\mathbb{F}_q$, and there exists $P \in \mathrm{GL}(2, q)$ such that $M = PRP^{-1}$ for the matrix $R = \begin{pmatrix} \lambda & 1 \\ 0 & \lambda \end{pmatrix}$. Then the m-th iteration $R^{(m)}(x)$ of $R(x)$ is obtained by

$$M^m = PR^mP^{-1} = P \begin{pmatrix} \lambda^m & m\lambda^{m-1} \\ 0 & \lambda^m \end{pmatrix} P^{-1}. \tag{16}$$

By Eq. (16), we conclude that $R^{(m)}(x) \equiv x$ if and only if $m = p$, where p is the characteristic of the field. □

Now let $f \in \mathscr{C}'_{1,n}$, and consider the corresponding rational transformation R_n. Then, the associated matrix $M \in \mathrm{GL}(2, q)$ is given by

$$M = \begin{pmatrix} \alpha_{n-1} & \beta_{n-1} \\ \alpha_n & \beta_n \end{pmatrix}. \tag{17}$$

Lemma 4 *Let $f \in \mathscr{C}'_{1,n}$ and M be the matrix representation of its associated rational transformation. Then the determinant of M is equal to $(-1)^n$. Hence, the characteristic polynomial of M is $h_M(T) = T^2 - (\alpha_{n-1} + \beta_n)T + (-1)^n$.*

Proof The proof is a simple computation of the determinant of M using the recurrence relations given in Eq. (3) and induction on n. □

The criteria given in Lemma 3 can be used to identify polynomials in $\mathscr{C}'_{1,n}$ with $R_n^{(m)}(x) \equiv x$. Lemma 5 below indicates the choices for $\mathrm{Tr}(M)$. Moreover, it enables us to construct permutations of prescribed Carlitz rank n with $R_n^{(m)}(x) \equiv x$; see Remark 5 and Example 1.

Lemma 5 *Let $h_1(T) = T^2 - \gamma T - 1$ and $h_2(T) = T^2 - \delta T + 1$ be in $\mathbb{F}_q[T]$.*

(i) Let λ_1, λ_2 be the roots of h_1 in $\mathbb{F}_{q^2}$. Then γ satisfies

$$\lambda_1^m - \lambda_2^m = (\lambda_1 - \lambda_2)H_m(\gamma), \text{ with } H_m(T) = \sum_{i=0}^{\lfloor \frac{m-1}{2} \rfloor} \binom{m-i-1}{i} T^{m-2i-1} \tag{18}$$

for $m \geq 1$.

(ii) Let λ_1, λ_2 be the roots of h_2 in $\mathbb{F}_{q^2}$. Then δ satisfies

$$\lambda_1^m - \lambda_2^m = (\lambda_1 - \lambda_2)G_m(\delta), \text{ with } G_m(T) = \sum_{i=0}^{m-1} \binom{m+i}{2i+1} (T-2)^i \tag{19}$$

for $m \geq 1$.

Proof If $\lambda_1 = \lambda_2$, then Eqs. (18) and (19) are clearly satisfied. Therefore we assume that $\lambda_1 \neq \lambda_2$. For $m \geq 1$, we define

$$L_m = \frac{\lambda_1^m - \lambda_2^m}{\lambda_1 - \lambda_2} .$$

(i) It can be seen easily that

$$L_1 = H_1(\gamma) = 1 \quad \text{and} \quad L_2 = H_2(\gamma) = \gamma . \tag{20}$$

Since λ_1, λ_2 are the roots of h_1, for $m \geq 1$, L_m satisfies

$$L_{m+2} - \gamma L_{m+1} - L_m = 0 .$$

By Eq. (20), it is enough to show that $H_m(T)$ satisfies the same recurrence for $T = \gamma$ and $m \geq 1$. For $m = 2k, 2k+1$, we can write $H_m(\gamma)$ as follows:

$$\begin{aligned} H_{2k}(\gamma) &= \sum_{i=0}^{k-1} \binom{2k-i-1}{i} \gamma^{2k-2i-1} = \sum_{i=0}^{k-1} \binom{k+i}{k-i-1} \gamma^{2i+1} \\ &= \sum_{i=0}^{k-1} \binom{k+i}{2i+1} \gamma^{2i+1}, \\ H_{2k+1}(\gamma) &= \sum_{i=0}^{k} \binom{2k-i}{i} \gamma^{2k-2i} = \sum_{i=0}^{k} \binom{k+i}{k-i} \gamma^{2i} \\ &= \sum_{i=0}^{k} \binom{k+i}{2i} \gamma^{2i} . \end{aligned}$$

Hence for $m = 2k$, we have

$$\begin{aligned} &H_{2k+2}(\gamma) - \gamma H_{2k+1}(\gamma) - H_{2k}(\gamma) \\ &= \sum_{i=0}^{k} \binom{k+i+1}{2i+1} \gamma^{2i+1} - \gamma \sum_{i=0}^{k} \binom{k+i}{2i} \gamma^{2i} - \sum_{i=0}^{k-1} \binom{k+i}{2i+1} \gamma^{2i+1} . \end{aligned} \tag{21}$$

Since the identity

$$\binom{k+i+1}{2i+1} - \binom{k+i}{2i} - \binom{k+i}{2i+1} = 0$$

holds, the coefficient of γ^{2i+1} in Eq. (21) is equal to zero for all $i = 0, \ldots, k$. The same argument holds for $m = 2k - 1$ as the coefficient of γ^{2i} in

$$H_{2k+1}(\gamma) - \gamma H_{2k}(\gamma) - H_{2k-1}(\gamma)$$

satisfies

$$\binom{k+i}{2i} - \binom{k+i-1}{2i-1} - \binom{k+i-1}{2i} = 0$$

for all $i = 0, \ldots, k$.

(ii) Similarly, we have $L_1 = G_1(\delta) = 1$ and $L_2 = G_2(\delta) = \delta$, and we replace Eq. (21) by

$$G_{m+2}(\delta) - (\delta - 2)G_{m+1}(\delta) - 2G_{m+1}(\delta) + G_m(\delta)\,. \tag{22}$$

Then, by similar calculations, we observe that the coefficient of $(\delta - 2)^i$ in Eq. (22) satisfies

$$\binom{m+2+i}{2i+1} - \binom{m+i}{2i-1} - 2\binom{m+1+i}{2i+1} + \binom{m+i}{2i+1} = 0$$

for all $i = 0, \ldots, m + 1$, which proves the desired result.

□

Corollary 3 *Let $f \in \mathscr{C}'_{1,n}$. We define*

$$A_m(T) = \begin{cases} H_m(T), & \text{if } n \text{ is odd,} \\ G_m(T), & \text{if } n \text{ is even,} \end{cases}$$

where H_m, G_m are given as in Lemma 5 . Then, $A_m(\alpha_{n-1} + \beta_n) = 0$ if and only if $R_n^{(m)}(x) \equiv x$. In particular, $\mathrm{ord}(R_n) = \min\{m : A_m(\alpha_{n-1} + \beta_n) = 0\}$.

Proof It follows from Lemmata 4 and 5. □

Remark 5 One can construct permutations f, represented as in Eq. (1), by an algorithm given in [3], when $R_n(x)$ and the poles $x_1, \ldots, x_n \in \mathbb{P}^1(\mathbb{F}_q)$ are prescribed.

Example 1 For $q = 29$, we fix $n = 4$ and choose $R_4(x) = (x - 5)/6x$, with $\alpha_3 + \beta_4 = 1$, being a root of $A_3(T)$ (and hence of $A_6(T)$). Hence, $\mathrm{ord}(R_4) = 3$ and $R_4^{(3)}(x) \equiv x \equiv R_4^{(6)}(x)$. We prescribe the poles as $(x_1, x_2, x_3, x_4) = (27, 16, 5, 0)$, and as in [3] we determine $(a_1, a_2, a_3, a_4) = (2, 8, 7, 14)$, so that $f(x) = ((((x + 2)^{27} + 8)^{27} + 7)^{27} + 14)^{27}$. It can also be checked in this case that $\mathrm{ord}_{S_{29}}(f) = m_f = 6$ and $29 > 4m_f + 2$.

Example 2 Consider the permutation polynomial $f = (((x+a)^{q-2}+b)^{q-2}+c)^{q-2}$ with $f(0) = 0$ and $a(b^2+4) \neq 0$. Then $R_3^{(m)} \equiv x$ if and only if b is a root of the polynomial

$$A_m(T) = \sum_{j=0}^{\lfloor \frac{m-1}{2} \rfloor} \binom{m-j-1}{j} T^{m-2j-1}. \tag{23}$$

In particular, $\mathrm{ord}(R_3) = \min\{m \mid A_m(b) = 0\}$.

The following result on $\mathrm{ord}_{S_q}(f)$ enables one to find many examples of polynomials of Carlitz rank 3 with a given order.

Proposition 1 *Let $f(x) = (((x+a)^{q-2}+b)^{q-2}+c)^{q-2} \in \mathbb{F}_q[x]$, with $f(0) = 0$ and $a(b^2+4) \neq 0$. Put $m_f = \mathrm{ord}_{S_q}(f)$ and $m_R = \mathrm{ord}(R_3)$. Suppose $q > 3m_f + 2$ and the order of $(b+\gamma(b))/2$ in the multiplicative group $\mathbb{F}_{q^2}^*$ is k, where $\gamma(b)^2 = b^2 + 4$. Then $m_R | m_f$, where*

$$m_R = \begin{cases} k/2, & \text{if } k \equiv 0 \mod 8, \\ k, & \text{if } k \equiv 2, 6 \mod 8, \\ k/4, & \text{if } k \equiv 4 \mod 8, \\ 2k, & \text{if } k \equiv 1, 3, 5, 7 \mod 8. \end{cases}$$

Proof Since $a \neq 0$, the assumption $f(0) = 0$ implies that $abc + a + c = 0$, i.e., $\beta_3 = 0$. Since $\alpha_4 = b$ and $b^2 + 4 \neq 0$, Lemma 3 implies that $R_3^{(m)} \equiv x$ if and only if the distinct eigenvalues satisfy $\lambda_1^m = \lambda_2^m$. Then the argument follows from $\lambda_1^{2m} = (-1)^m$ as $\lambda_1 \lambda_2 = -1$. □

Example 3

(1) Let $f(x) = (((x+7)^{71} + 14)^{71} + 25)^{71}$ be a permutation over $\mathbb{F}_{73}$. In this case, we have $b^2 + 4 = 54$. Then for $\gamma(b) = 28$, the order k of the element $(b+\gamma(b))/2$ is 24 and $\mathrm{ord}_{S_q}(f) = 12 = k/2$. Note that $b = 14$ is a root of $A_{12}(T)$ given in Eq. (23), but it is not root of $A_\ell(T)$ for any $\ell < 12$.
(2) Let $f(x) = (((x+13)^{41}+13)^{41}+28)^{41}$ be a permutation over $\mathbb{F}_{43}$. In this case, we have $b^2+4 = 1$. Then for $\gamma(b) = 1$, the order k of the element $(b+\gamma(b))/2$ is 6 and $\mathrm{ord}_{S_q}(f) = 6 = k$. We remark that $b = 13$ is not a root of A_ℓ for any $\ell < 6$.
(3) If we choose $\gamma(b) = -1$ in Example (2), then we have $\mathrm{ord}_{S_q}(f) = 6 = 2k$.
(4) Let $f(x) = (((x+8)^{35} - 6)^{35} + 23)^{35}$ be a permutation over $\mathbb{F}_{37}$. In this case, $b^2 + 4 = 3$ and $\gamma(b) = 22$ yield $k = 12$ and $m_R = 3 = k/4$. We also have $\mathrm{ord}_{S_q}(f) = 6 = 2m_R$.

Acknowledgements The initial work on this project began during "Women in Numbers Europe 2 (WIN-E2)" workshop, held in the Lorentz Center, Leiden, in September 2016. The authors are grateful to the Lorentz Center and all supporting institutions for making this conference and collaboration possible. They would especially like to thank the organizers of WIN-E2, Irene Bouw, Rachel Newton, and Ekin Ozman for all of their hard work, as this resulted in an extremely fruitful and enjoyable meeting.

The authors N.A., A.O., V.P., L.Q., and A.T. are partially supported by H.C. Ørsted COFUND Post-doc Fellowship from the project "Algebraic curves with many rational points" and the Austrian Science Fund FWF Project P 30405-N32; Federal Ministry of Education and Science, grant No.05-39-3663-1/14; an EPSRC studentship; CNPq, PDE grant number 200434/2015-2; and TUBITAK project number 114F432, respectively.

References

1. A. Akbary, D. Ghioca, Q. Wang, On permutation polynomials of prescribed shape. Finite Fields Appl. **15**, 195–206 (2009)
2. A. Akbary, D. Ghioca, Q. Wang, On constructing permutations of finite fields. Finite Fields Appl. **17**, 51–67 (2011)
3. E. Aksoy, A. Çeşmelioğlu, W. Meidl, A. Topuzoğlu, On the Carlitz rank of permutation polynomials. Finite Fields Appl. **15**, 428–440 (2009)
4. N. Anbar, A. Odžak, V. Patel, L. Quoos, A. Somoza, A. Topuzoğlu, On the difference between permutation polynomials. Finite Fields Appl. **49**, 132–142 (2018)
5. D. Bartoli, M. Giulietti, G. Zini, On monomial complete permutation polynomials. Finite Fields Appl. **41**, 132–158 (2016)
6. D. Bartoli, M. Giulietti, L. Quoos, G. Zini, Complete permutation polynomials from exceptional polynomials. J. Number Theory **176**, 46–66 (2017)
7. L. Carlitz, Permutations in a finite field. Proc. Am. Math. Soc. **4**, 538 (1953)
8. A. Çeşmelioğlu, W. Meidl, A. Topuzoğlu, Enumeration of a class of sequences generated by inversions. In *Proceedings of the First Int. Workshop on Coding and Cryptology, Fujian, China, June 2007*. Series in Coding Theory and Cryptology, vol. 4 (World Scientific, Singapore, 2008)
9. A. Çeşmelioğlu, W. Meidl, A. Topuzoğlu, On the cycle structure of permutation polynomials. Finite Fields Appl. **14**, 593–561 (2008)
10. A. Çeşmelioğlu, W. Meidl, A. Topuzoğlu, Permutations with prescribed properties. J. Comput. Appl. Math. **259**, 536–545 (2014)
11. S. Chowla, H. Zassenhaus, Some conjectures concerning finite fields. Nor. Vidensk. Selsk. Forh. (Trondheim) **41**, 34–35 (1968)
12. S.D. Cohen, Proof of a conjecture of Chowla and Zassenhaus on permutation polynomials. Can. Math. Bull. **33**, 230–234 (1990)
13. S.D. Cohen, G.L. Mullen, P.J.-S. Shiue, The difference between permutation polynomials over finite fields. Proc. Am. Math. Soc. **123**, 2011–2015 (1995)
14. L. Comtet, *Advanced Combinatorics: The Art of Finite and Infinite Expansions* (Reidel, Dordrecht, 1974)
15. D. Gomez-Perez, A. Ostafe, A. Topuzoğlu, On the Carlitz rank of permutations of $\mathbb{F}_q$ and pseudorandom sequences. J. Complexity **30**, 279–289 (2014)
16. X. Guangkui, X. Cao, Complete permutation polynomials over finite fields of odd characteristic. Finite Fields Appl. **31**, 228–240 (2015)
17. J.W.P. Hirschfeld, G. Korchmáros, F. Torres, *Algebraic Curves Over a Finite Field* (Princeton University Press, Princeton, 2013)
18. X. Hou, A class of permutation trinomials over finite fields. Acta Arith. **162**, 51–64 (2014)
19. X. Hou, A survey of permutation binomials and trinomials over finite fields, in *Topics in Finite Fields*. Contemp. Math., vol. 632 (Amer. Math. Soc., Providence, 2015), pp. 177–191
20. X. Hou, Determination of a type of permutation trinomials over finite fields, II. Finite Fields Appl. **35**, 16–35 (2015)

21. X. Hou, Permutation polynomials over finite fields – a survey of recent advances. Finite Fields Appl. **32**, 82–119 (2015)
22. L. Işık, A. Topuzoğlu, A note on value set of polynomials over finite fields (preprint)
23. L. Işık, A. Winterhof, Carlitz rank and index of permutation polynomials. Finite Fields Appl. **49**, 156–165 (2018)
24. L. Işık, A. Topuzoğlu, A. Winterhof, Complete mappings and Carlitz rank. Des. Codes Cryptogr. **85**, 121–128 (2017)
25. G. M. Kyureghyan, Constructing permutations of finite fields via linear translators. J. Comb. Theory Ser. A **118**, 1052–1061 (2011)
26. C.F. Laywine, G. Mullen, *Discrete Mathematics Using Latin Squares*. Wiley-Interscience Series in Discrete Mathematics and Optimization (Wiley, New York, 1998)
27. R. Lidl, H. Niederreiter, *Finite Fields* (Cambridge University Press, Cambridge, 1997)
28. H.B. Mann, The construction of orthogonal Latin squares. Ann. Math. Stat. **13**, 418–423 (1942)
29. G.L. Mullen, Permutation polynomials over finite fields, in *Finite Fields, Coding Theory, and Advances in Communications and Computing* (Las Vegas, NV, 1991). Lecture Notes in Pure and Applied Mathematics, vol. 141 (Taylor & Francis, Routledge, 1993), pp. 131–151
30. G.L. Mullen, H. Niederreiter, Dickson polynomials over finite fields and complete mappings. Can. Math. Bull. **30**, 19–27 (1987)
31. G.L. Mullen, D. Panario, *Handbook of Finite Fields* (Chapman and Hall, London, 2013)
32. A. Muratović-Ribić, E. Pašalić, A note on complete mapping polynomials over finite fields and their applications in cryptography. Finite Fields Appl. **25**, 306–315 (2014)
33. H. Niederreiter, K.H. Robinson, Complete mappings of finite fields. J. Aust. Math. Soc. A **33**, 197–212 (1982)
34. A. Ostafe, Iterations of rational functions: some algebraic and arithmetic aspects, in *Finite Fields and Their Applications*. Radon Series on Computational and Applied Mathematics, vol. 11 (De Gruyter, Berlin, 2013), pp. 197–231
35. F. Pausinger, A. Topuzoğlu, On the discrepancy of two families of permuted van der Corput 480 sequences. Unif. Distrib. Theory **13**(1), 47–64 (2018)
36. R.-H. Schulz, On check digit systems using anti-symmetric mappings, in *Numbers, Information and Complexity (Bielefeld, 1998)* (Kluwer Academic, Boston, 2000), pp. 295–310
37. R. Shaheen, A. Winterhof, Permutations of finite fields for check digit systems. Des. Codes Cryptogr. **57**, 361–371 (2010)
38. I.E. Shparlinski, *Finite Fields: Theory and Computation: The Meeting Point of Number Theory, Computer Science, Coding Theory and Cryptography*. Mathematics and Its Applications, vol. 477 (Kluwer Academic, Dordrecht, 1999)
39. N.J. Sloane, Online Encyclopedia of Integer Sequences, published electronically at http://www.research,att.com/~njas/sequences
40. P. Stănică, S. Gangopadhyay, A. Chaturvedi, A.K. Gangopadhyay, S. Maitra, Investigations on bent and negabent functions via the nega-Hadamard transform. IEEE Trans. Inf. Theory **58**, 4064–4072 (2012)
41. A. Topuzoğlu, Carlitz rank of permutations of finite fields: a survey. J. Symb. Comput. **64**, 53–66 (2014)
42. Z. Tu, X. Zeng, L. Hu, Several classes of complete permutation polynomials. Finite Fields Appl. **25**, 182–193 (2014)
43. A. Winterhof, Generalizations of complete mappings of finite fields and some applications. J. Symb. Comput. **64**, 42–52 (2014)
44. G. Wu, N. Li, T. Helleseth, Y. Zhang, Some classes of monomial complete permutation polynomials over finite fields of characteristic two. Finite Fields Appl. **28**, 148–165 (2014)
45. Z. Zha, L. Hu, X. Cao, Constructing permutations and complete permutations over finite fields via subfield-valued polynomials. Finite Fields Appl. **31**, 162–177 (2015)

Dynamical Belyi Maps

Jacqueline Anderson, Irene I. Bouw, Ozlem Ejder, Neslihan Girgin, Valentijn Karemaker, and Michelle Manes

Abstract We study the dynamical properties of a large class of rational maps with exactly three ramification points. By constructing families of such maps, we obtain $\mathcal{O}(d^2)$ conservative maps of fixed degree d defined over $\mathbb{Q}$; this answers a question of Silverman. Rather precise results on the reduction of these maps yield strong information on their $\mathbb{Q}$-dynamics.

Keywords Arithmetic dynamics · Conservative rational maps · Reduction

2010 Mathematics Subject Classification. Primary 37P05; Secondary 11G32, 37P15

J. Anderson
Bridgewater State University, Bridgewater, MA, USA
e-mail: jacqueline.anderson@bridgew.edu

I. I. Bouw
University of Ulm, Ulm, Germany
e-mail: irene.bouw@uni-ulm.de

O. Ejder
Colorado State University, Fort Collins, CO, USA
e-mail: ejder@usc.edu

N. Girgin
Boğaziçi University, Istanbul, Turkey

V. Karemaker
University of Pennsylvania, Philadelphia, PA, USA
e-mail: vkarem@math.upenn.edu

M. Manes (✉)
University of Hawaii, Honolulu, HI, USA
e-mail: mmanes@math.hawaii.edu

I. I. Bouw et al. (eds.), *Women in Numbers Europe II*, Association for Women in Mathematics Series 11, https://doi.org/10.1007/978-3-319-74998-3_5

1 Introduction

Let X be a smooth projective curve. A *Belyi map* is a finite cover $f : X \to \mathbb{P}^1$ that is branched exactly at 0, 1, and ∞. In this paper we restrict to the special case that $X = \mathbb{P}^1$, allowing for iteration of f and the study of dynamical behavior. We write f^n for the n-fold composition of f with itself. If we further insist that $f(\{0, 1, \infty\}) \subseteq \{0, 1, \infty\}$, then all iterates f^n are also Belyi maps. These maps have been called *dynamical Belyi maps* [13, 19].

Dynamical Belyi maps are the central objects of study in the present work. These maps are a special class of *post-critically finite* (PCF) maps; a map $f : \mathbb{P}^1 \to \mathbb{P}^1$ is PCF if each ramification point has a finite forward orbit. The study of PCF maps has a long history in complex and arithmetic dynamics, starting from Thurston's topological characterization of these maps in the early 1980s and continuing to the present day [1, 2, 6, 10].

1.1 Outline and Summary of Results

In Sect. 2, we describe a large class of dynamical Belyi maps with exactly three ramification points: the genus-0 single-cycle normalized Belyi maps. In addition, each of the three ramification points—0, 1, and ∞—is fixed. Maps $f : \mathbb{P}^1 \to \mathbb{P}^1$ for which every ramification point is fixed are called *conservative rational maps* or *critically fixed* and have been the subject of some recent study [3, 9, 16].

The genus-0 single-cycle Belyi maps are each naturally defined over $\mathbb{Q}$ (Proposition 1), a result which suffices to answer in the negative a question of Silverman [15, p. 110]: Is the number of PGL_2-conjugacy classes of conservative rational maps of degree d in $\mathbb{Q}[z]$ or $\mathbb{Q}(z)$ bounded independently of d?

In Sect. 3, we give explicit equations for two particular infinite families of maps in this class (Propositions 2 and 3), providing many new examples of maps of interest in complex and arithmetic dynamics.

Studying the reduction of the maps from Sect. 2 to characteristic p leads to one of our main results (Theorem 1): Given a genus-0 single-cycle normalized Belyi map f of degree $d \geq 3$, we describe a necessary and sufficient condition for the reduction of f modulo a prime p to be the monomial map x^d.

In Sect. 6, we use this result, together with standard local-global theorems in arithmetic dynamics, to study the dynamical behavior of the genus-0 single-cycle Belyi maps f. Our main result (Theorem 2) gives conditions for the set of $\mathbb{Q}$-rational preperiodic points to coincide with the set of $\mathbb{Q}$-rational fixed points of f. For one family of explicit Belyi maps, we determine the set of preperiodic points exactly (Proposition 8).

2 Background on Belyi Maps

A *Belyi map* is a finite cover $f : X \to \mathbb{P}^1$ of smooth projective curves defined over $\mathbb{C}$ that is branched exactly at $0, 1, \infty$. Belyi maps can be described topologically as *dessins d'enfants* [12] or combinatorially in terms of generating systems.

Definition 1 Fix an integer $d > 1$. A *generating system* of degree d is a triple $\boldsymbol{\rho} = (\rho_1, \rho_2, \rho_3)$ of permutations $\rho_i \in S_d$ that satisfy

- $\rho_1\rho_2\rho_3 = 1$,
- $G := \langle \rho_1, \rho_2, \rho_3 \rangle \subset S_d$ acts transitively on $\{1, 2, \ldots, d\}$.

The *combinatorial type* of $\boldsymbol{\rho}$ is a tuple $\underline{C} := (d; C(\rho_1), C(\rho_2), C(\rho_3))$, where d is the degree and $C(\rho_i)$ is the conjugacy class of ρ_i in S_d.

Two generating systems $\boldsymbol{\rho}$ and $\boldsymbol{\rho}'$ are *equivalent* if there exists a permutation $\tau \in S_d$ such that $\rho_i' = \tau\rho_i\tau^{-1}$ for $i = 1, 2, 3$. Furthermore, two Belyi maps $f_i : X_i \to \mathbb{P}^1$ are *isomorphic* if there exists an isomorphism $\iota : X_1 \to X_2$ making

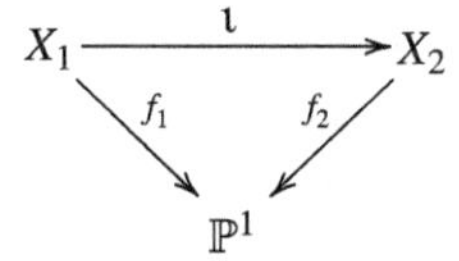

commutative. In particular, the f_i have the same branch locus. Riemann's Existence Theorem [17, Theorem 2.13] yields a bijection between equivalence classes of generating systems and isomorphism classes of Belyi maps $f : X \to \mathbb{P}^1$.

Let $\boldsymbol{\rho}$ be a generating system. The conjugacy class $C_i := C(\rho_i)$ of S_d corresponds to a partition $\sum_{j=1}^{r_i} n_j = d$ of d. The *length* $r_i = r(C_i)$ of C_i is the number of cycles of the elements of C_i. Note that we include the 1-cycles. The nonnegative integer $g := (d+2-r_1-r_2-r_3)/2$ is called the *genus* of the generating system. If $f : X \to \mathbb{P}^1$ is the Belyi map corresponding to $\boldsymbol{\rho}$, then $g = g(X)$ and $r_i = |f^{-1}(t_i)|$ is the cardinality of the inverse image of the ith branch point t_i.

In this paper we only consider the case that $g = g(X) = 0$. We write

$$f : \mathbb{P}^1_x \to \mathbb{P}^1_t, \qquad x \mapsto t = f(x);$$

the subscript indicates the coordinate of the corresponding projective line. Note that we write f both for the rational function $f(x) \in \mathbb{C}(x)$ and the cover defined by it.

We restrict, moreover, to the case that C_i is the conjugacy class of a single cycle, i.e., for each $i \in \{1, 2, 3\}$ the partition $\sum_j n_j$ corresponding to C_i contains a unique part n_j different from 1. We denote this part by e_i. Formulated differently, the corresponding Belyi map f has a unique ramification point above t_i, with ramification index e_i. The assumption that $g = g(X) = 0$ translates to the condition

$$2d + 1 = e_1 + e_2 + e_3.$$

We call this situation the *genus-0 single-cycle case*, and we write $(d; e_1, e_2, e_3)$ for the combinatorial type of f. More generally, given four integers d, e_1, e_2, and e_3 satisfying $2 \leq e_i \leq d$ and $2d + 1 = e_1 + e_2 + e_3$, we call $(d; e_1, e_2, e_3)$ an *abstract combinatorial type of genus* 0.

We say that a genus-0 single-cycle Belyi map f is *normalized* if its ramification points are 0, 1, and ∞, and if, moreover,

$$f(0) = 0, \quad f(1) = 1, \quad f(\infty) = \infty.$$

Since f has three ramification points, every isomorphism class of covers contains a unique normalized Belyi map.

Remark 1 In this paper we always assume that f has exactly three (as opposed to at most three) ramification points. We may allow some $e_i = 1$ without substantial change to our theoretical results. In this case the Belyi map f is Galois, and its Galois group is cyclic. Therefore f is linearly conjugate (in the sense of Definition 2 below) to the pure power map $f(x) = x^d$. This case is well understood, so we omit the details.

For instance, if we include the case $e_2 = 1$ (i.e., the combinatorial type $(d; d, 1, d)$) in our considerations, this only affects a few results: the count of normalized Belyi maps of each degree in Corollary 1 slightly changes, and the statement of Proposition 7 needs to be altered by including the assumption $\nu > 0$ for the conclusion to hold. If one of the other e_i is trivial, slightly different but equally straightforward adaptions are necessary.

The following proposition states that the triple of conjugacy classes corresponding to the combinatorial type $(d; e_1, e_2, e_3)$ is rigid and rational.

Proposition 1 *Let $\underline{C} := (d; e_1, e_2, e_3)$ be an abstract combinatorial type of genus 0. Then there exists a unique normalized Belyi map $f : \mathbb{P}^1_{\mathbb{C}} \to \mathbb{P}^1_{\mathbb{C}}$ of combinatorial type $\underline{C}$. Moreover, the rational map f may be defined over $\mathbb{Q}$.*

Proof We have already seen that the number of normalized Belyi maps of combinatorial type $\underline{C}$ is the cardinality of the finite set

$$\left\{(\rho_1, \rho_2, \rho_3) \in C_1 \times C_2 \times C_3 \text{ such that } \langle \rho_i \rangle \subset S_d \text{ transitive, } \prod_i \rho_i = e\right\} \Big/ \sim$$

of generating systems of combinatorial type $\underline{C}$ up to equivalence. Liu and Osserman prove in [5, Lemma 2.1] that in the genus-0 single-cycle case the cardinality of this set is 1. In other words, the triple (C_1, C_2, C_2) of conjugacy classes is rigid [17, Def. 2.15]. Moreover, the triple (C_1, C_2, C_3) of conjugacy classes in S_d, together with the choice of branch points $t_1 = 0, t_2 = 1$, and $t_3 = \infty$, is rational [17, Def. 3.7]. Corollary 3.13 of [17] therefore implies that $f : \mathbb{P}^1_x \to \mathbb{P}^1_t$ is defined over $\mathbb{Q}$. (In [17], Völklein states his results for Galois covers. Our statement follows from this, since we consider rational conjugacy classes in S_d.) □

Remark 2 Let $f : \mathbb{P}^1_x \to \mathbb{P}^1_t$ be a normalized Belyi map of combinatorial type $\underline{C} = (d; e_1, e_2, e_3)$, i.e., a genus-0 single-cycle map. Proposition 1 states that the rational function defining the Belyi map f satisfies $f(x) \in \mathbb{Q}(x)$.

Normalized Belyi maps are examples of *conservative rational maps*, i.e., rational maps $f : \mathbb{P}^1 \to \mathbb{P}^1$ such that every ramification point is fixed. In [15, top of p. 110], Silverman asked if the number of PGL_2-equivalence classes of conservative maps $\mathbb{P}^1 \to \mathbb{P}^1$ of degree d in $\mathbb{Q}[z]$ or $\mathbb{Q}(z)$ may be bounded independently of d. We use Proposition 1 to answer this question.

Definition 2 Two rational functions $f, g : \mathbb{P}^1_K \to \mathbb{P}^1_K$ are *linearly conjugate over a field* K if there is a $\phi \in \mathrm{Aut}(\mathbb{P}^1_K) \cong \mathrm{PGL}_2(K)$ such that $f^\phi := \phi^{-1} \circ f \circ \phi = g$.

Note that linear conjugacy respects iteration; that is, if $f^\phi = g$, then $(f^n)^\phi = \left(f^\phi\right)^n = g^n$. Note also that for normalized genus-0 single-cycle Belyi maps Proposition 1 implies that linear conjugacy over $\mathbb{C}$ is the same as linear conjugacy over $\mathbb{Q}$, so we may omit the mention of the field.

Remark 3 Definition 2 is the standard equivalence relation used in arithmetic and holomorphic dynamics. It is different from the notion of isomorphic covers introduced at the beginning of this section in which the target $\mathbb{P}^1$ is fixed. When we say that maps are conjugate, we mean linearly conjugate in the sense of this definition, where the same isomorphism is used on both the source and the target $\mathbb{P}^1$.

Lemma 1 *Let $f : \mathbb{P}^1_\mathbb{C} \to \mathbb{P}^1_\mathbb{C}$ be a normalized Belyi map of combinatorial type $(d; e_1, e_2, e_3)$ and let g be a normalized Belyi map with combinatorial type $(d; e'_1, e'_2, e'_3)$. Then f and g are linearly conjugate over $\mathbb{C}$ if and only if there is some permutation $\sigma \in S_3$ such that $e_i = e'_{\sigma(i)}$ for $i = 1, 2, 3$.*

Proof The ramification index of a point is a dynamical invariant in the following sense: for a nonconstant function $f : \mathbb{P}^1 \to \mathbb{P}^1$, any point $\alpha \in \mathbb{P}^1$, and any $\phi \in \mathrm{PGL}_2$, the ramification indices satisfy $e_\alpha(f) = e_{\phi^{-1}(\alpha)}(f^\phi)$.

In the case of normalized Belyi maps, the only points of $\mathbb{P}^1$ with ramification index greater than one are $t_1 = 0$, $t_2 = 1$, and $t_3 = \infty$. Assume that f and g are normalized Belyi maps with $f^\phi = g$ for some $\phi \in \mathrm{PGL}_2$. Then we may define a permutation $\sigma \in S_3$ by

$$e_i = \text{ramification index of } f \text{ at } t_i = \text{ramification index of } g \text{ at } \phi^{-1}(t_i) = e'_{\sigma(i)}.$$

Conversely, given a permutation $\sigma \in S_3$ and a normalized Belyi map g of combinatorial type $(d; e_{\sigma(1)}, e_{\sigma(2)}, e_{\sigma(3)})$, there exists a unique $\phi \in \mathrm{PGL}_2$ satisfying $\phi(t_i) = t_{\sigma(i)}$ for $i = 1, 2, 3$. Proposition 1 implies that $g = f^\phi$, since both normalized Belyi maps have the same combinatorial type. □

Lemma 1, together with the rationality result from Proposition 1, answers Silverman's question in the negative.

Corollary 1 *The number of* PGL_2*-conjugacy classes of conservative polynomials in* $\mathbb{Q}[z]$ *of degree* $d \geq 3$ *is at least* $\left\lfloor \frac{d-1}{2} \right\rfloor$. *The number of* PGL_2*-conjugacy classes of nonpolynomial conservative rational maps in* $\mathbb{Q}(z)$ *of degree* $d \geq 4$ *is at least*

$$\sum_{i=1}^{\left\lfloor \frac{d-1}{3} \right\rfloor} \left\lfloor \frac{d+1-3i}{2} \right\rfloor. \tag{1}$$

Proof We count the number of PGL_2-conjugacy classes of genus-0 single-cycle normalized Belyi maps in degree d, which serves as a lower bound for all conservative rational maps in the given degree, up to linear conjugacy. By Lemma 1, this equals the number of partitions of $2d+1$ into exactly three parts such that each part is at least 2 and none exceed d. The number of partitions equals the cardinality of

$$\{2 \leq e_1 \leq e_2 \leq e_3 \leq d \quad | \quad e_1 + e_2 + e_3 = 2d + 1\}. \tag{2}$$

If f is a polynomial of degree d, then the ramification index $e_3 = e_\infty(f) = d$. Hence, it is enough to count pairs (e_1, e_2) such that $2 \leq e_1 \leq e_2 \leq d-1$ and $e_1 + e_2 = d + 1$. We see that e_1 can take on $\left\lfloor \frac{d-1}{2} \right\rfloor$ distinct values, and e_2 is determined by e_1.

To count nonpolynomial maps, we use the same strategy for every possible value of $e_3 \leq d-1$. Fixing $e_3 = d - i$, we count pairs (e_1, e_2) such that

$$2 \leq e_1 \leq e_2 \leq d - i \quad \text{and} \quad e_1 + e_2 = d + i + 1.$$

These constraints give that $2i + 1 \leq e_1 \leq \lfloor \frac{d+i+1}{2} \rfloor$, yielding $\lfloor \frac{d+1-3i}{2} \rfloor$ distinct possibilities.

Finally, the constraints in (2) require that $d - 1 \geq e_3 \geq \left\lceil \frac{2d+1}{3} \right\rceil$. Writing $e_3 = d - i$ gives $1 \leq i \leq \lfloor \frac{d-1}{3} \rfloor$, and the result follows. □

Remark 4

1. The sum in (1) can be calculated explicitly. Let $N(d)$ be the number of PGL_2-conjugacy classes of genus-0 single-cycle normalized Belyi maps of degree d (including the polynomial maps). Then

$$N(d) = \frac{1}{12}(d^2 + 4d - c), \quad \text{where } c = \begin{cases} 5 & \text{if } d \equiv 1 \pmod 6 \\ 8 & \text{if } d \equiv 4 \pmod 6 \\ 9 & \text{if } d \equiv 3, 5 \pmod 6 \\ 12 & \text{if } d \equiv 0, 2 \pmod 6 \end{cases}$$

   This serves as a lower bound for the number of conservative rational maps of degree d in $\mathbb{Q}(z)$.

2. A result of Tischler [16] counts all monic conservative polynomials f in $\mathbb{C}[z]$ normalized by requiring that $f(0) = 0$. Tischler shows that there are exactly $\binom{2d-2}{d-1}$ such maps. This result is generalized in [3], where conservative rational maps rather than polynomials are counted. Note that in this more general situation counting is more difficult as one has to impose the condition that f is conservative. In our setting this condition is automatically satisfied.

3 Families of Dynamical Belyi Maps

In this section we determine some families of normalized dynamical Belyi maps explicitly. These results yield infinitely many explicit maps to which we can apply the dynamical system results from Sect. 6.

Proposition 2 *If a normalized Belyi map f has combinatorial type $(d; d-k, k+1, d)$, then $f(x)$ is given by*

$$f(x) = cx^{d-k}(a_0x^k + \ldots + a_{k-1}x + a_k),$$

where

$$a_i := \frac{(-1)^{k-i}}{(d-i)}\binom{k}{i} \quad \text{and} \quad c = \frac{1}{k!}\prod_{j=0}^{k}(d-j).$$

Proof It is clear that the ramification index e_3 is d, since f is a polynomial, and that e_1 is $d-k$. We need to show that the ramification index of e_2 is $k+1$. The derivative of f is given by

$$\begin{aligned} f'(x) &= c\sum_{i=0}^{k}\frac{(-1)^{k-i}}{(d-i)}\binom{k}{i}(d-i)x^{d-i-1} \\ &= cx^{d-k-1}\sum_{i=0}^{k}(-1)^{k-i}\binom{k}{i}x^{k-i} \\ &= (-1)^k cx^{d-k-1}(x-1)^k. \end{aligned}$$

Hence the only ramification points of f are 0, 1, and ∞, and the ramification index e_2 is equal to $k+1$. We are left to show that $f(1) = 1$ which is equivalent to showing that

$$\sum_{i=0}^{k}\left(a_i\prod_{j=0}^{k}(d-j)\right) = \sum_{i=0}^{k}\left((-1)^{k-i}\binom{k}{i}\prod_{\substack{j=0\\ j\neq i}}^{k}(d-j)\right) = k!.$$

We first show that in the above sum, the coefficients of d^l for each $1 \leq l \leq k$ are 0, and the constant term is $k!$. Notice that the coefficient of d^k is $\sum_{i=0}^{k} (-1)^{k-i}\binom{k}{i}$, which is 0 since this is the binomial expansion of $(x-1)^k$ evaluated at 1. Similarly, for the other terms d^l for $l \geq 1$, we obtain a sum $\sum_{i=0}^{k} (-1)^{k-i}\binom{k}{i}p(i)$ where $p(x)$ is a polynomial of degree less than k. This sum is also zero by [11, Corollary 2]. The constant coefficient is

$$\sum_{i=0}^{k}\left((-1)^i\binom{k}{i}\prod_{\substack{j=0\\ j\neq i}}^{k}(j)\right)=\binom{k}{0}k!=k!.$$

□

Remark 5 We now provide an alternative proof of Proposition 2. A variant of this proof can be found in the unpublished master's thesis of Michael Eskin; his PhD thesis [4, Proposition 5.1.2] contains a slightly weaker version of the result.
To have the correct ramification at 0, 1, and ∞, we see that f must be of the form

$$f(x) = x^{d-k} f_1(x) \tag{3}$$

for some $f_1(x) = \sum_{i=0}^{k} c_i(x-1)^i$, such that

$$f'(x) = (-1)^k c x^{d-k-1}(x-1)^k$$

for some $c \in \mathbb{C}^\times$. This implies that

$$c(x-1)^k = (d-k)f_1 + x f_1' = c_k d(x-1)^k + \sum_{i=0}^{k-1}\left((d-k+i)c_i + (i+1)c_{i+1}\right)(x-1)^i.$$

This yields a recursive formula for the c_i, from which it follows that

$$c_i = (-1)^i\binom{d-k+i-1}{i}$$

for all $i = 0, \ldots, k$, and

$$c = \binom{d-1}{k}d.$$

Substituting these values for c_i and c back into Eq. (3), the reader may check we obtain the claimed result.

Example 1 The unique normalized Belyi map f of combinatorial type $(d; d-1, 2, d)$ is given by

$$f(x) = -(d-1)x^d + dx^{d-1}.$$

Proposition 3 *The unique normalized Belyi map f of combinatorial type $(d; d-k, 2k+1, d-k)$ is given by*

$$f(x) = x^{d-k}\left(\frac{a_0x^k - a_1x^{k-1} + \ldots + (-1)^k a_k}{(-1)^k a_k x^k + \ldots - a_1 x + a_0}\right),$$

where

$$a_i := \binom{k}{i} \prod_{k+i+1 \le j \le 2k} (d-j) \prod_{0 \le j \le i-1} (d-j) = k!\binom{d}{i}\binom{d-k-i-1}{k-i}.$$

Proof The combinatorial type $(d; d-k, 2k+1, d-k)$ is characterized by the fact that the ramification at $x = 0, \infty$ is given by the same conjugacy class in the sense of Definition 1. This implies that the Belyi map f admits an automorphism in the following sense: Write $\psi(x) = 1/x$. Then $f^{\psi} = \psi^{-1} \circ f \circ \psi$ has the same combinatorial type as f. By Proposition 1, f is the unique normalized Belyi map of the given type, so $\psi^{-1} \circ f \circ \psi = f$.

From this, it immediately follows that we may write

$$f = x^{d-k} f_1/f_2, \qquad \text{with} \qquad f_2(x) = x^k f_1(1/x).$$

Let $f = g/h$ with

$$g(x) = x^{d-k} \sum_{i=0}^{k} (-1)^{k-i} a_{k-i} x^i \quad \text{and} \quad h(x) = \sum_{j=0}^{k} (-1)^j a_j x^j,$$

where the a_i are as in the statement of the proposition. It is clear that the ramification at $x = 0, \infty$ is as required. Moreover, we see that 0, 1, and ∞ are fixed points of f.

It therefore remains to determine the ramification at $x = 1$. More precisely, we need to show that the derivative satisfies

$$f'(x) = \frac{cx^{d-k-1}(x-1)^{2k}}{h(x)^2}, \tag{4}$$

for some nonvanishing constant c. Write $g'h - gh' = x^{d-k-1} \sum_l c_l x^l$. We have

$$c_l = (-1)^{k+l} \sum_{j=0}^{l} (d-k+l-2j) a_{k-l+j} a_j.$$

Here we have used the convention that $a_i = 0$ if $i > k$ or $i < 0$. Equation (4) on the a_i therefore translates to

$$
\begin{aligned}
c &= c_{2k} = (-1)^k (d-k) a_0 a_k, \\
c_l &= (-1)^l c \binom{2k}{l} = (-1)^{k+l} \binom{2k}{l} (d-k) a_0 a_k, \qquad l = 0, \ldots, 2k.
\end{aligned} \tag{5}
$$

Hence, to prove (5) it suffices to prove the following:

$$
\sum_{j=0}^{l} (d-k+l-2j) a_{k-l+j} a_j = \binom{2k}{l} (d-k) a_k a_0 \tag{6}
$$

for every $l \leq k$. (In fact, l runs from 0 to $2k$, but by symmetry it suffices to look at $l \leq k$.)

Here and for the rest of the proof, we use the convention that $\binom{n}{m} = 0$ if $m \leq 0$ or $m \geq n$. Hence, the right hand side of (6) translates to

$$
\begin{aligned}
\binom{2k}{l} (d-k) a_k a_0 &= \binom{2k}{l} (d-k) k! \binom{d}{k} k! \binom{d-k-1}{k} \\
&= d (k!)^2 \binom{2k}{l} \binom{d-1}{2k} \binom{2k}{k}.
\end{aligned}
$$

We write $d-k+l-2j$ as the difference of $d-k+l-j$ and j. Then the left hand side of (6) becomes

$$
\begin{aligned}
d(k!)^2 \sum_{j=0}^{l} \left(\binom{d}{j} \binom{d-1}{k-l+j} - \binom{d-1}{j-1} \binom{d}{k-l+j} \right) \cdot \\
\binom{d-2k+l-j-1}{l-j} \binom{d-k-j-1}{k-j}.
\end{aligned}
$$

Hence, dividing both sides of (6) by $d(k!)^2$, we find that we need to prove that the following equation holds for all integers d, k and l such that $d \geq 2k+1$ and $l \leq k$:

$$
\begin{aligned}
\sum_{j=0}^{l} \left(\binom{d}{j} \binom{d-1}{k-l+j} - \binom{d-1}{j-1} \binom{d}{k-l+j} \right) \cdot \\
\binom{d-2k+l-j-1}{l-j} \binom{d-k-j-1}{k-j} = \binom{2k}{l} \binom{d-1}{2k} \binom{2k}{k}.
\end{aligned} \tag{7}
$$

We fix k, l such that $l \leq k$ and define $F_{k,l}(d, j)$ as the quotient of the jth term in the sum on the left hand side of Eq. (7) by $\binom{2k}{k}\binom{d-1}{2k}\binom{2k}{l}$. Note that $F_{k,l}(d, j) = 0$ when $j > l$; this allows us to restate Eq. (7) in the following form:

$$\sum_{j=0}^{\infty} F_{k,l}(d,j) = 1. \tag{8}$$

To prove that Eq. (7), or equivalently (8), holds for every value of $d \geq 2k+1$, we first prove it for $d = 2k+1$, and then show that $\sum_{j=0}^{\infty} F_{k,l}(d+1,j) = \sum_{j=0}^{\infty} F_{k,l}(d,j)$ for any $d \geq 2k+1$.

So first suppose that $d = 2k+1$. Then Eq. (7) holds, since

$$\sum_{j=0}^{l}\left(\binom{2k+1}{j}\binom{2k}{k-l+j} - \binom{2k}{j-1}\binom{2k+1}{k-l+j}\right)$$
$$= \sum_{j=0}^{l}\binom{2k}{j}\binom{2k}{k-l+j} - \sum_{j=0}^{l-1}\binom{2k}{j}\binom{2k}{k-l+j} = \binom{2k}{l}\binom{2k}{k}.$$

Next, to show that $\sum_{j=0}^{\infty} F_{k,l}(d+1,j) = \sum_{j=0}^{\infty} F_{k,l}(d,j)$, we write

$$\sum_{j=0}^{\infty}\big(F_{k,l}(d+1,j) - F_{k,l}(d,j)\big)$$

as a telescoping series and use an algorithm due to Zeilberger [18] that proves hypergeometric identities involving infinite sums of binomial coefficients. More explicitly, running this algorithm in Maple produces an explicit function $G_{k,l}(d,j)$ which satisfies

$$F_{k,l}(d+1,j) - F_{k,l}(d,j) = G_{k,l}(d,j+1) - G_{k,l}(d,j).$$

One may check that $G_{k,l}(d,0) = 0$. Moreover, $G_{k,l}(d,j) = 0$ for all $j > l$ since the same is true for $F_{k,l}(d,j)$. Hence, $\sum_{j=0}^{\infty}(F_{k,l}(d+1,j) - F_{k,l}(d,j))$ equals

$$\sum_{j=0}^{\infty}(G_{k,l}(d,j+1) - G_{k,l}(d,j)) = G_{k,l}(d,l+1) - G_{k,l}(d,0) = 0.$$

□

Example 2 If a normalized Belyi map f has combinatorial type $(d; d-1, 3, d-1)$, then $f(x)$ is given by

$$f(x) = x^{d-1}\frac{(d-2)x - d}{-dx + (d-2)}.$$

(Note that necessarily $d \geq 3$ in this case.)

Example 3 If a normalized Belyi map f has combinatorial type $(d; d-2, 5, d-2)$, then $f(x)$ is given by

$$f(x) = x^{d-2}\left(\frac{(d-3)(d-4)x^2 - 2d(d-4)x + d(d-1)}{d(d-1)x^2 - 2d(d-4)x + (d-3)(d-4)}\right).$$

(Note that necessarily $d \geq 5$ in this case.)

4 Reduction Properties of Normalized Belyi Maps

Let

$$f : \mathbb{P}^1_x \to \mathbb{P}^1_t, \qquad x \mapsto f(x) := t$$

be a normalized Belyi map of combinatorial type $\underline{C} := (d; e_1, e_2, e_3)$. Proposition 1 implies that $f(x) \in \mathbb{Q}(x)$ is a rational function with coefficients in $\mathbb{Q}$. We start by defining the reduction of f at a rational prime p. Since we assume that the rational function f is normalized, we may write

$$f(x) = \frac{f_1(x)}{f_2(x)},$$

where $f_1, f_2 \in \mathbb{Q}[x]$ are polynomials that are relatively prime. Multiplying numerator and denominator by a common constant $c \in \mathbb{Q}_{>0}$, we may assume that $f_1, f_2 \in \mathbb{Z}[x]$.

For $k = 1, 2$, write

$$f_k = c_k \tilde{f}_k, \quad \text{with } \tilde{f}_k \in \mathbb{Z}[x] \text{ a polynomial of content 1} \tag{9}$$

Let $c = c_1/c_2$. The assumption that $f(1) = 1$ translates to

$$\tilde{f}_2(1) = c\tilde{f}_1(1) \in \mathbb{Z}.$$

Note that $\tilde{f}(x) := \frac{\tilde{f}_1(x)}{\tilde{f}_2(x)}$ need no longer be normalized. The ramification points of $\tilde{f}$ are still 0, 1, and ∞, but $\tilde{f}(1)$ need not be 1. Nonetheless, it makes sense to consider the reduction of $\tilde{f}$ modulo p. We denote the reduction of $\tilde{f}_k$ by $\overline{f}_k$ and put

$$\overline{f} = \frac{\overline{f}_1}{\overline{f}_2}, \qquad \overline{f}_k \in \mathbb{F}_p[x].$$

The definition of $\tilde{f}$ implies that $\overline{f} \neq 0$. We claim that in our situation $\overline{f}$ is not a constant. The proof below is inspired by a remark in [7, Section 4]; note however that Osserman works only with maps in characteristic p, while we consider reduction to characteristic p of maps in characteristic zero.

Proposition 4 *Let f be a normalized Belyi map of combinatorial type $\underline{C} := (d; e_1, e_2, e_3)$.*

(1) The reduction $\overline{f}$ is nonconstant.
(2) We have $\overline{f}(0) = 0$ and $\overline{f}(\infty) = \infty$.
(3) We have $\overline{f}(1) \neq 0, \infty$.

Proof Define $\tilde{f}_1$ and $\tilde{f}_2$ as in Eq. (9). Let i (resp. j) be maximal such that the coefficient of x^i in $\tilde{f}_1$ (resp. of x^j in $\tilde{f}_2$) is a p-adic unit. Note that the reduction $\overline{f}$ of f is constant if and only if $\tilde{f}_1 \equiv a\tilde{f}_2 \pmod{p}$ or $a\tilde{f}_1 \equiv \tilde{f}_2 \pmod{p}$ for some constant a. It follows that if $\overline{f}$ is constant, then $i = j$.

The definition of the combinatorial type implies that

$$e_1 \leq i \leq d = \deg(\tilde{f}_1), \qquad 0 \leq j \leq d - e_3 = \deg(\tilde{f}_2).$$

Since e_2 is a ramification index, we have that

$$e_2 = 2d + 1 - (e_1 + e_3) \leq d.$$

This implies that $e_1 + e_3 - d > 0$. It follows that

$$i \geq e_1 > d - e_3 \geq j. \tag{10}$$

This implies that $\overline{f}$ is nonconstant, and (1) is proved.

Equation (10) also implies that

$$\deg(\overline{f}_1) = i > j = \deg(\overline{f}_2).$$

This implies that $\overline{f}(\infty) = \infty$.

Applying the same argument to the minimal i' (resp. j') such that the coefficient of $x^{i'}$ in $\tilde{f}_1$ (resp. the coefficient of $x^{j'}$ in $\tilde{f}_2$) is a p-adic unit shows that

$$\mathrm{ord}_0(\overline{f}_1) = i' > j' = \mathrm{ord}_0(\overline{f}_2).$$

We conclude that $\overline{f}(0) = 0$, thus proving (2).

It remains to show that $\mu := \overline{f}(1) \neq 0, \infty$. We have

$$\mathrm{ord}_0(\overline{f}_1) \geq e_1, \qquad \mathrm{ord}_1(\overline{f}_1) \geq e_2.$$

We assume that $\mu = 0$, i.e. $\overline{f}(0) = \overline{f}(1) = 0$. This implies that

$$e_1 + e_2 \leq \deg(\overline{f}_1) \leq \deg(f_1) = d.$$

This yields a contradiction with $e_3 = 2d + 1 - (e_1 + e_2) \leq d$. We conclude that $\mu \neq 0$. Similarly, we conclude that $\mu \neq \infty$. This finishes the proof of (3). □

The following example shows that the reduction of f may be a constant if we omit the assumption on the combinatorial type of f.

Example 4 We consider the rational function

$$f(x) = \frac{px^4 + x^2}{x^2 + p}.$$

The corresponding cover $f : \mathbb{P}^1 \to \mathbb{P}^1$ is ramified at 6 points, each with ramification degree 2, so it is not a Belyi map as considered in this paper. Moreover, $x = 1$ is not a ramification point. Both the numerator and the denominator of f have content 1. Therefore with our definition of the reduction we obtain

$$\overline{f} = \frac{x^2}{x^2} = 1.$$

Remark 6 In [14, Section 2.3] Silverman gives a different definition of the reduction of f. The difference between Silverman's definition and ours is (roughly speaking) that he does not divide f by the constant c before reducing, as we do in passing from f to $\tilde{f}$. Instead, Silverman only multiplies f_1 and f_2 by a common constant to assume that at least one of the polynomials f_1 or f_2 has content 1.

We claim that in the case of a normalized Belyi map of ramification type $(d; e_1, e_2, e_3)$, Silverman's definition agrees with ours. To see this, let p be a prime. Recall that

$$1 = f(1) = c\tilde{f}(1) = c\frac{\tilde{f}_1(1)}{\tilde{f}_2(1)}.$$

Then Proposition 4.(3) implies that $\tilde{f}_1(1)$ and $\tilde{f}_2(1)$ have the same p-adic valuation, so $\tilde{f}(1)$ is a p-adic unit. Hence, $c = 1/\tilde{f}(1)$ is a p-adic unit, as well.

Note that $c = c_1/c_2$, where c_i is the content of the polynomial f_i, so in particular c is positive. We conclude that $c \in \mathbb{Q}_{>0}$ is a p-adic unit for all primes p and hence that $c = 1$.

Let $g \in \overline{\mathbb{F}}_p(x)$ be a rational function. We say that the map $\mathbb{P}^1 \to \mathbb{P}^1$ defined by g is *(in)separable* if g is (in)separable. Recall that $g \in \overline{\mathbb{F}}_p(x)$ is inseparable if and only if it is contained in $\overline{\mathbb{F}}_p(x^p)$.

Definition 3 Let $f : \mathbb{P}^1 \to \mathbb{P}^1$ have combinatorial type $(d; e_1, e_2, e_3)$. Let p be a prime. We say that f has *good reduction* if the reduction $\overline{f}$ also has degree d. If $\overline{f}$ is additionally (in)separable, we say that f has *good (in)separable reduction*. If f does not have good reduction, we say it has *bad reduction*.

In Corollary 2 we show that if f has bad reduction, then $\overline{f}$ is inseparable. In particular, we do not have to consider the case of bad separable reduction.

Definition 3 is the definition of good reduction used in the theory of arithmetic dynamics. From the point of view of Galois theory, one usually defines "good reduction" to mean good and separable reduction. In our terminology f has bad reduction if and only if $\deg(\overline{f}) < \deg(f)$.

Proposition 5 *Let $f : \mathbb{P}^1_{\mathbb{Q}} \to \mathbb{P}^1_{\mathbb{Q}}$ be a normalized Belyi map of combinatorial type $\underline{C} := (d; e_1, e_2, e_3)$. Assume that the reduction $\overline{f}$ of f to characteristic p is separable. Then*

(a) f has good reduction (i.e., $\overline{d} = d$), and
(b) $p \nmid e_i$ for all i.

Proof Our definition of the reduction of f, together with the assumption that f is normalized, implies that the points $x = 0, 1, \infty$ on the source $\mathbb{P}^1_{\mathbb{Q}}$ specialize to pairwise distinct points of $\mathbb{P}^1_{\mathbb{F}_p}$ (by Proposition 4.(2,3)). In particular, multiplying $\overline{f}$ by a constant (if necessary), we may assume that $\overline{f}$ is also normalized.

We write $f = f_1/f_2$ and $d_1 = \deg(f_1)$, $d_2 = \deg(f_2)$. We denote the degree of $\overline{f}_i$ by $\overline{d}_i$ and define $\overline{d} = \deg(\overline{f})$. The polynomials $\overline{f}_1$ and $\overline{f}_2$ are not necessarily relatively prime. Put $g = \gcd(\overline{f}_1, \overline{f}_2)$ and $\delta = \deg(g)$.

Let $\overline{e}_i$ be the ramification indices of $\overline{f}$ at $x = 0, 1, \infty$, respectively. Our first goal is to compare these to the ramification indices e_i of f. We start by considering what happens at $x = 0$. For this we write

$$\overline{f}_i = gh_i, \qquad i = 1, 2.$$

Since $\gcd(h_1, h_2) = 1$ it follows that

$$\overline{e}_1 = \text{ord}_0(\overline{f}) = \text{ord}_0(h_1).$$

The definition of the reduction implies that

$$\text{ord}_0(\overline{f}_1) = \text{ord}_0(g) + \text{ord}_0(h_1) \geq \text{ord}_0(f_1) = e_1.$$

For the right-most equality, we have used that $\gcd(f_1, f_2) = 1$. Defining $\varepsilon_1 := \text{ord}_0(g)$ we obtain

$$\overline{e}_1 + \varepsilon_1 \geq e_1. \tag{11}$$

Interchanging the roles of $x = 0$ and $x = 1$, we similarly obtain

$$\overline{e}_2 + \varepsilon_2 \geq e_2, \tag{12}$$

where $\varepsilon_2 := \mathrm{ord}_1(g)$. Note that interchanging the roles of $x = 0$ and $x = 1$ corresponds to conjugating $\overline{f}$ by $\varphi(x) = 1 - x$. From the definitions it follows immediately that

$$\varepsilon_1 + \varepsilon_2 \le \delta. \tag{13}$$

The definition of the reduction of f and our normalization implies that

$$d = d_1 \ge \overline{d}_1 = \overline{d} + \delta. \tag{14}$$

Finally, for the ramification index $\overline{e}_3$ of $\overline{f}$ at ∞, we have

$$d - e_3 = d_2 \ge \overline{d}_2 = \overline{d}_1 - \overline{e}_3 = \overline{d} + \delta - \overline{e}_3. \tag{15}$$

Since we assume that $\overline{f}$ is separable, the Riemann–Hurwitz formula applied to $\overline{f}$, together with the inequalities (11), (12), (13), (14), and (15), yields

$$\begin{aligned} -2 &\ge -2\overline{d} + (\overline{e}_1 - 1) + (\overline{e}_2 - 1) + (\overline{e}_3 - 1) \\ &= (\overline{e}_1 + \varepsilon_1 - 1) + (\overline{e}_2 + \varepsilon_2 - 1) + (\overline{e}_3 - \overline{d} - \delta - 1) + (\delta - \varepsilon_1 - \varepsilon_2) + (-\overline{d}) \\ &\ge -2d + (e_1 - 1) + (e_2 - 1) + (e_3 - 1) = -2. \end{aligned} \tag{16}$$

It follows that both inequalities are equalities. The fact that the last inequality is an equality implies that $\overline{d} = d$, and that the inequalities (11), (12), (13), (14), and (15) are also equalities. This proves Statement (a).

The first inequality in (16) is an equality if and only if all ramification of $\overline{f}$ is tame. Hence we have $p \nmid \overline{e}_i$ for all i. The statement $\overline{d} = d$ implies that $\varepsilon_1 = \varepsilon_2 = \delta = 0$. Hence $\overline{e}_i = e_i$ for all i. Statement (b) follows. □

The following is an immediate consequence of Lemma 5.

Corollary 2 *Let $f : \mathbb{P}^1_{\mathbb{Q}} \to \mathbb{P}^1_{\mathbb{Q}}$ be a normalized Belyi map of combinatorial type $\underline{C} := (d; e_1, e_2, e_3)$. Assume that f has bad reduction to characteristic p. Then the reduction $\overline{f}$ is inseparable.*

5 Good Inseparable Monomial Reduction

In Sect. 6 we determine the dynamical behavior of separable covers f of degree d (of a given combinatorial type), whose reduction modulo p satisfies $\overline{f}(x) = x^d$. Since 1 is a ramification point of $\overline{f}$, it follows that $\overline{f}$ is inseparable and hence that $p \mid d$. If this happens, we say that f has *good (inseparable) monomial reduction* to characteristic p. In Theorem 1 we prove necessary and sufficient conditions for this to occur.

Definition 4

1. A rational map ψ of degree d in characteristic p can be written uniquely as $\psi = \psi' \circ \phi^n$, where ϕ is the p-Frobenius map and ψ' is separable. Suppose that ψ' is a normalized Belyi map of combinatorial type $(d'; e'_1, e'_2, e'_3)$. Then we call the $\overline{e}_i = p^n e'_i$ for $i = 1, 2, 3$ the *generalized ramification indices* of ψ; we allow d' and each of the e'_i to be trivial.
2. Let $\underline{C} = (d; e_1, e_2, e_3)$ be a combinatorial type such that $e_1 + e_2 + e_3 = 2d + 1$. Then we define

$$S_{\underline{C},p} := \{\psi : \mathbb{P}^1_{\overline{\mathbb{F}}_p} \to \mathbb{P}^1_{\overline{\mathbb{F}}_p} \mid \psi \text{ satisfies the following combinatorial conditions}\}$$

   a. $\deg(\psi) := \overline{d} \leq d$, and
   b. there exist $\varepsilon_1, \varepsilon_2, \delta \geq 0$ such that

$$\varepsilon_1 + \varepsilon_2 \leq \delta \leq d - \overline{d}$$

   and the generalized ramification indices $\overline{e}_i$ $(i = 1, 2, 3)$ of ψ satisfy

$$\overline{e}_1 \geq e_1 - \varepsilon_1,$$
$$\overline{e}_2 \geq e_2 - \varepsilon_2,$$
$$\overline{e}_3 \geq e_3 - (d - \overline{d} - \delta).$$

The set $S_{\underline{C},p}$ may be considered as a characteristic-p analog of the set of normalized Belyi maps of combinatorial type $\underline{C}$. Lemma 2 and Proposition 6 below imply that this set consists of one element. Moreover, it follows that $\psi \in S_{\underline{C},p}$ is the reduction (in the sense of Sect. 4) of the (unique) normalized Belyi map of type $\underline{C}$ in characteristic zero. In particular, it follows that $\psi \in S_{\underline{C},p}$ may be defined over $\mathbb{F}_p$.

Lemma 2 *Let $f : \mathbb{P}^1 \to \mathbb{P}^1$ be a normalized cover in characteristic zero of combinatorial type $\underline{C} = (d; e_1, e_2, e_3)$. Its reduction $\overline{f}$ modulo p lies in $S_{\underline{C},p}$.*

Proof If f has good reduction at p, choose $\varepsilon_1 = \varepsilon_2 = \delta = 0$, and the result is immediate. If f has bad reduction, the result follows immediately from the proof of Proposition 5, for δ and ε_i $(i = 1, 2)$ as in that proof. □

The following proposition is a reformulation in our terminology of a result of Osserman.

Proposition 6 *[7, Theorem 4.2.(i)] For any combinatorial type $\underline{C}$ and prime number p, we have $|S_{\underline{C},p}| = 1$.*

We sketch the idea of Osserman's approach in his proof of Proposition 6. For details we refer to [7] and [8]. Osserman interprets a rational map $f : \mathbb{P}^1_K \to \mathbb{P}^1_K$ (up to automorphisms of the image) of degree d over a field K as a linear series by associating with the rational map $f = f_1/f_2$ the two-dimensional vector subspace $V := \langle f_1, f_2\rangle$ of the polynomials of degree less than or equal to d. This linear series may be considered as a point on the Grasmannian $G(1, d)$. The condition that the map has ramification index at least e_i at the point P_i defines a Schubert cycle $\Sigma_{e_i-1}(P_i)$ on $G(1, d)$. Base points of the linear series correspond to common zeros of f_1 and f_2.

Consider an arbitrary linear series in positive characteristic, which we denote by $\langle\psi_1, \psi_2\rangle$. The inequalities (a) and (b) in Definition 4 may be interpreted as conditions on the linear series. Note that we do not require the polynomials ψ_i to be relatively prime. The zeros of $g := \gcd(\psi_1, \psi_2)$ are base points of the linear series. (Compare to the proof of Proposition 5, where we denoted the orders of these zeros at 0, 1 by $\varepsilon_1, \varepsilon_2$, respectively.)

Assume that $\langle\psi_1, \psi_2\rangle$ is a linear series satisfying the inequalities (a) and (b) from Definition 4. The Riemann–Hurwitz formula, together with the condition that $2d + 1 = e_1 + e_2 + e_3$, implies that the linear series $\langle\psi_1, \psi_2\rangle$ does not have base points if $\psi = \psi_1/\psi_2$ is separable. (This follows as in the proof of Proposition 5.) However, the linear series associated with the reduction of a normalized Belyi map (as defined above) may have base points. Moreover, the base-point divisor $D := \varepsilon_1[0] + \varepsilon_2[1] - \delta[\infty]$ need not be unique. (See Example 5 below for an example.)

Our Proposition 1 states that in characteristic zero, the intersection of the three Schubert cycles $\Sigma_{e_1-1}(0) \cap \Sigma_{e_2-1}(1) \cap \Sigma_{e_3-1}(\infty)$ has dimension 0, and the intersection product is 1. In other words, the intersection consists of one point. Osserman gives an intersection-theoretic argument to prove the same statement in arbitrary characteristic. More precisely, he proves that the intersection product of the three Schubert cycles in positive characteristic is scheme-theoretically a point. The underlying point is the unique map $\psi \in S_{\underline{C},p}$.

In this paper, we are mainly interested in the case of good inseparable reduction. In this case, we have $\varepsilon_1 = \varepsilon_2 = \delta = 0$; hence, the base-point divisor is unique in this situation.

We give an example of a combinatorial type $\underline{C}$ for which the reduction of the normalized dynamical Belyi map of this type has base points. Moreover, we will see in that in this case the linear series satisfying the inequalities (a) and (b) is not unique, even though the underlying rational function is.

Example 5 This example is taken from Osserman [7, §2] (two paragraphs above Proposition 2.1). Let p be a prime and $\underline{C} = (d; e_1, e_2, e_3)$ be a type such that $d > p$ and $e_i < p$ for $i = 1, 2, 3$. Then

$$x^p \in S_{\underline{C},p}$$

as one may verify directly.

We therefore have that $\overline{d} = p = \overline{e}_i$ for all i. The inequalities (a) and (b) from Definition 4 become

$$\varepsilon_1 \geq 0, \qquad \varepsilon_2 \geq 0, \qquad \delta \leq d - e_3, \qquad \varepsilon_1 + \varepsilon_2 \leq \delta.$$

We conclude that for a given combinatorial type $\underline{C}$ and prime p, the base-point divisor $D = \varepsilon_1[0] + \varepsilon_2[1] - \delta[\infty]$ need not be unique. The linear series corresponding to a solution $(\varepsilon_1, \varepsilon_2, \delta)$ of the inequalities is

$$\langle x^p g, g\rangle, \qquad \text{with } g = x^{\varepsilon_1}(x-1)^{\varepsilon_2}.$$

Dynamical Belyi maps as considered in this example do exist; see [8, Cor. 2.5]. Here is a concrete instance. Choose $p \geq 7$ and d with $p < d < 3(p-1)/2$ and k such that $d - p < k < (p-1)/2$. Let $\underline{C} = (d; d-k, 2k+1, d-k)$. The normalized Belyi map of this combinatorial type is given in Proposition 3. The expression for the coefficients a_i in that lemma shows both that $p|a_i$ for $d - p + 1 \leq i \leq d$, and that $a_i \equiv (-1)^k a_{d-p-i} \bmod p$ for $0 \leq i \leq d - p$ (these are non-zero modulo p). From this it follows that

$$\overline{f} = x^p.$$

Moreover, it follows that

$$g = \gcd(\overline{f}_1, \overline{f}_2) = (-1)^{d-p} a_{d-p} x^{d-p} + \cdots + a_0 \in \mathbb{F}_p[x]$$

has degree $d - p$. The roots of the polynomial g correspond to base points of the linear series.

Theorem 1 *Suppose that $f\colon \mathbb{P}^1 \to \mathbb{P}^1$ is a normalized dynamical Belyi map of combinatorial type $\underline{C} = (d = p^n d'; e_1, e_2, e_3)$, where $p \nmid d'$. Then the reduction $\overline{f}$ of f modulo p satisfies $\overline{f}(x) = x^d$ if and only if $e_2 \leq p^n$.*

Proof In the good monomial reduction case, i.e., where $\overline{f}(x) = x^d$, we have $\overline{d} = d$, and the generalized ramification indices are $\overline{e}_1 = \overline{e}_3 = d$ and $\overline{e}_2 = p^n$. Hence, $e_2 \leq p^n$ is a necessary condition for good inseparable monomial reduction.

Conversely, let f be of combinatorial type $\underline{C} = (d = p^n d', e_1, e_2, e_3)$ as in the statement of the theorem and assume that $e_2 \leq p^n$. We claim that the map $g(x) = x^d$ lies in $S_{\underline{C},p}$.

As before, we write ψ as the composition of the purely inseparable map of degree p^n and the separable map $\psi'(x) = x^{d'}$, and we write e_1', e_2', e_3' for the ramification indices of ψ' at $x = 0, 1, \infty$, respectively. Clearly, $\deg(\psi) =: \overline{d} = d$ satisfies $\overline{d} \leq d$. Moreover, the ramification indices $\overline{e}_i$ of g satisfy

$$\begin{aligned}
\overline{e}_1 &:= p^n e_1' = d \geq e_1,\\
\overline{e}_2 &:= p^n \geq e_2, \qquad \text{(by assumption)},\\
\overline{e}_3 &:= p^n e_3' = d \geq e_3.
\end{aligned}$$

Choosing $\varepsilon_1 = \varepsilon_2 = \delta = 0$, we see that ψ satisfies the combinatorial conditions in Definition 4, so indeed $\psi \in S_{\underline{C},p}$. By Lemma 2 and Proposition 6, we obtain that $\overline{f} = g$, i.e., that f has good inseparable monomial reduction modulo p. □

Remark 7 Theorem 1 can be viewed as a special case of [8, Theorem 2.4], which proves the existence of a (necessarily unique) (in)separable cover for any combinatorial type $(d; e_1, e_2, e_3)$ by studying the combinatorial properties of the e_i. One can prove variants of the statement of Theorem 1 by specifying different possibilities for the reduction $\overline{f}$ of f. This would amount to formulating conditions on the e_i for reduction types other than the good inseparable monomial one.

Example 6 Consider the combinatorial type $\underline{C} = (d = 15; e_1, e_2, e_3 = d = 15)$. The equation for the associated cover is given in [4, Proposition 5.1.2] and can alternatively be determined from Proposition 2. Computing the reduction of f modulo the primes $p = 2, 3, 5$, and 7 yields the following table. We immediately see that the results of the table are in accordance with Theorem 1.

e_2	$\overline{f}(x)$ at $p = 2$	**Reduction**
2	x^{14}	Bad
3	$x^{15} + x^{14} + x^{13}$	Good sep.
4	x^{12}	Bad
5	$x^{15} + x^{12} + x^{11}$	Good sep.
6	$x^{14} + x^{12} + x^{10}$	Bad
7	$x^{15} + x^{14} + x^{13} + x^{12} + x^{11} + x^{10} + x^9$	Good sep.
8	x^8	Bad
9	$x^{15} + x^8 + x^7$	Good sep.
10	$x^{14} + x^8 + x^6$	Bad
11	$x^{15} + x^{14} + x^{13} + x^8 + x^7 + x^6 + x^5$	Good sep.
12	$x^{12} + x^8 + x^4$	Bad
13	$x^{15} + x^{12} + x^{11} + x^8 + x^7 + x^4 + x^3$	Good sep.
14	$x^{14} + x^{12} + x^{10} + x^8 + x^6 + x^4 + x^2$	Bad
e_2	$\overline{f}(x)$ at $p = 3$	**Reduction**
$e_2 \leq p = 3$	x^{15}	Good insep.
$p = 3 < e_2 \leq 2p = 6$	$2x^{15} + 2x^{12}$	Good insep.
$2p = 6 < e_2 \leq 3p = 9$	x^9	Bad
$3p = 9 < e_2 \leq 4p = 12$	$2x^{15} + x^9 + x^6$	Good insep.
$4p = 12 < e_2 < 5p = d = 15$	$x^{15} + x^{12} + x^9 + 2x^6 + 2x^3$	Good insep.
e_2	$\overline{f}(x)$ at $p = 5$	**Reduction**
$e_2 \leq p = 5$	x^{15}	Good insep.
$p = 5 < e_2 \leq 2p = 10$	$3x^{15} + 3x^{10}$	Good insep.
$2p = 10 < e_2 < 3p = d = 15$	$x^{15} + 2x^{10} + 3x^5$	Good insep.
e_2	$\overline{f}(x)$ at $p = 7$	**Reduction**
$e_2 \leq 7$	x^{14}	Bad
$e_2 = 8$	$5x^{15} + x^{14} + 2x^8$	Good sep.
$8 < e_2 \leq 14$	$6x^{14} + 2x^7$	Bad

6 Dynamics

Let $f : \mathbb{P}^1 \to \mathbb{P}^1$ be a rational map, and let f^n denote the nth iterate of f. The *(forward) orbit* of a point P under f is the set $\mathcal{O}_f(P) = \{f^n(P) : n \geq 0\}$. The *backward orbit* of a point P under f is the set $\bigcup_{n=1}^{\infty}\{Q \in \mathbb{P}^1 : f^n(Q) = P\}$. We say a point $P \in \mathbb{P}^1$ is *periodic* if $f^n(P) = P$ for some positive integer n. The smallest such n is called the *exact period* of P. For a point P of exact period n, we define the *multiplier* of f at P to be the nth derivative of f evaluated at P, denoted by $\lambda_P(f)$. A point P is *preperiodic* if $f^n(P) = f^m(P)$ for some positive integers $n \neq m$. If P is preperiodic but not periodic, we say it is *strictly preperiodic*. Let PrePer($f, \mathbb{Q}$) denote the set of all rational preperiodic points for f. Our goal is to determine PrePer($f, \mathbb{Q}$) for an interesting class of Belyi maps.

Theorem 2 *Let f be a normalized Belyi map of combinatorial type $(d; e_1, e_2, e_3)$, where d satisfies at least one of the following conditions:*

1. *$p = 2$ is a divisor of d with valuation $\ell = \nu_2(d)$*
2. *$p = 3$ is a divisor of d with valuation $\ell = \nu_3(d)$*
3. *$d = p^\ell$ for some prime p*

Assume that $e_2 \leq p^\ell$. Then PrePer($f, \mathbb{Q}$) consists entirely of all rational fixed points for f and their rational preimages.

Recall that the condition $e_2 \leq p^\ell$ implies that f has good monomial reduction modulo p (Theorem 1). To prove Theorem 2, we will use the following well-known theorem.

Theorem 3 ([14, Theorem 2.21]) *Let $f : \mathbb{P}^1_K \to \mathbb{P}^1_K$ be a rational function of degree $d \geq 2$ defined over a local field K with residue field k of characteristic p. Assume that f has good reduction and that $P \in \mathbb{P}^1(K)$ is a periodic point for f of exact period n. Let m denote the exact period of $\overline{P}$ under the reduced map $\overline{f}$, and let r denote the order of the multiplier $\lambda_{\overline{f}}(\overline{P})$ in k^*. Then one of the following holds:*

$$n = m$$

$$n = mr$$

$$n = mrp^e, e \in \mathbb{Z}, e > 0.$$

Proof (Proof of Theorem 2) Let p be a prime in one of the three cases of the statement. To apply Theorem 3, we consider f as element of $\mathbb{Q}_p(x)$.

First suppose that $d = p^\ell$. When we reduce f modulo p, we get $\overline{f}(x) = x^d$ (Theorem 1). All points in $\mathbb{P}^1(\mathbb{F}_p)$ are fixed points for $\overline{f}$. Moreover, they are all critical points because the derivative of $\overline{f}$ is identically zero, so the multiplier of any point in $\mathbb{F}_p$ is zero. In the language of Theorem 3, for any $\alpha \in \mathbb{Q}$ that is periodic under f, we have $m = 1$ and $r = \infty$. Therefore, $n = 1$, so any rational periodic point for f must be a fixed point.

If $2 \mid d$, reduce f modulo 2 to get $\overline{f}(x) = x^d$. All points in $\mathbb{P}^1(\mathbb{F}_2)$ are fixed and critical, so Theorem 3 implies that any periodic point for f in $\mathbb{Q}$ must also be fixed.

Now assume that $3 \mid d$. In the case that d is even, the points in $\mathbb{P}^1(\mathbb{F}_3)$ are all fixed under the reduction $\overline{f}$ of f modulo 3. In the case that d is odd, the points $0, 1, \infty$ are fixed and $\overline{f}(-1) = 1$. This implies that -1 is strictly preperiodic. In either case, the only periodic points for $\overline{f}$ are fixed and critical, so once again Theorem 3 implies that all rational periodic points for f must also be fixed points.

In all cases, the only periodic rational points for f are fixed points. Thus, $\mathrm{PrePer}(f, \mathbb{Q})$ consists solely of rational fixed points and their rational preimages.

□

Remark 8 Each of the three conditions on primes dividing d in Theorem 2 ensures that all periodic points for the reduced map $\overline{f}$ are fixed points. This is not always true for arbitrary d and p. For example, if $d = 35$ and we reduce modulo 5, the resulting map $\overline{f}(x) = x^{35}$ on $\mathbb{F}_5$ contains a periodic cycle of length two: $\overline{f}(2) = 3$ and $\overline{f}(3) = 2$. If we instead reduce modulo 7, we see that $\overline{f}$ on $\mathbb{F}_7$ also has a 2-cycle: $\overline{f}(2) = 4$ and $\overline{f}(4) = 2$. Thus in this case, we cannot use Theorem 3 to deduce a statement analogous to that of Theorem 2 because it is possible that f contains a rational periodic point of exact period 2.

The following proposition gives a slightly stronger statement than Theorem 2 in the first case of that theorem.

Proposition 7 *Let f be the unique normalized Belyi map of combinatorial type $(d; d-k, k+1, d)$. Write $\nu := \nu_2(d)$ for the 2-adic valuation of d. Assume that $k+1 \le 2^\nu$. Then the only fixed points of f in $\mathbb{P}^1(\mathbb{Q})$ are $x = 0, 1, \infty$.*

Proof Recall from Theorem 1 that the condition $k+1 \le 2^\nu$ implies that f has good monomial reduction to characteristic 2. As in Remark 5, we write

$$f(x) = x^{d-k}\left(\sum_{i=0}^{k} c_i (x-1)^i\right), \qquad \text{with } c_i = (-1)^i \binom{d-k+i-1}{i}.$$

In particular, $c_0 = 1$. One easily checks that

$$h(x) := \frac{f(x) - x}{x(x-1)} = \left(\sum_{i=0}^{d-k-2} x^i + x^{d-k-1} \sum_{i=0}^{k-1} c_{i+1}(x-1)^i\right).$$

Since f is branched at 3 points, we have that $d - k \ge 2$. It follows that

$$h(0) \equiv 1 \pmod 2, \qquad h(1) = d - k - 1 - \binom{d-k}{1} \equiv 1 \pmod 2.$$

Therefore, the reduction $\overline{h}(x)$ of $h(x)$ modulo 2 does not have any roots in $\mathbb{F}_2$, and hence h does not have any roots in $\mathbb{Q}$. Here we have used that h has good reduction to characteristic 2, i.e., $\deg(h) = \deg(\overline{h})$. This implies that h does not have any rational roots that specialize to ∞ when reduced modulo 2. □

We will now look at one particular family of normalized Belyi maps and use Theorem 2 to determine PrePer(f, $\mathbb{Q}$). Let $d \geq 3$ be the degree of f. Consider the following family:

$$f(x) = -(d-1)x^d + dx^{d-1}. \tag{17}$$

Recall from Example 1 that this is the unique normalized Belyi map of combinatorial type $(d; d-1, 2, d)$.

Proposition 8 *Let f be defined as in Eq. (17). Then:*

1. *The only fixed points for f in $\mathbb{P}^1(\mathbb{Q})$ are 0, 1, and ∞ (for all d) and $\frac{1}{2}$ (for $d = 3$).*
2. *The only additional rational points in the backward orbits of these fixed points are $\frac{d}{d-1}$ (for all d) and $-\frac{1}{2}$ (for $d = 3$).*

Proof

1. The fixed points of f are the roots of $f(x) - x = -(d-1)x^d + dx^{d-1} - x$, which factors as follows:

$$f(x) - x = x(x-1)(-(d-1)x^{d-2} + x^{d-3} + x^{d-4} + \ldots + x + 1).$$

By the rational root theorem, any non-zero rational zero of the above polynomial is of the form $\frac{1}{b}$, where b divides $d-1$. If $\frac{1}{b}$ is a root of $f(x) - x$, then b satisfies:

$$\frac{b^{d-1} - 1}{b - 1} = b^{d-2} + b^{d-3} + \ldots + b + 1 = d. \tag{18}$$

Claim Equation (18) does not have any integer solutions for $d \geq 4$.

Statement (1) immediately follows from the claim.

By inspection, it follows that $b \notin \{0, \pm 1\}$, so we may assume $|b| \geq 2$. Note that we must have $b \leq -2$ because if $b > 1$, the left hand side of Eq. (18) is strictly greater than d. Moreover, since b is negative, d must be even, since the left hand side of Eq. (18) is positive. Since $d \geq 4$ and $b \leq -2$, we have:

$$\sum_{i=0}^{d-2} b^i > b^{d-2} + b^{d-3} = (-b)^{d-3}(-b+1) \geq 3 \cdot 2^{d-3} > d.$$

The claim follows.

2. We have the following by direct calculation:

$$f^{-1}(0) = \left\{0, \frac{d}{d-1}\right\}$$

If $d = 3$, $f^{-1}(1) = \{1, -\frac{1}{2}\}$. Otherwise, if $d > 3$, an argument similar to that in Part 1 shows that $f^{-1}(1) \bigcap \mathbb{Q} = \{1\}$: Suppose that $f(\frac{1}{b}) = 1$, where $b \in \mathbb{Z}$. (By the rational root theorem, any such rational preimage is of this form.) Then,

$f(\frac{1}{b}) - 1 = 0$, which, after factoring $(x-1)$ from the left hand side, gives the following equation:

$$\sum_{i=0}^{d-1} b^i = d.$$

Note that $b = 1$ is one solution to this equation. Any other solution for b would require $b < 0$ and in particular, $b \le -2$. Therefore, d must be odd for the sum to be positive. If $d \ge 5$, we have the following:

$$\sum_{i=0}^{d-1} b^i \ge b^{d-1} + b^{d-2} \ge |b|^{d-2} \ge 2^{d-2} > d.$$

Thus, $f^{-1}(1) \bigcap \mathbb{Q} = \{1\}$.

A direct calculation also shows that if $d = 3$, then $-\frac{1}{2}$ has no rational preimages, and $\frac{1}{2}$ has no rational preimages except itself. It remains to show that $\frac{d}{d-1}$ has no rational preimages. Suppose $f(\frac{a}{b}) = \frac{d}{d-1}$ for some relatively prime integers a and b. The rational root theorem implies that $a|d$ and $b|(d-1)^2$. After clearing denominators, we have the following equation:

$$-(d-1)^2 a^d + d(d-1)a^{d-1}b - db^d = 0. \tag{19}$$

Reducing modulo $d-1$ yields $-b^d \equiv 0$, so $(d-1)|b^d$. Reducing modulo d yields $-a^d \equiv 0$, so $d|a^d$. Let p be a prime dividing d. Suppose that the valuation $\nu_p(d) = k \ge 1$ and $\nu_p(a) = \ell$, for $1 \le \ell \le k$. Then $\nu_p(-(d-1)^2 a^d + d(d-1)a^{d-1}b - db^d) = k$ because $\nu_p(db^d) = k$ and $\nu_p(-(d-1)^2 a^d + d(d-1)a^{d-1}b) \ge \max\{\ell d, k + \ell^{d-1}\} > k$. This contradicts Eq. (19).

□

Corollary 3 *Let f be the polynomial of degree d in the family defined in Eq.* (17), *where either $2 \mid d, 3 \mid d$, or $d = p^\ell$ for some prime p. Then:*

1. *PrePer$(f, \mathbb{Q}) = \{0, 1, \frac{3}{2}, \frac{1}{2}, -\frac{1}{2}, \infty\}$ if $d = 3$.*
2. *PrePer$(f, \mathbb{Q}) = \left\{0, 1, \frac{d}{d-1}, \infty\right\}$ if $d \neq 3$.*

Proof Theorem 2 states that PrePer$(f, \mathbb{Q})$ consists solely of fixed points for f and their rational preimages. Proposition 8 then completely describes all rational preperiodic points for f. □

Remark 9 The statement of Proposition 8.(2) may be partially generalized. For simplicity we restrict to the case that f is the unique normalized Belyi map of combinatorial type $(d; d-k, k+1, d)$. An explicit formula for f was determined in Proposition 2. We use the terminology of that result.

In the proof of Proposition 2, we showed that the derivative of f satisfies

$$f'(x) = (-1)^k c x^{d-k-1}(x-1)^k, \qquad \text{with } c > 0.$$

Distinguishing four cases depending on whether k and d are even or odd and considering the sign of f' yields the following statement for the real elements in the fibers $f^{-1}(0)$ and $f^{-1}(1)$.

1. Suppose that d and k are both even. Then $f^{-1}(0) \cap \mathbb{R} = \{0\}$ and $f^{-1}(1) \cap \mathbb{R} = \{1, \beta\}$ for some $\beta < 0$.
2. Suppose that d is odd and k is even. Then $f^{-1}(0) \cap \mathbb{R} = \{0\}$ and $f^{-1}(1) \cap \mathbb{R} = \{1\}$.
3. Suppose that d is even and k is odd. Then $f^{-1}(0) \cap \mathbb{R} = \{0, \gamma\}$ for some $\gamma > 1$ and $f^{-1}(1) \cap \mathbb{R} = \{1\}$.
4. Suppose that d and k are both odd. Then $f^{-1}(0) \cap \mathbb{R} = \{0, \gamma\}$ for some $\gamma > 1$ and $f^{-1}(1) \cap \mathbb{R} = \{1, \beta\}$ for some $\beta < 0$.

In particular, this determines the rational values in $f^{-1}(0)$ and $f^{-1}(1)$ in the case that d is odd and k is even. In the other cases, in principle it is possible to analyze when the real roots β, γ are rational, similarly to the proof of Lemma 8. In Proposition 2, we showed that the leading coefficient of $f(x)$ is $ca_0 = (-1)^k \binom{d-1}{k}$. It follows that if $\beta < 0$ is a rational root of $f(x) - 1$, then we have

$$\beta = \frac{-1}{b} \qquad \text{with } b \in \mathbb{N} \text{ such that } b \mid \binom{d-1}{k}.$$

Similarly, assume that $\gamma > 1$ is a rational root of $f(x)$. We use the expression $f(x) = x^{d-k} \sum_{i=0}^{k} c_i (x-1)^i$ from Remark 5. Since $c_0 = 1$ and $c_k = \pm\binom{d-1}{k}$, we find

$$\gamma = 1 + \frac{1}{c} \qquad \text{with } c \in \mathbb{N} \text{ such that } c \mid \binom{d-1}{k}.$$

Acknowledgements This project began at the Women in Numbers Europe 2 conference at the Lorentz Center. We thank the Lorentz Center for providing excellent working conditions, and we thank the Association for Women in Mathematics for supporting WIN-E2 and other research collaboration conferences for women through their NSF ADVANCE grant. We also thank the referee for numerous helpful comments, all of which greatly improved the paper.

MM partially supported by NSF-HRD 1500481 (AWM ADVANCE grant) and by the Simons Foundation grant #359721.

References

1. R. Benedetto, P. Ingram, R. Jones, A. Levy, Attracting cycles in p-adic dynamics and height bounds for postcritically finite maps. Duke Math. J. **163**(13), 2325–2356 (2014)
2. E. Brezin, R. Byrne, J. Levy, K. Pilgrim, K. Plummer, A census of rational maps. Conform. Geom. Dyn. **4**, 35–74 (2000)
3. K. Cordwell, S. Gilbertson, N. Nuechterlein, K.M. Pilgrim, S. Pinella, On the classification of critically fixed rational maps. Conform. Geom. Dyn. **19**, 51–94 (2015)

4. M. Eskin, Stable reduction of three-point covers. Ph.D. thesis, Ulm University, 2015. https://doi.org/10.18725/OPARU-3280
5. F. Liu, B. Osserman, The irreducibility of certain pure-cycle Hurwitz spaces. Am. J. Math. **130**(6), 1687–1708 (2008)
6. D. Lukas, M. Manes, D. Yap, A census of quadratic post-critically finite rational functions defined over $\mathbb{Q}$. LMS J. Comput. Math. **17**(Suppl. A), 314–329 (2014)
7. B. Osserman, Rational functions with given ramification in characteristic p. Compos. Math. **142**(2), 433–450 (2006)
8. B. Osserman, Linear series and the existence of branched covers. Compos. Math. **144**(1), 89–106 (2008)
9. F. Pakovich, Conservative polynomials and yet another action of $\mathrm{Gal}(\overline{\mathbb{Q}}/\mathbb{Q})$ on plane trees. J. Théor. Nombres Bordeaux **20**(1), 205–218 (2008)
10. K.M. Pilgrim, An algebraic formulation of Thurston's characterization of rational functions. Ann. Fac. Sci. Toulouse Math. (6) **21**(5), 1033–1068 (2012)
11. S.M. Ruiz, An algebraic identity leading to Wilson's theorem. Math. Gaz. **80**(489), 579–582 (1996)
12. L. Schneps (ed.), *The Grothendieck Theory of Dessins D'enfants*. London Mathematical Society Lecture Note Series, no. 200 (Cambridge University Press, Cambridge, 1994)
13. J. Sijsling, J. Voight, *On Computing Belyi Maps*, Numéro consacré au trimestre "Méthodes arithmétiques et applications", automne 2013. Publ. Math. Besançon Algèbre Théorie Nr., vol. 2014/1 (Presses Univ. Franche-Comté, Besançon, 2014), pp. 73–131
14. J.H. Silverman, *The Arithmetic of Dynamical Systems*. Graduate Texts in Mathematics, vol. 241 (Springer, New York, 2007)
15. J.H. Silverman, *Moduli Spaces and Arithmetic Dynamics*. CRM Monograph Series, vol. 30 (American Mathematical Society, Providence, 2012)
16. D. Tischler, Critical points and values of complex polynomials. J. Complexity **5**(4), 438–456 (1989)
17. H. Völklein, *Groups as Galois Groups*. Cambridge Studies in Advanced Mathematics, vol. 53 (Cambridge University Press, Cambridge, 1996)
18. D. Zeilberger, A fast algorithm for proving terminating hypergeometric identities. Discrete Math. **306**(10–11), 1072–1075 (2006)
19. A. Zvonkin, Belyi functions: examples, properties, and applications. http://www.labri.fr/perso/zvonkin/Research/belyi.pdf

Discriminant Twins

Alyson Deines

Abstract The conductor and minimal discriminant are two invariants that measure the bad reduction of an elliptic curve. The conductor of an elliptic curve E over $\mathbb{Q}$ is an arithmetic invariant. It is an integer N that measures the ramification in the extensions $\mathbb{Q}(E[p^\infty])/\mathbb{Q}$. The minimal discriminant Δ is a geometric invariant. It counts the number of irreducible components of $\tilde{E}(\mathbb{F}_p)$. When two elliptic curves have the same conductor and discriminant, we call them **discriminant twins**. In this paper, we explore when discriminant twins occur. In particular, we prove there are only finitely many semistable isogenous discriminant twins.

Keywords Elliptic curves · Discriminant · Conductor · p-Adic Uniformization · Isogenies

2010 Mathematics Subject Classification. 11G05, 11G07, 11Y40

1 Statement of Theorem

The conductor and minimal discriminant are two invariants that measure the bad reduction of an elliptic curve. The conductor of an elliptic curve E over $\mathbb{Q}$ is an arithmetic invariant. It is an integer N that measures the ramification in the extensions $\mathbb{Q}(E[p^\infty])/\mathbb{Q}$. The minimal discriminant Δ is a geometric invariant. It counts the number of irreducible components of $\tilde{E}(\mathbb{F}_p)$.

Definition 1 We say two curves E and E' are **discriminant twins** if they have the same minimal discriminant and conductor.

In the 1990s, Ribet and Takahashi gave a method to compute the degree of parameterization of a semistable elliptic defined over $\mathbb{Q}$ curve by a Shimura curve [9]. Within each isogeny class, there is one elliptic curve such that this map has

A. Deines (✉)
CCR La Jolla, San Diego, CA, USA

I. I. Bouw et al. (eds.), *Women in Numbers Europe II*, Association for Women in Mathematics Series 11, https://doi.org/10.1007/978-3-319-74998-3_6

a connected kernel. This curve is called the optimal quotient. When computing the Shimura degree, it is not known before which curve is the optimal quotient. However, a by-product of the computation is the minimal discriminant of the optimal quotient. Thus if we do not have any semistable, isogenous discriminant twins, we can always compute the optimal quotient.

Our main theorem is as follows:

Theorem 1 (Main Theorem) *Over $\mathbb{Q}$, there are only finitely many semistable, isogenous discriminant twins. In terms of LMFDB labels [13], they are* $11a.1, 11a.3$, $17a.1, 17a.4$, $19a.1, 19a.3$, *and* $37b.1, 37b.3$.

If we do not restrict to the semistable case, there are infinitely many discriminant twins. Via twisting curves of conductor 37, we show there are infinitely many isogenous and non-isogenous discriminant twins. Since we are twisting curves, this gives us a family of discriminant twins with additive reduction and all with the same j-invariants. Even though data hints that there are infinitely many semistable, non-isogenous discriminant twins, we do not yet know how to show this. Further, the results of Ribet and Takahashi on modular parameterization of elliptic curves by Shimura curves can be generalized to elliptic curves over totally real number fields, so we can ask about discriminant twins in this case as well. Unfortunately, while our results generalize to totally real number fields, due to a plethora of units, we cannot yet reach the same conclusion.

2 Elliptic Curves of Prime Conductor

The simplest case of semistable elliptic curves is those with prime conductor. Such curves have been studied since the 1970s [11] and remain mysterious objects. We don't even know if there are infinitely many of them [4]. Due to Mestre's method of graphs, computing modular elliptic curves of prime degree has been considerably easier than the general case.

Elliptic curves of prime conductor fall into three categories based on isogeny classes and parameterization. First, there are those with singleton isogeny classes; secondly, Neumann-Setzer curves, i.e., curves with a 2-isogeny and a rational parameterization; and finally, the rest specifically curves with conductors 11, 17, 19, and 37.

Definition 2 A Neumann-Setzer curve is a curve of the form:

$$E : y^2 + xy = x^3 - \frac{u+1}{4}x^2 + 4x - u$$

where $u \equiv 3 \pmod 4$ such that $p = u^2 + 64$ is prime.

The discriminant is Δ_E of E is $-(u^2 + 64)^2$ and E is 2-isogenous to

$$E' : y^2 + xy = x^3 - \frac{u+1}{4}x^2 - x$$

with discriminant $\Delta_{E'} = u^2+64 = p$. Both curves then have conductor $N = p$. Up to conductor 1000, the only Neumann-Setzer curves are those with conductors 73, 113, and 593. In fact there are only 83 Neumann-Setzer isogeny classes of curves in the Cremona database which currently has largest conductor 379998, while there are 3765 prime conductors.

In particular, we have the following theorem of Setzer.

Theorem 2 (Setzer [11]) *Neumann-Setzer curves are the only curves with prime conductor and non-singleton isogeny class except for the curves with conductors* 11, 17, 19, *and* 37.

The following table gives the conductor, N, and the discriminants, Δ, of the curves in the non-singleton isogeny classes.

N	Δ_{a_1}	Δ_{a_2}	Δ_{a_3}	Δ_{a_4}
11	11	11^5	11	–
17	17	17^2	-17^2	17
19	-19	-19^3	-19	–
N	Δ_{b_1}	Δ_{b_2}	Δ_{b_3}	–
37	37	37^3	37	–

Corollary 3 ([11]) *All elliptic curves with prime conductor that are not Neumann-Setzer curves and not listed in the table have discriminant* $\pm p$.

Corollary 4 *The only isogenous discriminant twins with prime conductor are those listed in the table.*

How far does this generalize? A natural generalization of curves with prime conductor is semistable curves, i.e., those with square-free conductor. Searching the Cremona database for semistable, isogenous discriminant twins yield only those in the table. However, it's easy to see that by twisting any of the discriminant twins, we get infinitely many more. Searching Cremona's database, we also find examples of isogenous discriminant twins with additive reduction that are not twists of one of these known semistable isogenous discriminant twins.

The rest of this paper will be structured as follows. Necessary background will be given in the next section. The proof of Theorem 1 will then be broken up into the different isogeny degree cases, all of which are proved in an analogous manner: pick an isogeny graph, use Kubert's parameterizations of torsion to write out a parameterization of the curves with given torsion, use Velu's isogeny formula to write out the other curves in the isogeny class, write down discriminants in terms of this parameterization, and run Tate's algorithm to find restrictions on parameterizations and discriminants. Finally, we discuss generalizations of discriminant twins, in particular to the non-isogenous and number field cases.

All elliptic curves will be given in terms of their LMFDB labels. LMFDB labels are similar to Cremona labels and allow one to search for the curve in the LMFDB. See http://www.lmfdb.org/knowledge/show/ec.q.lmfdb_label for how the labels are defined.

3 p-Adic Uniformization of Elliptic Curves and Isogenies

Let K be a field and E an elliptic curve over K. Following Silverman [12], we can write E as

$$y^2 + a_1xy + a_3y = x^3 + a_2x^2 + a_4x + a_6$$

$a_1, a_2, a_3, a_4, a_6 \in K$, with the usual invariants:

$$\begin{aligned} b_2 &= a_1^2 + 4a_2 & b_4 &= 2a_4 + a_1a_3 \\ b_6 &= a_3^2 + 4a_4 & b_8 &= a_1^2a_6 + 4a_2a_6 - a_1a_3a_4 \\ & & &\quad +a_2a_3^2 - a_4^2 \\ c_4 &= b^2 - 24b_4 & c_6 &= -b_2^3 + 36b_2b_4 - 216b_6 \\ \Delta &= -b_2b_8 - 8b_4^3 - 27b_6^2 + 9b_2b_4b_6 & j &= c_4^3/\Delta \\ &= (c_4^3 - c_6^2)/1728 \end{aligned}$$

As the discriminant depends on the choice of model, it is useful to work with a fixed model. Over $\mathbb{Q}$, we can find a curve isomorphic to E with discriminant valuation minimal at each prime. We take this to be the minimal discriminant. By the Laska-Kraus algorithm, we know the minimal discriminant of E over $\mathbb{Q}$ is unique [6].

Since we will frequently change models, it is useful to have the following notation. Let E, E' be isomorphic elliptic curves defined over a field K with isomorphism $\tau : E \to E'$, with E given by

$$y^2 + a_1xy + a_3y = x^3 + a_2x^2 + a_4x + a_6$$

and E' by

$$y'^2 + a_1'x'y' + a_3'y' = x'^3 + a_2'x'^2 + a_4'x' + a_6',$$

then we denote $\tau = [u, r, s, t]$. We can write τ as

$$\begin{aligned} x &= u^2x' + r \\ y &= u^3y' + su^2x' + t \\ \Delta &= u^{12}\Delta' \end{aligned}$$

Notice that τ changes the discriminant by 12th powers of u.

If all of $u, r, s, t \in K$, then we say τ is defined over K, and thus E and E' are isomorphic over K or simply isomorphic. In general, τ need not be defined over K. The isomorphism τ could instead be defined over some extension of K. When this happens, E and E' have the same j-invariants, $j = j'$, and we say E and E' are twists. In particular, we will be interested in quadratic twists of elliptic curves over $\mathbb{Q}$. The quadratic twist of E by d given in short Weierstrass form, denoted by E^d, is

$$dy^2 = x^3 + Ax + B.$$

This isomorphism is defined over $K(\sqrt{d})$. When the conductors N and d are coprime, we see that it changes the discriminant by 6th powers of D where D is the discriminant of $\mathbb{Q}(\sqrt{d})$. Similarly, it changes the conductor from N to ND^2 [2].

Let $E/\mathbb{Q}$ be a minimal model with discriminant Δ. For each prime p dividing Δ, we can examine $\tilde{E}$, the reduction of $E \pmod{p}$. The primes p that divide the minimal discriminant Δ are exactly the primes where E is not smooth mod p, i.e., the primes of bad reduction. Further, the primes dividing Δ can give a model with bad reduction in two different ways. Either $\tilde{E}$ can have a singularity and one tangent line, a cusp, in which case we say $\tilde{E}$ has additive reduction, or $\tilde{E}$ can have a singularity and two tangent lines, a node. If the slopes of the tangent lines are defined over the residue field, we say $\tilde{E}$ has split multiplicative reduction. Otherwise, if the slopes are defined over some extension of the residue field, we say $\tilde{E}$ has non-split multiplicative reduction.

The conductor N of an elliptic curve E can be written explicitly as

$$N = \prod_{p|\Delta} p^{f_p}$$

where $f_p = 1$ if E has a node (i.e., semistable reduction) at p, $f_p = 2$ if $p \geq 5$ and E has a cusp (i.e., additive reduction), and $f_p = 2 + \delta$ if $p = 2, 3$ E has a cusp at p, with $\delta \in \mathbb{Z}_+$. The conductor and minimal discriminant are both computable via Tate's algorithm [12]. An elliptic curve E is semistable if it has only multiplicative reduction, i.e., if N is square-free. Note that the conductor is an isomorphism class invariant; in fact, it does not depend on the choice of model. Further, the conductor is actually an isogeny class invariant and does not depend on the choice of curve in the isogeny class.

If K is a field of characteristic zero, then we can write E in short Weierstrass form

$$y^2 = x^3 + Ax + B, \qquad A, B \in K.$$

The isogeny defined by the composition $\tau = [1, -b_2/12, 0, 0] \circ [1, 0, -a_1/2, -a_3/2]$ sends E to this form. Note that for both isogenies in the composition of τ, we have $u = 1$, so the discriminant does not change. Thus if $E/\mathbb{Q}$ has a minimal discriminant, so does the short Weierstrass form.

Let K be an algebraically closed field of characteristic zero. If E and E' are isogenous elliptic curves defined over K, then there is a nonconstant map that sends identities to identities: $\mathcal{O} \mapsto \mathcal{O}'$. Let $\mu : E \to E'$ be an isogeny, then we can write μ as a rational map. In particular, if E and E' are in short Weierstrass form and $(x, y) \in E$, $\mu(x, y) = \left(f_\mu(x), \frac{1}{\gamma_\mu} y \frac{df_\mu(x)}{dx}\right)$, with $f_\mu \in K(x)$ and $\gamma_\mu \in K^\times$. If we know the kernel of the isogeny, we can write down f_μ explicitly. Let C be a finite, nontrivial subgroup of $E(K)$ and write E in short Weierstrass form

$y^2 = x^3 + Ax + B$. If $Q \in E(K)$, denote the coordinates of Q by $(x(Q), y(Q))$. Following González's reformulation of Vélu's formulas [3], we can write the isogenous curve explicitly in terms of C. Let $\mu_C : E \to E_C$ denote the isogeny given by $\mu_C(x, y) = \left(f_C(x), y\frac{df_C(x)}{dx}\right)$ with

$$f_C(x) = x + \sum_{Q \in C \setminus \{\mathcal{O}\}} \left(\frac{t(Q)}{x - x(Q)} + \frac{u(Q)}{(x - x(Q))^2}\right),$$

where $t(Q) = 3x(Q)^2$ and $u(Q) = 2(x(Q)^3 + Ax(Q) + B)$. Then take $A_C = A - 5t$, $t = \sum_{Q \in C \setminus \{\mathcal{O}\}} t(Q)$ and $B_C = B - 7w$ with $w = \sum_{Q \in C \setminus \{\mathcal{O}\}} (u(Q) + x(Q)t(Q))$, we can write E_C as $y^2 = x^3 + A_C x + B_C$.

4 Infinitely Many Discriminant Twin Pairs

Proposition 5 *There are infinitely many pairs of isogenous discriminant twins and infinitely many pairs of non-isogenous discriminant twins over* $\mathbb{Q}$.

Proof Quadratic twisting preserves isogenies, i.e., E and E' are isogenous if and only if the twists E^d and E'^d are isogenous [2]. We can prove both the isogenous and non-isogenous cases at the same time. We begin by examining the curves of conductor 37, as there are three curves in two isogeny classes with the same conductor and discriminant giving us our initial curves to twist. These curves are $37a.1$, $37b.1$, and $37b.3$, and all have conductor and discriminant 37. Let $p \neq 37$, $p \equiv 1 \pmod 4$ so that the discriminant of $\mathbb{Q}(\sqrt{p})$ is p. Then the quadratic twists of $37a.1$, $37b.1$, and $37b.3$ by p all have conductor $37p^2$ and discriminant $37p^6$. Moreover, twisting commutes with isogenies [2], so the twists of $37a.1$ and $37b.1$ remain non-isogenous, while the twists of $37b.1$ and $37b.3$ remain isogenous. Thus twisting $37a.1$ and $37b.2$ gives a family of non-isogenous discriminant twins, and twisting $37b.1$ and $37b.3$ gives a family of isogenous discriminant twins.

With the exception of the original curves, all curves in either family will have additive reduction at the prime p. While this gives infinitely many discriminant twin pairs, it does nothing to address whether there are infinitely many semistable, non-isogenous discriminant twins. Further, we do not even know if infinitely many j-invariants occur among the discriminant twins.

Question 1 Do infinitely many j-invariants occur among all discriminant twin pairs?

5 Semistable Isogenous Curves

For semistable curves, we use the Tate curve parameterization over $\mathbb{Q}_p$ to examine how discriminants change under isogeny. For background and proofs, see Chapter V, Sections 3–6 of [12]. Let K be a number field, fix a prime ideal $\mathfrak{p}$ in the ring of integers $\mathbb{Z}_K$, and let k be the residue field of $\mathfrak{p}$. Take $K_\mathfrak{p}$ to be the completion of K at $\mathfrak{p}$ and $\nu_\mathfrak{p} = \nu$ to be the valuation map.

If E has multiplicative reduction, then there exists a unique parameter $q \in \mathbb{Z}_{K_\mathfrak{p}}$ with $\nu(q) \geq 1$ such that $E\left(\overline{K}_\mathfrak{p}\right) \cong \overline{K}_\mathfrak{p}^\times/q^\mathbb{Z}$. Let $\gamma(E) = -c_4/c_6 \in K_\mathfrak{p}^\times/K_\mathfrak{p}^{\times 2}$. If $\sqrt{\gamma(E)} \in K_\mathfrak{p}$, then E has split multiplicative reduction at $\mathfrak{p}$ and $E(K_\mathfrak{p}) \cong K_\mathfrak{p}^\times/q^\mathbb{Z}$. If E has non-split multiplicative reduction at $\mathfrak{p}$, then for the extension $L_\mathfrak{p} = K_\mathfrak{p}\left(\sqrt{\gamma(E)}\right)$ of $K_\mathfrak{p}$, $E(L_\mathfrak{p}) \cong L_\mathfrak{p}^\times/q^\mathbb{Z}$. Additionally, we can write the discriminant and j-invariant of E in terms of its Tate parameter q: $\Delta(E) = q\prod_{n\geq 1}(1-q^n)^{24}$ with $\nu(\Delta(E)) = \nu(q)$, and $j(E) = \frac{1}{q} + \sum_{n\geq 0} c_n q^n$, where the c_n can be explicitly described, so that $\nu(j) = -\nu(q)$.

We now use Tate curves to study how discriminants change under isogeny. Let ℓ be a rational prime and $\mathfrak{p}$ a prime of the field K. Let E and E' be two ℓ-isogenous elliptic curves defined over K with split multiplicative reduction at $\mathfrak{p}$. Let E have parameter q and E' have parameter q'. The ℓ-isogeny and its dual can be written explicitly:

$$\begin{array}{ccc} K_\mathfrak{p}^\times/q^\mathbb{Z} \to K_\mathfrak{p}^\times/(q')^\mathbb{Z} & \text{and} & K_\mathfrak{p}^\times/(q')^\mathbb{Z} \to K_\mathfrak{p}^\times/q^\mathbb{Z} \\ u \mapsto u^n & & v \mapsto v^m \end{array}$$

for some integers n, m such that $q^n = (q')^m$ and $nm = \ell$. Since ℓ is prime, pick $n = 1$ and $m = \ell$ and $q = (q')^\ell$. Thus we can write precisely how the discriminant changes, $\nu(\Delta) = \ell\nu(\Delta')$.

If E has non-split multiplicative reduction, then E' does as well [12]. Then $E(L) \cong L_\mathfrak{p}^\times/q^\mathbb{Z}$, $E'(L) \cong L_\mathfrak{p}^\times/q'^\mathbb{Z}$, and the same argument applies. Thus, if an isogeny class has only two curves, as the isogeny between the curves must be prime degree, we get the following corollary:

Corollary 6 *Isogeny classes of size two with at least one prime of multiplicative reduction cannot have discriminant twins.*

When examining curves with only additive reduction, we quickly see that curves with prime isogeny can be discriminant twins. The first such example is for the curves $32.a1, 32.a2, 32.a3, and 32.a4$ which all have discriminants $2^9, 2^9, 2^6, and\ -2^{12}$, respectively, and have the following isogeny graph where each edge represents a 2-isogeny between curves.

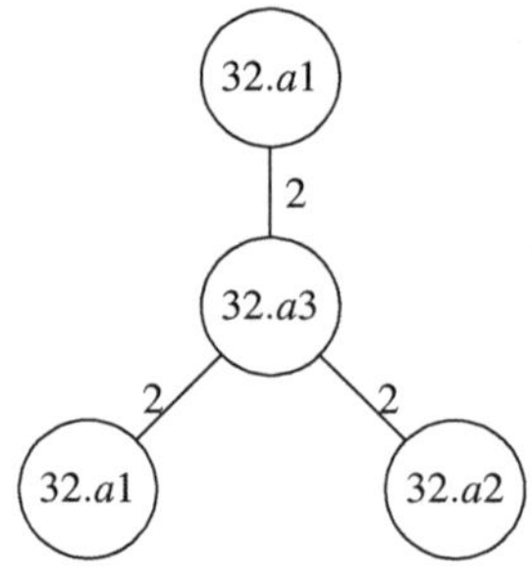

Again assuming a prime of semistable reduction, note that it follows from the above discussion that if E has two prime isogenies ℓ and ℓ', then the curves E_ℓ and $E_{\ell'}$ cannot have the same discriminant. Thus we are left with the prime power isogeny case.

Now let's examine isogeny classes with three non-isomorphic curves, E, E', E'', where there is a degree ℓ isogeny from $E \to E'$ and another degree ℓ isogeny from $E' \to E''$. Again we are assuming split multiplicative reduction at $\mathfrak{p}$. Then we get the same Tate parameterization as before, q, q', q'', respectively, and get the following sequence:

$$K_{\mathfrak{p}}^{\times}/q^{\mathbb{Z}} \to K_{\mathfrak{p}}^{\times}/(q')^{\mathbb{Z}} \to K_{\mathfrak{p}}^{\times}/(q'')^{\mathbb{Z}}.$$

If we start as before with $q = (q')^{\ell}$, then there are two cases for the second isogeny:

1. $q' = (q'')^{\ell}$ and then $\nu(\Delta) = \ell\nu(\Delta') = \ell^2\nu(\Delta'')$.
2. $q'' = (q')^{\ell}$, but this implies that $q'' = q$. As E and E'' are non-isomorphic, over K, they have distinct j-invariants; thus, they also do over $K_{\mathfrak{p}}$ and so $q'' \neq q$ and this case cannot occur.

Now instead assume $q' = q^{\ell}$. If $q'' = (q')^{\ell}$, we are in analogous case 1, and $\nu(\Delta'') = \ell\nu(\Delta') = \ell^2\nu(\Delta)$. If instead $q' = (q'')^{\ell}$ we have $(q'')^{\ell} = q^{\ell}$. If $K_{\mathfrak{p}}$ does not contain ℓ-th roots of unity, this implies that $q'' = q$ and again this cannot occur. If, however, $K_{\mathfrak{p}}$ does contain the ℓ-th roots of unity, this can and does occur. Over $\mathbb{Q}_p$ this occurs when $p \equiv 1 \pmod{\ell}$. Analyzing the discriminants, $\nu(\Delta) = \nu(\Delta'')$ and $\nu(\Delta') = \ell\nu(\Delta) = \ell\nu(\Delta'')$. In this case, we say that E and E'' are ℓ-discriminant twins (Fig. 1).

Using the same notation as before, we summarize this in the following table. Let q be the Tate parameter.

Corollary 7 *If E' is a semistable elliptic curve with two ℓ-isogenies, such that*

$$E \to E' \to E''$$

then $\nu_p(\Delta(E')) \equiv 0 \pmod{\ell}$ for all $p \mid \Delta(E')$.

Proof Following the above table of all possible q-parameters for the Tate curves, in the first two rows, q' is already an ℓth power. The third case cannot occur. In the fourth case, as $\nu(q'') \geq 1$, $q' = q^{1/\ell}$ must be an ℓth power so that the ℓth root is

Fig. 1 Possible q' and q'' given a Tate parameter q and $E \to E' \to E''$

q'	q''	Behavior
q^ℓ	$(q^\ell)^\ell$	not disc. twins
q^ℓ	$(q^\ell)^{1/\ell}$	ℓ-disc. twins
$q^{1/\ell}$	$(q^{1/\ell})^\ell = q$	not disc. twins
$q^{1/\ell}$	$(q^{1/\ell})^{1/\ell}$	not disc. twins

in $K_\mathfrak{p}$. As E' is semistable, we do this for all primes dividing $\Delta(E')$. Thus, modulo units, $\Delta(E')$ is an ℓ-th power.

6 Main Theorem

The goal of this section is to show that there are only finitely many semistable, isogenous, discriminant twin pairs of elliptic curves over $\mathbb{Q}$. Locally, we see that ℓ-discriminant twins occur frequently but only when the two curves have prime power isogenies. Using results of Mazur [8] and Kenku [5], we cut down on possible isogeny degrees as follows: The only prime power isogeny degrees that occur for semistable elliptic curves defined over $\mathbb{Q}$ are $\ell = 2^2, 2^3, 2^4, 3^2$, or 5^2 and have the following prime degree isogeny graphs:

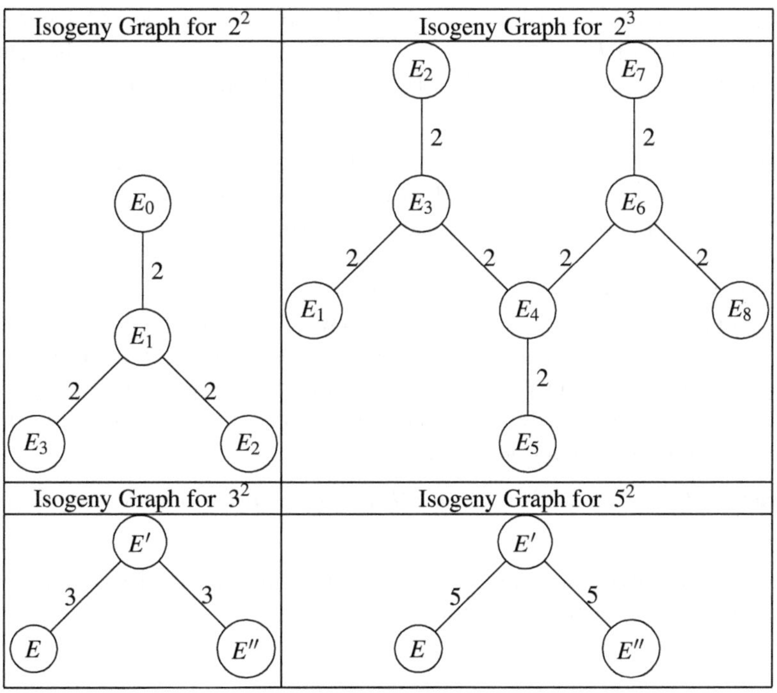

Kubert's [7] parameterizations of curves with rational torsion allow us to explicitly write the curves with prescribed torsion. Narrowing to semistable curves, this allows us to examine isogenies of a given degree. We then study how the discriminants of these curves change under isogeny. In this section, we go through each case, $2^n, n = 2, 3, 4$, and then $3^2, 5^2$, and find all possible semistable, isogenous, discriminant twin pairs of each isogeny degree. To set notation, let E, E', E'' be isogenous elliptic curves with isogenies of degree ℓ between $E \to E'$ and $E' \to E''$ and a degree ℓ^2 isogeny between $E \to E''$, so we have the following sequence:

$$E \to E' \to E''.$$

When $\ell = 3$ or 5, we need the following lemma.

Lemma 8 *If $K = \mathbb{Q}$ and $\ell = 3$ or 5, E and E' have rational ℓ torsion and E'' does not.*

Proof For any two ℓ-isogenous curves defined over $\mathbb{Q}$, $E \to E'$, the kernel is either $\mathbb{Z}/\ell\mathbb{Z}$, rational torsion in E or μ_ℓ, a Galois invariant subgroup. Thus one of E or E' must have rational ℓ-torsion [10]. By Mazur's theorem [8], the torsion subgroup of an elliptic curve over $\mathbb{Q}$ is only divisible by ℓ^2 if $\ell = 2$. Thus for $\ell \geq 3$ and ℓ-isogenous elliptic curves

$$E \to E' \to E'',$$

either E and E' both have ℓ-rational torsion, E' and E'' both have ℓ-rational torsion, or E and E'' have ℓ-rational torsion. As E' has two ℓ isogenies $E' \to E''$ and $E' \to E$, the dual of $E \to E'$, and $\ell \geq 3$, then the kernel of one of the isogenies is $\mathbb{Z}/\ell\mathbb{Z}$. Thus E' has ℓ-rational torsion. By switching E and E'' if necessary, we get the first statement.

Now we can start the case-by-case work for $\ell = 2, 3, 5$. All code for the computations in the proofs is found in the appendix and can also be found on Github: https://github.com/adeines/DiscriminantTwins. All three cases follow a similar pattern of proof. First, we use the parameterization of curves with given torsion as found in Kubert [7]. Next, we use the parameterization to find all the torsion points. Once we have the torsion points, we write the curves in a form which allows us to use Vélu's formulas to find the other curves in the isogeny class and write their discriminants in terms of the initial parameters. Finally, we assume semistability and again use Tate curves to find the only cases in which discriminant twins with semistable reduction can occur.

Theorem 9 *There are no 5-isogenous discriminant twins outside the pair $11a.2$ and $11a.3$.*

We start by proving a slightly stronger statement. Kubert gives the following parameterization for curves with rational 5-torsion:

$$E_{s,r} : y^2 + (s - r)xy - rs^2y = x^3 - rsx^2$$

for some $r, s \in \mathbb{Z}$, r, s coprime and a rational point $(0, 0)$ of order 5.

Proposition 10 *The only discriminant twins with 5-torsion occur when* $rs = \pm 1$ *or are twists of said curves.*

In fact what we show is that there are no semistable p-discriminant twins for $p \mid rs$.

Proof As above, let E, E', E'' be elliptic curves over $\mathbb{Q}$ such that there are $\ell = 5$ isogenies $E \to E'$ and $E' \to E''$, so we have the following sequence:

$$E \to E' \to E''$$

and we can assume both E and E' both have rational 5-torsion (and that E'' does not). In terms of Kubert's parameterization E has the form

$$y^2 + (s - r)xy - rs^2 y = x^3 - rsx^2$$

for some $r, s \in \mathbb{Z}$ coprime and a rational point $(0, 0)$ of order 5. Then the discriminant is

$$\Delta(E) = \Delta = r^5 s^5 (r^2 - 11rs - s^2) \in \mathbb{Z}.$$

Using SageMath [14], we can compute all the points in $E(\mathbb{Q})[5]$ in terms of r and s. They are the point at infinity $\mathcal{O}$, $(0, 0)$, $(rs, 0)$, $(0, rs^2)$, and $(rs, r^2 s)$. Since we know all the points in $E(\mathbb{Q})[5]$, it is possible to use Vélu's formula to explicitly compute E' [15].

$$\begin{aligned} y^2 + (s - r)xy - rs^2 y &= x^3 + rsx^2 \\ &\quad - 5rs(r^2 - 2rs - s^2)x - rs(r^4 + 10r^3 s - 5r^2 s^2 + 15rs^3 - s^4) \end{aligned}$$

Doing so we find

$$\Delta(E') = \Delta' = rs(r^2 - 11rs - s^2)^5.$$

Note that $\Delta(E)$ is already minimal at $p \mid rs$. To see this, pick any prime p dividing rs. Note that $c_4(E) = r^4 - 12r^3 s + 14r^2 s^2 + 12rs^3 + s^4$ is not divisible by p, Examining the parameterization of E by r, s, we can reduce mod p to get either:

$$y^2 + sxy = x^3 \text{ or } y^2 - rxy = x^3$$

where p divides r or s, respectively. Using Tate's algorithm, as $p \nmid c_4$, E has multiplicative reduction. As $T^2 - a_1 T - a_2 \equiv T(T + s)$ or $T(T - r) \pmod{p}$, again where p divides r or s, respectively, both tangents are in $\mathbb{F}_p$ and E has split multiplicative reduction at p. Thus E' and E'' also have split multiplicative reduction at p and we get the following sequence of Tate curve parameterizations:

$$\mathbb{Q}_p^\times / q^{\mathbb{Z}} \to \mathbb{Q}_p^\times / (q^{1/5})^{\mathbb{Z}} \to \mathbb{Q}_p^\times / (q'')^{\mathbb{Z}}.$$

The possibilities for q'' are as follows. If $q'' = (q^{1/5})^5 = q$, then we have $q = q''$ and $E = E''$ over $\mathbb{Q}_p$. Since we assumed that E and E'' are isogenous and not isomorphic, this cannot occur. Thus we have $q'' = q^{1/25}$ and $25v_p(\Delta'') = v_p(\Delta)$. So 5-discriminant twins can only occur if ab is a unit or if we take a twist of E, which will then have additive reduction.

Over $\mathbb{Q}$, if $ab = \pm 1$, then E is the elliptic curve with LMFDB label $11a.3$ and discriminant 11, E' is $11a.2$ with discriminant 11^5, and E'' is $11a.1$ with discriminant 11 [13], giving us all the semistable 5-discriminant twins.

Theorem 11 *The only rational* 3*-discriminant twins are the* $19a_2$, $19a_3$, $37b_2$, *and* $37b_3$.

Proof This proof follows similarly to the $\ell = 5$ case. If E has rational 3-torsion, we can write E as

$$y^2 + a_1xy + a_3y = x^3$$

with a point of order 3 at $(0, 0)$. Its discriminant is $\Delta = \Delta(E) = a_3^3(a_1^3 - 27a_3)$. By using Tate's algorithm, we see that Δ is minimal at all $p \mid \Delta$ unless $p \mid a_1$ and $p \mid a_3^3$. If $p \mid a_1$ and $p \mid a_3^3$,

$$\mathrm{ord}_p(\Delta) = 9\mathrm{ord}_p(a_3) + 3\max\{\mathrm{ord}_p(a_1) + \mathrm{ord}_p(a_3/3)\} \geq 12.$$

In this case $p^{12} \mid \Delta$ and thus we can take an isomorphism $[u, r, s, t] = [p, 0, 0, 0]$ that changes the model of E to the form $y^2 + a_1p^{-1}xy + a_3p^{-3}y = x^3$. Assume that E is minimal at $p \mid \Delta$. By Kraus's algorithm, and noting that $c_4(E) = a_1(a_1^3 - 24a_3)$, E has semistable reduction unless $p \mid a_1$ and $p \mid a_3$. As we are interested in semistable curves, we can assume $(a_1, a_3) = 1$.

Again we use Sage to find all torsion points of E. We then write E in the form

$$y^2 = x^3 + A(a_1, a_3)x + B(a_1, a_3),$$

and push the points onto this form As before, we proceed to use Vélu's formulas to find the curve $E' = E/E(\mathbb{Q})[3]$. Then $\Delta' = \Delta(E') = a_3(a_1^3 - 27a_3)^3$. Making again the assumption that E' has rational 3-torsion, equivalently, E' is isogenous to E'' with E'' and E non-isomorphic; Δ' must be a cube by Corollary 7. The only case where E and E'' can be discriminant twins is if

$$\mathbb{Q}_p^\times/q^\mathbb{Z} \to \mathbb{Q}_p^\times/(q^3)^\mathbb{Z} \to \mathbb{Q}_p^\times/(\xi q)^\mathbb{Z}$$

where ξ is a 3rd root of unity in $\mathbb{Q}_p$. If $p \mid a_3$, then $\mathrm{ord}_p(\Delta) = 3n$ for some $n \in \mathbb{Z}$ and $\mathrm{ord}_p(\Delta') = n$, so this case does not occur, and we can only have a discriminant twin in a_3 as a unit.

Again assume that E and E'' are discriminant twins. Then $a_3 = \pm 1$ and E' has 3-torsion, so we can write E' as

$$y^2 + a_1'xy + a_3'y = x^3$$

with a point of order 3 at $(0, 0)$. Its discriminant is $\Delta' = -a_3'^3(-a_1'^3 + 27a_3') = -a_3(-a_1^3 + 27a_3)^3$. Further, $\Delta'' = \Delta(E'') = -a_3'(-a_1'^3 + 27a_3')^3$. As Δ' is a cube and $v_p(\Delta') = 3v_p(\Delta'')$, $-a_1'^3 + 27a_3' = \pm 1$ and $a_3' = -a_1^3 + 27a_3$. The only integers (a_1, a_3), (a_1', a_3') satisfying these relations are $(-4, -1)$, $(-10, -37)$ corresponding to $37b.3$ and $37b.2$, $(-3, -1)$, $(-1, 0)$ (which give singular curves) and $(-2, -1)$, $(-8, -19)$ corresponding to $19a.3$ and $19a.2$. Thus we get the only 3-isogenous discriminant twins which are $37b.3$, $37b.1$ and $19a.3$, $19a.1$.

Finally, we examine $\ell = 2$. If E has rational 2-torsion, then we can write E in the form

$$y^2 = x(x - s)(x - r) = x^3 - (r + s)x^2 + rsx$$

for $r, s \in \mathbb{Q}$. Via the isomorphism $x \mapsto u^{-2}x$, $y \mapsto u^{-3}y$, we can take $r, s \in \mathbb{Z}$. Then $\Delta = 16r^2s^2(r - s)^2 \in \mathbb{Z}$. As before, we examine when this gives a minimal model.

Lemma 12 *For $p = 2$ and $p \mid \Delta$, if $p \mid r$ and $p \mid s$, then either the curve is not minimal or it has additive reduction.*

Proof If $p = 2$, then $\tilde{E}$ can be written as $y^2 = x^3$ and

$$\mathrm{ord}_2(\Delta) = 4 + 2\mathrm{ord}_2(r) + 2\mathrm{ord}_2(s).$$

If either $4 \mid s$ or $4 \mid r$, then via the isomorphism $[2, 0, 0, 0]$ we can reduce the model. So assume $\mathrm{ord}_2(s) = 1$ and $\mathrm{ord}_2(r) = 1$. Then $c_4 = 16(a^2 - ab + b^2)$. Then we fly through Tate's algorithm, and every outcome tells us that E has additive reduction.

Thus we can assume $p = 2$ only divides one of r, s, or $r - s$. Without loss of generality, assume $2 \mid r$. Note that under the isomorphism $[1, s, 0, 0]$, the model of E changes to $y^2 = x^3 - (s + (s - r))x^2 + s(s - r)x$, and the torsion is parameterized by the points s and $s - r$ instead of s and r. So we can assume $2 \mid r$ without any loss of generality. Then we have the following lemma.

Lemma 13 *Assume $2 \mid r$ and $2 \nmid s$. If $2^4 \parallel r$, then E has good reduction; if $2^n \mid r$ for $n > 4$, E has multiplicative reduction; and if $2^n \mid r$ for $n < 4$, E has additive reduction.*

Proof This follows from applying Tate's algorithm.

Now, let p be any prime not equal to 2. As before, without loss of generality, let $p \mid r$. Notice that then $p \mid r$ if and only if $p \mid s + r$.

Lemma 14 *If $p \mid r$ and $(r, s) = 1$, then E has multiplicative reduction. Otherwise, E has additive reduction or is not minimal.*

Proof As $c_4 = 16(r^2 - rs + s^2)$, this follows from Kraus's algorithm.

Now we repeat the process used for $\ell = 3, 5$ for $\ell = 2^2$; we find the minimal discriminants (via the previous lemmas) of the isogenous curves using Vélu's formulas. Again, the SageMath code is in the appendix. These are

$$\begin{aligned}
\Delta_0 &= 2^{-8}s^2r^2(r-s)^2\\
\Delta_1 &= 2^{-4}sr(r-s)^4\\
\Delta_2 &= 2^{-4}sr^4(-r+s)\\
\Delta_3 &= 2^{-4}s^4r(r-s)
\end{aligned}$$

where E_1, E_2, and E_3 are all 4-isogenous and factor through $E = E_0$. There are three cases to check, $\Delta_1 = \Delta_2$, $\Delta_1 = \Delta_3$, and $\Delta_2 = \Delta_3$. These are the following relation:

$$r^3 = (2^4)^3(s-r)^3.$$

As r and $s-r$ are coprime integers, this forces $s - r = 1$, and the only solution is $s = 17, r = 16$. This solution gives us the discriminant twin pair $17a.3$ and $17a.4$.

Thus the only case we have left to cover is the isogeny class of eight curves, all connected by 2-isogenies. In particular, the last case is if E and E' are 16-isogenous curves. Let $E \to A \to B \to C \to E'$ be the chain of 2-isogenies. Then we can write the discriminants of A and B in terms of C. Without loss of generality, let $t = r - s$, then we can write

$$\begin{aligned}
\Delta_A &= -2^{-4}s^2r^8t^2\\
\Delta_B &= 2^{-8}s^4r^4t^4\\
\Delta_C &= 2^{-4}s^8r^2t^2
\end{aligned}$$

As $E \to A$ is a 2-isogeny, for each $p \mid \Delta_A$, the power to twice $p \mid \Delta_E$ must change by a multiple of 2 and similarly with $B \to E'$. Thus the only way $\Delta_E = \Delta_{E'}$ is if $\Delta_E = \Delta_B = \Delta_{E'}$, which we've already shown only happens for $17a.1$ and $17a.4$, and this isogeny class only has four curves.

7 Number Fields

With a bit of care, the above proofs hold for semistable elliptic curves over some number fields as well, by work of Anni and Siksek [1] where we can bound the isogeny degrees. What we mean by a bit of care is that one must recognize that the conductor is now defined as an ideal of the number field and that a global minimal discriminant can only be defined if the number field has class number one. Thus we are now only working up to units. Unfortunately, even with this machinery, we cannot draw the same conclusions. For example, when $\ell = 5$, we see that in the parameterization, rs must be a unit for a discriminant twin to occur. As the only units over $\mathbb{Q}$ are ± 1, we find our bad case and move on. Over totally real number fields, unit groups have infinite order, and thus we cannot draw the same conclusions. However, over $\mathbb{Q}(\sqrt{d})$ for $d = 2, 3, 5, 7, 13$, the parameterization $t = u^n$ with u a generator of the unit group and $-50 \leq n \leq 50$, $n \neq 0$ does not yield any discriminant twin examples. As expected, we do get an isogeny class of at least three 5-isogenous curves exactly when $5 \mid n$, so it is not due to isogeny class structure. This leads to the following question.

Question 2 Does the 5-torsion parameterization yield infinitely many discriminant twin pairs over number fields with infinite unit group?

Further, we examine instances of discriminant twins from the tables over $\mathbb{Q}(\sqrt{5})$ in LMFDB. This number field is particularly nice as it has narrow class number one, so the elliptic curves defined over this field have particularly nice global minimal models, as when working over $\mathbb{Q}$. Examining these tables, we examine instances of discriminant twins from the tables over $\mathbb{Q}(\sqrt{5})$ as in LMFDB, we find that out of all curves up to and including norm conductor 5000, there are 4605 isogeny classes and 20 pairs of discriminant twins (including conjugates), 16 of which have additive reduction and 4 with only multiplicative reduction. Most interestingly, all semistable isogenous discriminant twin pairs came from those over $\mathbb{Q}$.

Question 3 Are there finitely many semistable isogenous discriminant twin pairs over $\mathbb{Q}(\sqrt{5})$?

Question 4 What about over number fields in general? Or restricting to totally real or totally imaginary number fields?

Question 5 Can we get more semistable isogenous discriminant twin pairs from number fields that allow larger isogeny classes?

8 Semistable Non-isogenous Discriminant Twins

The data seems to support that there are in fact infinitely many semistable, non-isogenous discriminant twin pairs over $\mathbb{Q}$. We have computed all such pairs for all curves up to conductor 299998 in the Cremona database [2]. More precisely, for small values of X, we can easily compute the following functions:

$$\kappa(X) = \frac{\#\{N \leq X : \exists \text{ discriminant twins of conductor } N\}}{\#\{N \leq X : \exists E \text{ of conductor } N\}}$$

and similarly for semistable N, κ_{ss} and N with additive reduction, κ_{ad}. In Fig. 2, κ, κ_{ss}, and κ_{ad} have been computed for all curves in the Cremona database up to conductor 299998.

$$\kappa(X) = \frac{\#\{N \leq X : \exists \text{ discriminant twins of conductor } N\}}{\#\{N \leq X : \exists E \text{ of conductor } N\}}$$

κ_{ad} non-square-free (top) - red, κ All Curves (middle) - purple, κ_{ss} square-free (bottom) - blue

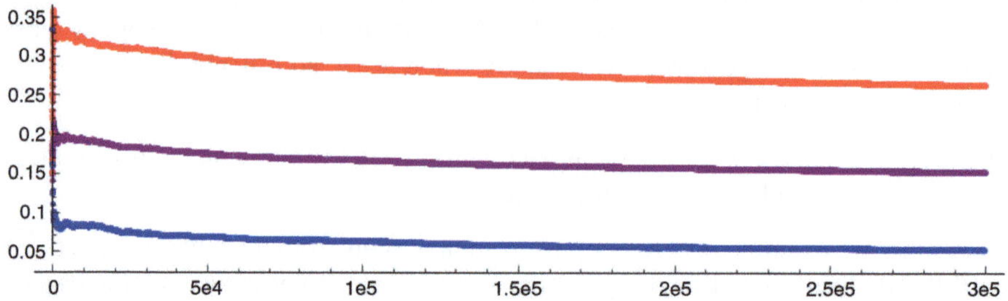

Fig. 2 Ratio of number of discriminant twin conductors up to X by number of conductors up to X

The final values:

$\kappa(299998)$	$\kappa_{ss}(299998)$	$\kappa_{ad}(299998)$
0.15189	0.26314	0.05223

Further, we have a computational method that seems to generate large numbers of such discriminant twin pairs. This method is inspired by work of Howe and Joshi [4]. By examining non-torsion, integral points on the family of curves of the form $y^2 = x^3 - 1728\Delta$, for Δ square-free, we can find curves with conductor $N = \Delta$. If $P = (x, y)$ is an integral point on $y^2 = x^3 - 1728\Delta$, then (x, y) correspond to the c_4, c_6 pair of a curve with discriminant Δ, and as Δ is square-free, the curve has conductor $N = \Delta$. Thus if we could make a statement about some subfamily always having two non-torsion, integral points, we would be able to show there are infinitely many non-isogenous, semistable, discriminant twins.

Conjecture 15 *There are infinitely many semistable non-isogenous discriminant twin pairs.*

With the data from graph Fig. 2, we can also ask the following questions.

Question 6 Do discriminant twin pairs occur with positive density?

Question 7 Do semistable non-isogenous discriminant twin pairs occur with positive density?

Acknowledgements The author would like to thank Ben Lundell for many helpful discussions and the referee for pointing out a hole in Proposition 6.3, noting simpler Weierstrass equations, and other useful comments.

Appendix

All of the discriminant twins code is written in SageMath [14]. It works with Sage 6.0 or higher. It should work with much older versions of Sage as well, but I have not tested this.

5-Isogenies

```
R.<t>=QQ[]
Q = R.fraction_field()
E = EllipticCurve([(1-t),-t,-t,0,0])
print ''Original Discriminant''
print E.discriminant().factor()

def Point(E,xcoord,ycoord):
    P = E.defining_polynomial()
    if P(x = xcoord,y = ycoord,z=1) == 0:
        pass
    else:
        print P(x = xcoord,y = ycoord,z=1)
        raise ValueError, ''Point must lie on curve''
    return [xcoord,ycoord]

def add_points(E,P,Q):
    '' '' ''
    INPUT: curve and two points
    OUTPUT: lambda and nu from silverman
    '' '' ''
    x1,y1 = P
    x2,y2 = Q
    a1 = E.a1()
    a2 = E.a2()
    a3 = E.a3()
    a4 = E.a4()
    a6 = E.a6()
    if x1 != x2:
        lam = (y2-y1)/(x2-x1)
        nu = (y1*x2-y2*x1)/(x2-x1)
    if x1 == x2:
        lam = (3*x1^2+2*a2*x1+a4-a1*y1)/(2*y1+a1*x1+a3)
        nu = (-x1^3+a4*x1+2*a6-a3*y1)/(2*y1+a1*x1+a3)
    x3 = lam^2+a1*lam-a2-x1-x2
```

```
    y3 = -(lam+a1)*x3-nu-a3
    return Point(E,x3,y3)

def neg_point(E,P):
    '''
    INPUT: a curve and a point
    '''
    x0,y0 = P
    x1 = x0
    y1 = -y0-E.a1()*x0-E.a3()
    return Point(E,x1,y1)
P1 = Point(E,0,0)
#P2 = point at infinity
P3 = Point(E,t,0)
P4 = Point(E,0,t)
P5 = Point(E,t,t^2)

u = 1
r = -(E.a1()^2+4*E.a2())/12
s = -E.a1()/2
tt = -(E.a3()+r*E.a1())/2

EAB = E.change_weierstrass_model([u,r,s,tt])

def Pnew(Enew,P):
    x0,y0 = P
    xnew = x0-r
    ynew = y0-s*x0+s*r-tt
    return Point(Enew,xnew,ynew)

def t_tilde(EABform,Q):
    x0,y0 = Q
    A = EABform.a4()
    return 3*x0^2+A
def u_tilde(EABform,Q):
    x0,y0 = Q
    A = EABform.a4()
    B = EABform.a6()
    return 2*(x0^3+A*x0+B)

A = EAB.a4()
B = EAB.a6()
t_total = 0
w_total = 0
```

```
C = [Pn1,Pn3,Pn4,Pn5]
for Q in C:
    t_total = t_total + t_tilde(EAB,Q)
    w_total = w_total + (u_tilde(EAB,Q)+Q[0]*t_tilde
        (EAB,Q))
AC = A - 5*t_total
BC = B - 7*w_total
EC = EllipticCurve([0,0,0,AC,BC])
print ''5-isogenous Discriminant''
print EC.discriminant().factor()
```

3-Isogenies

```
R.<A1,A3>=QQ[]
Q = R.fraction_field()
E = EllipticCurve([A1,0,A3,0,0]);
print ''Original Discriminant,''
print E.discriminant().factor()

def Point(E,xcoord,ycoord):
    P = E.defining_polynomial()
    if P(x = xcoord,y = ycoord,z=1) == 0:
        pass
    else:
        print P(x = xcoord,y = ycoord,z=1)
        raise ValueError, ''Point must lie on curve''
    return [xcoord,ycoord]

def add_points(E,P,Q):
    '' '' ''
    INPUT: curve and two points
    OUTPUT: lambda and nu from silverman
    '' '' ''
    x1,y1 = P
    x2,y2 = Q
    a1 = E.a1()
    a2 = E.a2()
    a3 = E.a3()
    a4 = E.a4()
    a6 = E.a6()
    if x1 != x2:
        lam = (y2-y1)/(x2-x1)
        nu = (y1*x2-y2*x1)/(x2-x1)
    if x1 == x2:
        lam = (3*x1^2+2*a2*x1+a4-a1*y1)/(2*y1+a1*x1+a3)
```

```
        nu = (-x1^3+a4*x1+2*a6-a3*y1)/(2*y1+a1*x1+a3)
    x3 = lam^2+a1*lam-a2-x1-x2
    y3 = -(lam+a1)*x3-nu-a3
    return Point(E,x3,y3)

def neg_point(E,P):
    '''
    INPUT: a curve and a point
    '''
    x0,y0 = P
    x1 = x0
    y1 = -y0-E.a1()*x0-E.a3()
    return Point(E,x1,y1)

P1 = Point(E,0,0)
P2 = add_points(E,P1,P1)

u = 1
r = -(E.a1()^2+4*E.a2())/12
s = -E.a1()/2
t = -(E.a3()+r*E.a1())/2

EAB = E.change_weierstrass_model([u,r,s,t])

def Pnew(Enew,P):
    x0,y0 = P
    xnew = x0-r
    ynew = y0-s*x0+s*r-t
    return Point(Enew,xnew,ynew)

Pn1 = Pnew(EAB,P1)
Pn2 = Pnew(EAB,P2)

def t_tilde(EABform,Q):
    x0,y0 = Q
    A = EABform.a4()
    return 3*x0^2+A
def u_tilde(EABform,Q):
    x0,y0 = Q
    A = EABform.a4()
    B = EABform.a6()
    return 2*(x0^3+A*x0+B)

A = EAB.a4()
```

```
B = EAB.a6()
t_total = 0
w_total = 0
C = [Pn1,Pn2]
for Q in C:
    t_total = t_total + t_tilde(EAB,Q)
    w_total = w_total + (u_tilde(EAB,Q)+Q[0]*t_tilde
        (EAB,Q))
AC = A - 5*t_total
BC = B - 7*w_total

EC = EllipticCurve([0,0,0,AC,BC])

r = A1^2/12
s = A1/2
t = -9*A3/2

EC1 = EC.change_weierstrass_model([1,r,s,t])
print ''3-isogenous discriminant''
print EC1.discriminant().factor()
```

2-Isogenies

```
R.<r,s,u>=QQ[]
Q = R.fraction_field()
E = EllipticCurve([0,-(r+s),0,r*s,0])

def Point(E,xcoord,ycoord):
    P = E.defining_polynomial()
    if P(x = xcoord,y = ycoord,z=1) == 0:
        pass
    else:
        print P(x = xcoord,y = ycoord,z=1)
        raise ValueError, ''Point must lie on curve''
    return [xcoord,ycoord]

def add_points(E,P,Q):
    '' '' ''
    INPUT: curve and two points
    OUTPUT: lambda and nu from silverman
    '' '' ''
    x1,y1 = P
    x2,y2 = Q
    a1 = E.a1()
    a2 = E.a2()
```

```
    a3 = E.a3()
    a4 = E.a4()
    a6 = E.a6()
    if x1 != x2:
        lam = (y2-y1)/(x2-x1)
        nu = (y1*x2-y2*x1)/(x2-x1)
    if x1 == x2:
        lam = (3*x1^2+2*a2*x1+a4-a1*y1)/(2*y1+a1*x1+a3)
        nu = (-x1^3+a4*x1+2*a6-a3*y1)/(2*y1+a1*x1+a3)
    x3 = lam^2+a1*lam-a2-x1-x2
    y3 = -(lam+a1)*x3-nu-a3
    return Point(E,x3,y3)

def neg_point(E,P):
    '''
    INPUT: a curve and a point
    '''
    x0,y0 = P
    x1 = x0
    y1 = -y0-E.a1()*x0-E.a3()
    return Point(E,x1,y1)

P1 = Point(E,0,0)
P2 = Point(E,s,0)
P3 = Point(E,r,0)

uu = 1
rr = -(E.a1()^2+4*E.a2())/12
ss = -E.a1()/2
tt = -(E.a3()+rr*E.a1())/2

EAB = E.change_weierstrass_model([uu,rr,ss,tt])

def Pnew(Enew,P):
    x0,y0 = P
    xnew = x0-rr
    ynew = y0-ss*x0+ss*rr-tt
    return Point(Enew,xnew,ynew)

Pn1 = Pnew(EAB,P1)
Pn2 = Pnew(EAB,P2)
Pn3 = Pnew(EAB,P3)

def t_tilde(EABform,Q):
```

```
    x0, y0 = Q
    A = EABform.a4()
    return 3*x0^2+A
def u_tilde(EABform,Q):
    x0, y0 = Q
    A = EABform.a4()
    B = EABform.a6()
    return 2*(x0^3+A*x0+B)

A = EAB.a4()
B = EAB.a6()
t_total = 0
w_total = 0
C1 = [Pn1]
for Q in C1:
    t_total = t_total + t_tilde(EAB,Q)
    w_total = w_total + (u_tilde(EAB,Q)+Q[0]*t_tilde
        (EAB,Q))
AC1 = A - 5*t_total
BC1 = B - 7*w_total
EC1 = EllipticCurve([0,0,0,AC1,BC1])

t_total = 0
w_total = 0
C2 = [Pn2]
for Q in C2:
    t_total = t_total + t_tilde(EAB,Q)
    w_total = w_total + (u_tilde(EAB,Q)+Q[0]*t_tilde
        (EAB,Q))
AC2 = A - 5*t_total
BC2 = B - 7*w_total
EC2 = EllipticCurve([0,0,0,AC2,BC2])

t_total = 0
w_total = 0
C3 = [Pn3]
for Q in C3:
    t_total = t_total + t_tilde(EAB,Q)
    w_total = w_total + (u_tilde(EAB,Q)+Q[0]*t_tilde
        (EAB,Q))
AC3 = A - 5*t_total
BC3 = B - 7*w_total
EC3 = EllipticCurve([0,0,0,AC3,BC3])
```

```
print ''Discriminants''
print 'D0', E.discriminant().factor()
print 'D1', EC1.discriminant().factor()
print 'D2', EC2.discriminant().factor()
print 'D3', EC3.discriminant().factor()
```

References

1. S. Anni, S. Siksek, On Serre's uniformity conjecture for semistable elliptic curves over totally real fields. arXiv e-prints (2016)
2. J.E. Cremona, *Algorithms for Modular Elliptic Curves* (Cambridge University Press, Cambridge, 1997)
3. J. González, On the division polynomials of elliptic curves. Rev. R. Acad. Cienc. Exactas Fís. Nat. (Esp.) **94**, 377–381 (2000)
4. S. Howe, K. Joshi, Elliptic curves of almost-prime conductor. arXiv e-prints (2012)
5. M.A. Kenku, On the number of **Q**-isomorphism classes of elliptic curves in each **Q**-isogeny class. J. Number Theory. **15**, 199–202 (1982)
6. A. Kraus, Quelques remarques à propos des invariants c_4, c_6 et Δ d'une courbe elliptique. Acta Arith. **54**, 75–80 (1989)
7. D.S. Kubert, Universal bounds on the torsion of elliptic curves. Compos. Math. **38**, 121–128 (1979)
8. B. Mazur, Rational isogenies of prime degree. Invent. Math. **44**, 129–162 (1978)
9. K.A. Ribet, S. Takahashi, Parametrizations of elliptic curves by Shimura curves and by classical modular curves. Proc. Natl. Acad. Sci. USA **94**, 11110–11114 (1997)
10. J.-P. Serre, Propriétés galoisiennes des points d'ordre fini des courbes elliptiques. Invent. Math. **15**, 259–331 (1972)
11. B. Setzer, Elliptic curves of prime conductor. J. Lond. Math. Soc. (2) **10**, 367–378 (1975)
12. J.H. Silverman, *Advanced Topics in the Arithmetic of Elliptic Curves*. Graduate Texts in Mathematics (Springer, New York, 1994)
13. The LMFDB Collaboration: LMFDB: The L-functions and Modular Forms Database (2017). http://www.lmfdb.org
14. W.A. Stein et al, The Sage Development Team: Sage Mathematics Software (Version 7.3) (2017). http://www.sagemath.org
15. J. Vélu, Isogénies entre courbes elliptiques. C. R. Acad. Sci. Paris Sér. A-B. **273**, A238–A241 (1971)

The a-Number of Hyperelliptic Curves

Sarah Frei

Abstract It is known that for a smooth hyperelliptic curve to have a large a-number, the genus must be small relative to the characteristic of the field, $p > 0$, over which the curve is defined. It was proven by Elkin that for a genus g hyperelliptic curve C to have $a_C = g - 1$, the genus is bounded by $g < \frac{3p}{2}$. In this paper, we show that this bound can be lowered to $g < p$. The method of proof is to force the Cartier-Manin matrix to have rank 1 and examine what restriction that places on the affine equation defining the hyperelliptic curve. We then use this bound to summarize what is known about the existence of such curves when $p = 3, 5$ and 7.

Keywords Hyperelliptic · Curve · a-Number · p-Rank · Cartier operator · Cartier-Manin matrix

2010 Mathematics Subject Classification. 11G20, 14H45, 14H40, 15A04, 15B33

1 Introduction

Associated with an algebraic curve defined over a field of positive characteristic p are a number of invariants used to better understand the structure of the curve, such as p-rank, Newton polygon, Ekedahl-Oort type, and a-number. Knowing if and when certain properties of a curve exist gives information about the moduli space of smooth projective curves of genus g over a field k. Studied here is the a-number of hyperelliptic curves of genus g. The a-number a_C of a hyperelliptic curve C defined over an algebraically closed field k of characteristic $p > 0$

S. Frei (✉)
Department of Mathematics, University of Oregon, Eugene, OR, USA
e-mail: sfrei@uoregon.edu

I. I. Bouw et al. (eds.), *Women in Numbers Europe II*, Association for Women in Mathematics Series 11, https://doi.org/10.1007/978-3-319-74998-3_7

is $a_C = \dim_k \mathrm{Hom}(\alpha_p, \mathrm{Jac}(C)[p])$, where α_p is the kernel of the Frobenius endomorphism on the additive group scheme $\mathbb{G}_a$. While the a-number of a curve is easily computable, there are still many open questions about this invariant.

For an algebraic curve of genus g defined over $\mathbb{C}$, its Jacobian will have p^{2g} p-torsion points. However, for a curve in characteristic p, the number of p-torsion points drops to p^{f_C}, where $0 \leq f_C \leq g$. We define f_C to be the p-rank of the curve. A generic curve of genus g will have $f_C = g$. It must also be that the a-number is bounded above by $g - f_C$, so a typical curve of genus g will have $a_C = 0$. This means curves with larger a-numbers do not occur as often, and in fact curves with $a_C = g$ are very rare. An algebraic curve with $a_C = g$, called a superspecial curve, has the property that its Jacobian is isomorphic to a product of supersingular elliptic curves [7]. Because superspecial curves are as far from ordinary as possible, they are a popular topic for research.

For a curve to have a large a-number, the genus of that curve must be small relative to the characteristic $p > 0$ of the field over which the curve is defined. It is a result of Ekedahl [1] that for any curve with $a_C = g$, the genus is bounded by $g \leq \frac{p(p-1)}{2}$. If the curve is hyperelliptic and $a_C = g$, then $g \leq \frac{p-1}{2}$.

If superspecial curves occur the least, then the next most infrequently occurring type of curve should be one with $a_C = g - 1$. The next question that can be asked then is what kind of bound exists on the genus when $a_C = g - 1$, and for any known bound, is that bound attained? It should be that the genus must still be small relative to the characteristic of the field. For a curve with $a_C = g - 1$, it was shown by Re [8] that $g \leq p^2$. In fact, Re's results were more general, giving the bound $g \leq (g - a_C + 1)\frac{p(p-1)}{2} + p(g - a_C)$ on the genus of a curve with any a-number.

Further results by Elkin [2] show that for a hyperelliptic curve with $a_C = g - 1$, the bound on the genus is even lower: $g < \frac{3p}{2}$. Elkin's bound was also proven more generally, showing that if $g - a \leq \frac{2g}{p} - 2$, then there are no hyperelliptic curves of genus g with $a_C \geq a$. Work by Johnston [4] confirms Elkin's bound of $g < \frac{3p}{2}$.

While these general results are useful, it is not clear whether the bound is optimal for a given a-number. The goal of this paper is to explore this bound when $a_C = g - 1$ and show that it can be lowered even further. The following result is proven in Sect. 3.

Theorem 1 *Let $g \geq p$ where p is an odd prime. Then there are no smooth hyperelliptic curves of genus g defined over a field of characteristic p with a-number equal to $g - 1$.*

These results show that for a hyperelliptic curve with $a = g - 1$, the bound on the genus is even lower than was previously known. We must actually have $g < p$ for such a curve to exist. Section 4 summarizes what this bound looks like for small fields.

Based on computations for $p = 5$, $p = 7$, and $p = 11$, it seems possible that this bound may be even lower when $p > 3$. When $g = p - 1$, for a genus g hyperelliptic curve to have $a = g - 1$, its affine equation $y^2 = f(x)$ must take on a particular form. This is discussed in Sect. 5.

2 Background Information

2.1 The Cartier Operator

Let $K = k(x, y)$ be the algebraic function field of a hyperelliptic curve C given by $y^2 = f(x)$, and let $d : K \to \Omega^1(K)$ be the canonical derivation of elements in K. For a holomorphic 1-form $\omega \in H^0(C, \Omega_C^1)$, we can write it as $\omega = d\phi + \eta^p x^{p-1} dx$ with $\phi, \eta \in K$.

Definition 1 The modified Cartier operator $C' : H^0(C, \Omega_C^1) \to H^0(C, \Omega_C^1)$ is defined for ω given as above by $C'(\omega) = \eta dx$.

For a full discussion on the Cartier operator as well as the modified Cartier operator, see [10].

A canonical basis for $H^0(C, \Omega_C^1)$ is given by

$$\left\{ \omega_i = \frac{x^{i-1} dx}{y} : 1 \leq i \leq g \right\}.$$

We want to consider what the modified Cartier operator does to these basis elements. Recall that C is given by $y^2 = f(x)$, and if we let $f(x)^{(p-1)/2} = \sum_{j=0}^{N} \kappa_j x^j$ where $N = \frac{p-1}{2}(2g+1)$, then we can rewrite ω_i as follows:

$$\begin{aligned} \omega_i =& x^{i-1} y^{-p} y^{p-1} dx = y^{-p} x^{i-1} \sum_{j=0}^{N} \kappa_j x^j dx \\ =& y^{-p} \left(\sum_{\substack{j \\ i+j \not\equiv 0 (\text{mod } p)}} \kappa_j x^{i+j-1} dx \right) + \sum_{l} \kappa_{(l+1)p-i} \frac{x^{lp}}{y^p} x^{p-1} dx. \end{aligned}$$

The highest possible power of x is $N + i - 1$, so $lp + p - 1 \leq N + i - 1$, which forces

$$0 \leq l \leq \frac{N+i}{p} - 1 = g - \frac{1}{2} - \left(\frac{2g-2i+1}{2p}\right) < g - \frac{1}{2}.$$

This means the sum in the second term is over $0 \leq l \leq g-1$. Thus we can now see that

$$C'(\omega_i) = \sum_{l=0}^{g-1} \kappa_{(l+1)p-i}^{1/p} \frac{x^l}{y} dx.$$

This shows that C' is a map on $H^0(C, \Omega_C^1)$ and we can represent its action on the basis with a matrix. If we write $\bar{\omega} = (\omega_1, \ldots, \omega_g)$, then

$$C'(\bar{\omega}) = A^{(1/p)}\bar{\omega}$$

where A is a $g \times g$ matrix $[a_{ij}]$ with $a_{ij} = \kappa_{pi-j}$.

Definition 2 The matrix A described above is the Cartier-Manin matrix of the hyperelliptic curve C of genus g defined over k.

2.2 p-Rank and a-Number

The group scheme $\mu_p \cong \mathrm{Spec}(k[x]/(x-1)^p)$ is the kernel of the Frobenius endomorphism on the multiplicative group $\mathbb{G}_m = \mathrm{Spec}(k[x, x^{-1}])$. The group scheme $\alpha_p \cong \mathrm{Spec}(k[x]/x^p)$ is the kernel of the Frobenius endomorphism on the additive group $\mathbb{G}_a = \mathrm{Spec}(k[x])$. For more on group schemes, see [9].

The p-rank of a hyperelliptic curve C is $f_C = \dim_k \mathrm{Hom}(\mu_p, \mathrm{Jac}(C)[p])$. An equivalent definition of the p-rank is that it is the positive integer f_C such that $\mathrm{Jac}(C)[p](k) \cong (\mathbb{Z}/p\mathbb{Z})^{f_C}$, so $\#\mathrm{Jac}(C)[p](k) = p^{f_C}$. We see that $0 \leq f_C \leq g$, where $g = \dim(\mathrm{Jac}(C))$. A curve is called ordinary if $f_C = g$ and non-ordinary otherwise.

The a-number of C is $a_C = \dim_k \mathrm{Hom}(\alpha_p, \mathrm{Jac}(C)[p])$. We also have $0 \leq a_C \leq g$, and in fact $a_C \leq g - f_C$. Curves with $a_C = g$ are called superspecial and do not occur often, due to the fact that a typical curve of genus g has $f_C = g$. Curves with $a_C = g-1$ are forced to have $f_C = 0$ or $f_C = 1$ which limits their occurrences.

The a-number is also related to the rank of the Cartier-Manin matrix introduced above. For an abelian variety X of dimension g, such as the Jacobian of a genus g hyperelliptic curve, the Frobenius operator $F : X \to X^{(p)}$ is the p-th power map on X, and the Verschiebung operator $V : X^{(p)} \to X$ is the map such that $V \circ F = [p]$, the multiplication-by-p map. The a-number is also defined [5] as the dimension of the kernel of the action of V on $H^0(X, \Omega_X^1)$. If we let $v = \dim V H^0(X, \Omega_X^1)$, this gives us that $a_C = g - v$. It is also known for a smooth projective curve C, such as a hyperelliptic curve, that the action of the Cartier operator on $H^0(C, \Omega_C^1)$ agrees with the action of V on $H^0(\mathrm{Jac}(C), \Omega_{\mathrm{Jac}(C)}^1) \cong H^0(C, \Omega_C^1)$ [6]. Since we can express the action of the Cartier operator on $H^0(C, \Omega_C^1)$ with the Cartier-Manin matrix A, we see that $a_C = g - \mathrm{rank}(A)$.

It turns out that associated with any abelian variety X of dimension g is a short exact sequence

$$0 \to H^0(X, \Omega_X^1) \to H_{dR}^1(X) \to H^0(X, \Omega_X^1) \to 0.$$

The Frobenius operator acts on $H^0(X, \Omega_X^1)$ in this sequence, and the Verschiebung operator acts on $H_{dR}^1(X)$ so $H^0(X, \Omega_X^1) = VH_{dR}^1(X)$.

For the sake of notation, we will let $a_C = a$ for the rest of this paper. In studying hyperelliptic curves with $a = g - 1$, we will thus be looking for curves with a Cartier-Manin matrix of rank one. We will utilize the fact that for a matrix of rank 1, there is at least one nonzero entry, and every 2×2 minor has determinant 0.

3 Results

In this section, we will use the following notation. Let C be a hyperelliptic curve given by the equation $y^2 = f(x)$ where $f(x) = \sum_{i=1}^{2g+1} c_i x^i$ with $c_i \in k$ where k is an algebraically closed field of characteristic $p > 0$. Note that by a change of variables, we can assume $c_0 = 0$ and $c_{2g+1} = 1$. We will assume that C has $a = g - 1$. Then we will define the coefficients κ_i as follows:

$$f(x)^{(p-1)/2} = \sum_{i=0}^{\left(\frac{p-1}{2}\right)(2g+1)} \kappa_i x^i$$

and $\kappa_i = 0$ if $i < \frac{p-1}{2}$ or $i > (\frac{p-1}{2})(2g+1)$. The Cartier-Manin matrix A associated with C is the $g \times g$ matrix $[a_{ij}]$ where $a_{ij} = \kappa_{pi-j}$. We will denote row m of A by A_m. For C to have a-number equal to $g - 1$, A must have rank one.

Theorem 2 *Let $g \geq p$ where p is an odd prime. Then there are no smooth hyperelliptic curves of genus g defined over an algebraically closed field of characteristic p with a-number equal to $g - 1$.*

Proof We will proceed by considering two separate cases: first when $g > p$ and then when $g = p$.

Case 1: Let $g > p$ where p is an odd prime. We consider the entries $a_{i,j} = \kappa_{pi-j}$ of the Cartier-Manin matrix A. Since $\kappa_i = 0$ for $0 \leq i \leq \frac{p-3}{2}$, $a_{1,j}$ is possibly nonzero for $1 \leq j \leq \frac{p+1}{2}$ and $a_{1,j} = 0$ for $\frac{p+3}{2} \leq j \leq g$. The largest nonzero term of $f(x)^{(p-1)/2}$ is $x^{g(p-1)+(p-1)/2}$, so $\kappa_{g(p-1)+(p-1)/2} = \kappa_{gp-(g-(p-1)/2)} = 1$ and any larger-indexed coefficient is zero. This means $a_{g,j} = 0$ for $1 \leq j \leq g - \frac{p+1}{2}$, and $a_{g,j}$ is possibly nonzero for $g - \frac{p-1}{2} \leq j \leq g$.

Now let us suppose that $g = p + m$ for some integer $m \geq 1$. We have

$$a_{1,(p+1)/2} = \kappa_{(p-1)/2} = c_1^{(p-1)/2},$$

and $a_{1,(p+1)/2+m} = 0$, since $a_{1,(p+1)/2}$ is the last nonzero entry in A_1. Also, $a_{g,(p+1)/2} = 0$, since $a_{g,j} = 0$ for $1 \leq j \leq g - \frac{p+1}{2} = \frac{p-1}{2} + m$ and $m \geq 1$. Hence $a_{g,(p+1)/2}$ is possibly the last zero term in A_g, if $m = 1$. Lastly, $a_{g,(p+1)/2+m} = 1$, since $g - \frac{p-1}{2} = p + m + \frac{p-1}{2} = \frac{p+1}{2} + m$, which is the first nonzero term in A_g. Using this 2×2 minor, we get

$$a_{1,(p+1)/2} \cdot a_{g,(p+1)/2+m} - a_{g,(p+1)/2} \cdot a_{1,(p+1)/2+m} = 0,$$

which means $c_1^{(p-1)/2} \cdot 1 - 0 \cdot 0 = 0$. This forces $c_1 = 0$. But then $f(x) = \sum_{i=2}^{2g+1} c_i x^i = x^2 \sum_{i=2}^{2g+1} c_i x^{i-2}$ is not squarefree and C is not a smooth curve. Therefore, when $g > p$, there are no smooth hyperelliptic curves of genus g defined over a field of characteristic p with a-number equal to $g - 1$.

Case 2: Let $g = p$ where p is an odd prime. We again consider the $a_{i,j}$ in the Cartier-Manin matrix. There will be $g - \frac{p+1}{2}$ zeros in A_1 and A_g. For $g = p$, this means the last $\frac{p-1}{2}$ entries of A_1 are zeros and the first $\frac{p-1}{2}$ entries of A_g are zeros. As above, $\kappa_{\frac{p-1}{2}} = c_1^{(p-1)/2}$ and $\kappa_{(2g+1)(p-1)/2} = \kappa_{(2p^2-p-1)/2} = 1$. We will assume $c_1 \neq 0$ so that C is not singular at $x = 0$. This gives us an idea of what A looks like:

$$\begin{pmatrix} \kappa_{\frac{p-1}{2}+\frac{p-1}{2}} & \cdots & \kappa_{\frac{p-1}{2}+1} & c_1^{(p-1)/2} & 0 & \cdots & 0 \\ & \cdots & & \kappa_{\frac{p-1}{2}+p} & \kappa_{\frac{p-1}{2}+(p-1)} & \cdots & \kappa_{\frac{p-1}{2}+\frac{p+1}{2}} \\ \vdots & \ddots & \vdots & \vdots & \vdots & \ddots & \vdots \\ \kappa_{\frac{2p^2-p-1}{2}-\frac{p+1}{2}} & \cdots & \kappa_{\frac{2p^2-p-1}{2}-(p-1)} & \kappa_{\frac{2p^2-p-1}{2}-p} & & \cdots & \\ 0 & \cdots & 0 & 1 & \kappa_{\frac{2p^2-p-1}{2}-1} & \cdots & \kappa_{\frac{2p^2-p-1}{2}-\frac{p-1}{2}} \end{pmatrix}$$

We can again consider the 2×2 minors of A, or we can simply use the fact that because $\text{rk}A = 1$, every column of A is a scalar multiple of the middle column. The columns to the left of the middle column must be zero since the last entry of the index $\frac{p-1}{2}$ column is 1 while the last entry of the previous columns is zero. The columns to the right of the middle column must also be zero since the first entry of the index $\frac{p-1}{2}$ column is $c_1^{(p-1)/2} \neq 0$ while the first entry of the following columns is zero. This means $f(x)^{(p-1)/2}$ has the following form:

$$f(x)^{(p-1)/2} = \sum_{i=0}^{p-1} \kappa_{\frac{p-1}{2}+ip} x^{(p-1)/2+ip} = x^{(p-1)/2} h(x^p) = x^{(p-1)/2} \widetilde{h}(x)^p,$$

where $h(x) = \sum_{i=0}^{p-1} \kappa_{\frac{p-1}{2}+ip} x^i$ and where the last equality is a consequence of the multinomial theorem in characteristic $p > 0$. Thus, since $f(x) = x(c_1 + c_2 x + \ldots + x^{2p}) = x\widetilde{f}(x)$, we see that $\widetilde{f}(x)^{(p-1)/2} = \widetilde{h}(x)^p$. Then we see that any root of $\widetilde{h}$ is a root of $\widetilde{f}^{(p-1)/2}$ with multiplicity p, making it a root of $\widetilde{f}$ with multiplicity greater than 1. Thus f is not squarefree and hence C is a singular hyperelliptic curve. Therefore, when $g = p$ there are no smooth hyperelliptic curves of genus g defined over a field of characteristic p with a-number equal to $g - 1$.

4 Computations and Examples for Small Primes

4.1 For $p = 3$

We see from Elkin's bound that hyperelliptic curves defined over $\overline{\mathbb{F}}_3$ with $a = g - 1$ will only occur when $g < 5$. By Theorem 2, in fact such a curve will only occur for $g < 3$. Genus 3 hyperelliptic curves have been studied extensively, and it was previously known that curves with $a = 2$ do not exist [3]. It is also known that genus 2 hyperelliptic curves with $a = 1$ exist for all $p \geq 3$. Hence for $p = 3$, genus 2 hyperelliptic curves are the only hyperelliptic curves with $a = g - 1$.

4.2 For $p = 5$

According to Elkin's bound, hyperelliptic curves with $a = g - 1$ will only occur when $g < \frac{15}{2}$. For $p = 5$, it is known that such hyperelliptic curves exist with genus 2 and with genus 3 [3]. When $g = 3$, they in fact occur with both p-rank 0 and 1.

It is next worth investigating $g = 4, 5, 6$, and 7, but Theorem 2 in Sect. 3 shows that for $g = 5, 6$, and 7, there are no smooth hyperelliptic curves of such a genus with $a = g - 1$. It can be shown that if we assume C is a genus 4 hyperelliptic curve with $a = 3$ defined by $y^2 = f(x)$, then $f(x) = x(x + 2c_8)^3(x + \sqrt[5]{c_4})^5$. This means there are no smooth hyperelliptic curves of $g = 4$ with $a = 3$ defined over a field of characteristic 5. Hence, the case $p = 5$ is completely determined, with curves having $a = g - 1$ only existing when $g = 2$ and $g = 3$.

4.3 For $p = 7$

Elkin's bound for $p = 7$ gives that for a hyperelliptic curve with $a = g - 1$, we must have $g < \frac{21}{2}$, so we are interested in looking for curves with genus up to 10. Theorem 2 shows that such a curve will not exist with $g \geq p$, so in fact we only need to study $g = 2, 3, 4, 5$, and 6. It was previously shown that genus 2 curves exist with $a = 1$ in characteristic 7.

Hyperelliptic curves of genus 3 with $a = 2$ exist, and as occurs for $p = 5$, they exist with p-rank both 0 and 1. In this case, as expected, there are far more such curves with p-rank 1 than p-rank 0 defined over $\mathbb{F}_7$.

It is still unknown whether or not curves of genus 4 exist with $a = 3$. The Sage code shown in Sect. 6 was used to determine that genus 4 hyperelliptic curves with a defining polynomial of the form $f(x) = c_1x + c_2x^2 + \ldots + c_8x^8 + x^9$ do not exist over $\mathbb{F}_7$. We note that there could still exist a curve with either $c_0 \neq 0$, $c_9 \neq 1$ or $c_{10} \neq 0$ and the desired $a = 3$ defined over $\mathbb{F}_7$, so this was not an exhaustive search. After checking 1,000,000 hyperelliptic curves defined over $\mathbb{F}_{49}$ with branch points fixed at $x = 0, 1$, and ∞, none were found to have $a = 3$. This code can also be seen in Sect. 6. However, this is a very small portion of the total number of curves defined over $\mathbb{F}_{49}$, and it is possible that such a curve could exist over a larger extension of $\mathbb{F}_7$.

When $g = 5$, we see similar results. It is still open whether or not curves of genus 5 exist with $a = 4$. It has been checked in Sage that there are no such hyperelliptic curves over $\mathbb{F}_7$ with defining polynomial of the form $f(x) = c_1x + \ldots + c_{10}x^{10} + x^{11}$ (again a non-exhaustive search). We next checked for curves branched at 0 and ∞ defined over $\mathbb{F}_{49}$. In this case, we use information from the Cartier-Manin matrix, again forcing the matrix to have rank one, to further shrink the search space. After checking 30,000,000 random curves under these restrictions, none were found to have $a = 4$. We note again that this is only a small portion of the curves defined over $\mathbb{F}_{49}$, and those checked were only curves in a restricted search space, since there could exist a genus 5 hyperelliptic curve defined over $\mathbb{F}_{49}$ with no rational ramification points having $a = 4$.

For genus 6 curves, it can be shown that if we assume C is a genus 6 hyperelliptic curve with $a = 5$ defined by $y^2 = f(x)$, then $f(x) = x(x + 3c_{12})^5(x + \sqrt[7]{c_6})^7$. Thus, there are no smooth hyperelliptic curves of genus 6 with $a = 5$ when $p = 7$.

5 Further Lowering the Bound

Without any known examples of algebraic curves of genus $g > 3$ with $a = g - 1$, it is unclear whether or not it is possible to lower the bound on the genus any further. Future work in this area could include exploring the cases of $g = p - 1$ and $g = p - 2$.

As stated in Sects. 4.2 and 4.3, neither smooth hyperelliptic curves of genus 4 with $a = 3$ nor smooth hyperelliptic curves of genus 6 with $a = 5$ exist when $p = 5$ or $p = 7$, respectively. It can also be shown that if we assume C is a genus 10 hyperelliptic curve with $a = 9$ defined by $y^2 = f(x)$ in characteristic 11, then $f(x) = x(x+5c_{20})^9(x+\sqrt[11]{c_{10}})^{11}$, and hence C is not smooth. These cases suggest that curves with $a = g - 1$ likely do not exist when $g = p - 1$. In fact, we have the following result.

Proposition 1 *Let C be a hyperelliptic curve defined over a field of characteristic $p > 3$ of genus $g = p - 1$, where C is defined above. If C has $a = g - 1$, then $f(x) \in k[x, c_g, c_{2g-g/2}, c_{2g}]$.*

Thus, for a hyperelliptic curve C with $a = g - 1$ to exist when $g = p - 1$, its affine equation $y^2 = f(x)$ must take on a very specific form; the polynomial $f(x)$ is completely determined by only three of its $2g$ coefficients. Proposition 1 is proven using the same methods employed in Sect. 3, where the associated Cartier-Manin matrix is assumed to have rank 1, and the relationships forced on the coefficients of $f(x)$ are studied.

As shown in Sect. 4.3, it seems possible that curves of genus 5 with $a = 4$ do not exist in characteristic 7. It would be worth generating data for $p = 11$ and $g = 9$ to explore the existence of hyperelliptic curves with $a = 8$. From there, an attempt could be made to make a general statement about the existence of hyperelliptic curves of genus $g = p - 2$ and $a = g - 1$ when $p > 5$.

6 Sage Code

The following is a sample of some of the codes used to obtain results discussed in Sect. 4.3. In both examples listed here, the returned output was $N = 0$.

```
F=GF(7)
R.<x>=PolynomialRing(F)
N=0
V=VectorSpace(F, 8)
for m in V:
    f=m[1]*x+m[2]*x^2+m[3]*x^3+m[4]*x^4+m[5]*\
        x^5+m[6]*x^6+m[7]*x^7+m[0]*x^8+x^9
    if f.is_squarefree()==True:
        C=HyperellipticCurve(f)
        B=C.Cartier_matrix()
        if B.rank()==1:
            N=N+1;
            C

N
```

```
F=GF(49, 'a')
R.<x>=PolynomialRing(F)
N=0
for i in range(1000000):
    m=random_vector(F, 7)
    f=(x-1)*(m[0]*x+m[1]*x^2+m[2]*x^3+m[3]*x^4\
          +m[4]*x^5+m[5]*x^6+m[6]*x^7+x^8
    if f.is_squarefree()==True:
        C=HyperellipticCurve(f)
        B=C.Cartier_matrix()
        if B.rank()==1:
            N=N+1;
            C
N
```

Acknowledgements I would like to thank my advisor Rachel Pries for her many helpful comments and suggestions on this paper, as well as for guiding me on this project while I was a graduate student at Colorado State University.

References

1. T. Ekedahl, On supersingular curves and abelian varieties. Math. Scand. **60**, 151–178 (1987)
2. A. Elkin, The rank of the cartier operator on cyclic covers of the projective line. J. Algebra **327**(1), 1–12 (2011)
3. A. Elkin, R. Pries, Hyperelliptic curves with a-number 1 in small characteristic. Albanian J. Math. **1**(4), 245–252 (2007)
4. O. Johnston, A note on the a-numbers and p-ranks of kummer covers (2007). arXiv preprint arXiv:0710.2120
5. K.-Z. Li, F. Oort, *Moduli of Supersingular Abelian Varieties*, vol. 1680 (Springer, Berlin, 1998)
6. T. Oda, The first de rham cohomology group and dieudonné modules. Ann. Sci. École Norm. Super. **4**(2), 63–135 (1969)
7. F. Oort, Which abelian surfaces are products of elliptic curves? Math. Ann. **214**(1), 35–47 (1975)
8. R. Re, The rank of the cartier operator and linear systems on curves. J. Algebra **236**(1), 80–92 (2001)
9. J. Tate, Finite flat group schemes, in *Modular Forms and Fermat's Last Theorem* (Springer, Berlin, 1997), pp. 121–154
10. N. Yui, On the jacobian varieties of hyperelliptic curves over fields of characteristic p > 2. J. Algebra **52**(2), 378–410 (1978)

Non-ordinary Curves with a Prym Variety of Low p-Rank

Turku Ozlum Celik, Yara Elias, Burçin Güneş, Rachel Newton, Ekin Ozman, Rachel Pries, and Lara Thomas

Abstract If $\pi : Y \to X$ is an unramified double cover of a smooth curve of genus g, then the Prym variety P_π is a principally polarized abelian variety of dimension $g - 1$. When X is defined over an algebraically closed field k of characteristic p, it is not known in general which p-ranks can occur for P_π under restrictions on the p-rank of X. In this paper, when X is a non-hyperelliptic curve of genus $g = 3$, we analyze the relationship between the Hasse-Witt matrices of X and P_π. As an application, when $p \equiv 5 \bmod 6$, we prove that there exists a curve X of genus 3 and p-rank $f = 3$ having an unramified double cover $\pi : Y \to X$ for which P_π has p-rank 0 (and is thus supersingular); for $3 \leq p \leq 19$, we verify the same for each $0 \leq f \leq 3$. Using theoretical results about p-rank stratifications of moduli spaces, we prove, for small p and arbitrary $g \geq 3$, that there exists an unramified double cover $\pi : Y \to X$ such that both X and P_π have small p-rank.

Keywords Curve · Jacobian · Prym variety · Abelian variety · p-Rank · Supersingular · Moduli space · Kummer surface

---

T. O. Celik
Laboratoire IRMAR, UMR CNRS, Rennes, France
e-mail: turku-ozlum.celik@univ-rennes1.fr

Y. Elias
Max Planck Institute for Mathematics, Bonn, Germany
e-mail: yara.elias@mail.mcgill.ca

B. Güneş
Sabanci University, Faculty of Engineering and Natural Sciences, Tuzla, Istanbul, Turkey
e-mail: bgunes@sabanciuniv.edu

R. Newton
Department of Mathematics and Statistics, University of Reading, Reading, UK
e-mail: r.d.newton@reading.ac.uk

E. Ozman
Bogazici University, Faculty of Arts and Sciences, Bebek, Istanbul, Turkey
e-mail: ekin.ozman@boun.edu.tr

I. I. Bouw et al. (eds.), *Women in Numbers Europe II*, Association for Women in Mathematics Series 11, https://doi.org/10.1007/978-3-319-74998-3_8

- when $g \geq 3$ and $f' = g - 1$, as a special case of [15, Theorem 1.1(1)];
- when $g \geq 3$ and $f' = g - 2$ (with $f \geq 2$ when $p = 3$), by [15, Theorem 7.1];
- when $p \geq 5$ and $g \geq 4$ and $\frac{g}{2} - 1 \leq f' \leq g - 3$, by [15, Corollary 7.3].

In this paper, we study the first open case of Question 1, which occurs when has genus $g = 3$ and P_π has p-rank $f' = 0$. We focus on the case that X is a smoo plane quartic or, equivalently, that X is not hyperelliptic. (If X has genus $g = 3$, Riemann-Hurwitz formula implies that the genus of Y is 5 and hence the dimens of P_π is 2.)

Given an unramified double cover $\pi : Y \to X$ of a smooth plane quartic X Lemma 2 we use work of Bruin [4] to analyze the Hasse-Witt matrices of X and simultaneously, in terms of the quadratic forms that determine π. As an applicat we verify that the answer to Question 1 is yes when $g = 3$ and $3 \leq p \leq 1$ Proposition 5.

Given a genus 2 curve $Z : z^2 = D(x)$, one can describe all smooth plane qu curves X having an unramified double cover $\pi : Y \to X$ whose Prym va P_π is isomorphic to Jac(Z). Specifically, the Kummer variety $K = \text{Jac}(Z)/\langle -$ Jac(Z) is a quartic surface in $\mathbb{P}^3$. Each smooth plane quartic X having an unram double cover $\pi : Y \to X$ such that $P_\pi \simeq \text{Jac}(Z)$ arises as the intersection V for some plane $V \subset \mathbb{P}^3$ [5, 13, 19]. Building on work of Kudo and Harashita we provide a method to determine the Hasse-Witt matrix of X from V and Proposition 6.

Suppose that $\pi : Y \to X$ is an unramified double cover with $P_\pi \simeq \text{Jac}($ in the previous paragraph. In Sect. 5, we first choose $Z : z^2 = x^6 - 1$ and ve Proposition 7 that the answer to Question 1 is yes when $(g, f, f') = (3, 3,$ $p \equiv 5 \bmod 6$.

In the second part of Sect. 5, for an arbitrary smooth curve Z of genus 2, commutative algebra to analyze the condition that (*) X is non-ordinary and p-rank 0. In Proposition 9, we prove that condition (*) is equivalent to the va of 4 homogeneous polynomials of degree $(p + 1)(p - 1)/2$ in the coeffici $D(x)$ and the vanishing of one homogeneous polynomial of degree $6(p - 1$ coefficients of $D(x)$ and V. As an application, when $p = 3$, we give an characterization of the curves Z and planes V for which $\pi : Y \to X$ condition (*); see Sect. 5.6.2.

Finally, we apply the new results described above for genus 3 curves characteristic to study the p-ranks of Prym varieties of smooth curves of genus $g \geq 3$. For this, we use an inductive method developed in [1]. Th Corollary 4, which extends [15, Corollary 7.3] for small p and gives the f application.

Corollary 1 *Let $3 \leq p \leq 19$. The answer to Question 1 is yes, for an under the following conditions on (f, f'):*

1. If $g = 3r$ and (f, f') is such that $2r \leq f \leq g$ and $r - 1 \leq f' \leq g - 1$

1.1 *Outline of the Paper*

Here is the material contained in each section.

Section 2: Definitions and background material.
Section 3: The Hasse-Witt matrices of X and P_π in terms of quadratic forms. Examples of unramified double covers $\pi : Y \to X$ with given p-ranks (f, f') for small p.
Section 4: The Hasse-Witt matrix of a smooth plane quartic X defined as the intersection of a plane and quartic surface in $\mathbb{P}^3$.
Section 5: The Kummer surface of an abelian surface $\mathrm{Jac}(Z)$, information about the determinant of the Hasse-Witt matrix from Sect. 4, the nonemptiness of $\mathscr{R}_3^{(3,0)}$ when $p \equiv 5 \bmod 6$, and information about the geometry of $\mathscr{R}_3^{(2,0)}$ when $p = 3$.
Section 6: An auxiliary result about the number of points on the Kummer surface of a supersingular curve of genus 2 defined over a finite field.
Section 7: Results about p-ranks of Prym varieties of unramified double covers of curves of arbitrary genus for small p proven inductively from results in Sect. 3.

2 Background

2.1 *Notation*

Let k be an algebraically closed field of characteristic $p > 0$. Unless stated otherwise, every curve is a smooth projective connected k-curve. Suppose that C is a curve of genus $g \geq 1$.

2.2 *The Cartier-Manin Matrix*

Let L be the function field of C/k. Since k is perfect, there exists a separating variable $x \in L \setminus k$ such that $L/k(x)$ is algebraic and separable. It follows that $L = L^p(x)$ and hence every element $z \in L$ can be written uniquely in the form

$$z = z_0^p + z_1^p x + \cdots + z_{p-1}^p x^{p-1}$$

with $z_0, \ldots, z_{p-1} \in L$. The *Cartier operator* $\mathscr{C}$ is defined on differentials of the first kind by

$$\mathscr{C}((z_0^p + z_1^p x + \cdots + z_{p-1}^p x^{p-1})dx) = z_{p-1}dx.$$

The Cartier operator is $\frac{1}{p}$-linear, meaning that

$$\mathscr{C}(a^p\omega_1 + b^p\omega_2) = a\mathscr{C}(\omega_1) + b\mathscr{C}(\omega_2)$$

for all $a, b \in L$ and all $\omega_1, \omega_2 \in \Omega^1(L)$. It is independent of the choice of separating variable and hence gives a well-defined map on the k-vector space of regular differentials on C, $\mathscr{C} : \mathrm{H}^0(C, \Omega^1_C) \to \mathrm{H}^0(C, \Omega^1_C)$.

Definition 1 Let $\omega_1, \ldots, \omega_g$ be a basis for $\mathrm{H}^0(C, \Omega^1_C)$ over k and write $\mathscr{C}(\omega_j) = \sum_{i=1}^{g} c_{ij}\omega_i$ with $c_{ij} \in k$. The Cartier-Manin matrix of C with respect to the basis $\omega_1, \ldots, \omega_g$ is the matrix $(c_{ij}^p)_{i,j}$.

Remark 1 The Cartier-Manin matrix depends on the choice of basis. Let $\omega'_1, \ldots, \omega'_g$ be another k-basis for $\mathrm{H}^0(C, \Omega^1_C)$, and let $T = (t_{ij})$ be the change of basis matrix so that $\omega_j = \sum_{i=1}^{g} t_{ij}\omega'_i$. Then the Cartier-Manin matrix with respect to the basis $\omega'_1, \ldots, \omega'_g$ is $T^{(p)}(c_{ij}^p)T^{-1}$, where $T^{(p)}$ denotes the matrix obtained from T by taking the pth power of each of its entries.

2.2.1 The Cartier-Manin Matrix of a Hyperelliptic Curve

Let p be odd and let Z be a hyperelliptic curve of genus g. Then Z has an equation of the form $z^2 = f(x)$ for a separable polynomial $f(x) \in k[x]$ having degree $2g+1$ or $2g+2$. Write $\omega_i = \frac{x^{i-1}}{z}dx$, so that $\{\omega_1, \ldots, \omega_g\}$ is a basis for $\mathrm{H}^0(Z, \Omega^1_Z)$.

Proposition 1 ([22, Proposition 2.3]) *Let c_s denote the coefficient of x^s in the expansion of $f(x)^{(p-1)/2}$. Then the Cartier-Manin matrix of Z is $A_0 = (c_{ip-j})_{i,j}$.*

2.3 *The Hasse-Witt Matrix*

The *(absolute) Frobenius* F of C is the morphism of schemes given by the identity on the underlying topological space and $f \mapsto f^p$ on $\mathscr{O}_C$. We write F^* for the induced endomorphism of $\mathrm{H}^1(C, \mathscr{O}_C)$. It is a p-linear map, meaning that

$$F^*(\lambda\xi) = \lambda^p F^*\xi$$

for all $\lambda \in k$ and all $\xi \in \mathrm{H}^1(C, \mathscr{O}_C)$.

Proposition 2 ([16, Proposition 9]) *Serre duality gives a perfect pairing*

$$\langle\ ,\ \rangle : \mathrm{H}^1(C, \mathscr{O}_C) \times \mathrm{H}^0(C, \Omega^1_C) \to k$$

such that

$$\langle F^*\xi, \omega\rangle = \langle \xi, \mathscr{C}\omega\rangle^p$$

for all $\xi \in \mathrm{H}^1(C, \mathscr{O}_C)$ and all $\omega \in \mathrm{H}^0(C, \Omega^1_C)$.

Definition 2 Let $\xi_1, \dots \xi_g$ be a k-basis of $\mathrm{H}^1(C, \mathcal{O}_C)$. Write $F^*(\xi_j) = \sum_{i=1}^{g} a_{ij}\xi_i$ with $a_{ij} \in k$. The Hasse-Witt matrix of C with respect to the basis $\xi_1, \dots \xi_g$ is the matrix $(a_{ij})_{i,j}$.

Remark 2 The Hasse-Witt matrix depends on the choice of basis. Let $\xi'_1, \dots, \xi'_g$ be another k-basis for $\mathrm{H}^1(C, \mathcal{O}_C)$, and let $S = (s_{ij})$ be the change of basis matrix so that $\xi'_j = \sum_{i=1}^{g} s_{ij}\xi_i$. Then the Hasse-Witt matrix with respect to the basis $\xi'_1, \dots, \xi'_g$ is $S^{-1}(a_{ij})S^{(p)}$, where $S^{(p)}$ denotes the matrix obtained from S by taking the pth power of each of its entries.

Remark 3 If the basis $\xi_1, \dots \xi_g$ of $\mathrm{H}^1(C, \mathcal{O}_C)$ is the dual basis for the basis $\omega_1, \dots, \omega_g$ of $\mathrm{H}^0(C, \Omega^1_C)$, then the Hasse-Witt matrix is the transpose of the Cartier-Manin matrix.

2.4 The p-Rank

If A is an abelian variety of dimension g over k, its p-rank is the number f_A such that $\#A[p](k) = p^{f_A}$. If C is a curve of genus g over k, its p-rank is the p-rank of $\mathrm{Jac}(C)$. We write f_A (resp. f_C) for the p-rank of A (resp. C).

Here is another definition of the p-rank. The k-vector space $\mathrm{H}^1(C, \mathcal{O}_C)$ has a direct sum decomposition into F^*-stable subspaces as

$$\mathrm{H}^1(C, \mathcal{O}_C) = \mathrm{H}^1(C, \mathcal{O}_C)_s \oplus \mathrm{H}^1(C, \mathcal{O}_C)_n$$

where F^* is bijective on $\mathrm{H}^1(C, \mathcal{O}_C)_s$ and nilpotent on $\mathrm{H}^1(C, \mathcal{O}_C)_n$. The dimension of $\mathrm{H}^1(C, \mathcal{O}_C)_s$ is equal to the rank of the composition of F^* with itself g times, and this rank is called the *stable* rank of Frobenius on $\mathrm{H}^1(C, \mathcal{O}_C)$.

Proposition 3 *The p-rank of C is equal to the stable rank of the Frobenius endomorphism $F^* : \mathrm{H}^1(C, \mathcal{O}_C) \to \mathrm{H}^1(C, \mathcal{O}_C)$.*

Proof See [16]. □

The p-rank of the Jacobian of C can be determined from either the Cartier-Manin or Hasse-Witt matrix. For a matrix M, we write $M^{(p^i)}$ for the matrix obtained from M by raising each of its entries to the power p^i.

Proposition 4 *Let C be a curve of genus g with Hasse-Witt matrix H and Cartier-Manin matrix M. Then the p-rank of C is*

$$f_C = \mathrm{rk}(HH^{(p)} \dots H^{(p^{g-1})}) = \mathrm{rk}(M^{(p^{g-1})} \dots M^{(p)}M).$$

Proof The first equality follows from Proposition 3 and the fact that Frobenius is p-linear. The second equality is a consequence of Serre duality; see Proposition 2. □

Remark 4 When $g = 1$ or $g = 2$, then C has p-rank 0 if and only if C is supersingular. When $g \geq 3$, there exist curves of p-rank 0 which are not supersingular.

2.5 Prym Varieties

Suppose that p is odd. If X is a curve of genus g defined over k, then $\mathrm{Jac}(X)$ is a principally polarized abelian variety of dimension g. There is a bijection between 2-torsion points on $\mathrm{Jac}(X)$ and unramified double covers $\pi : Y \to X$. Without further comment, we require that Y is connected, which is equivalent to the 2-torsion point being nontrivial. Also, we note that Y is smooth if X is smooth.

Let $\pi : Y \to X$ be an unramified double cover of X. By the Riemann-Hurwitz formula, Y has genus $2g - 1$. Also $\mathrm{Jac}(X)$ is isogenous to a sub-abelian variety of $\mathrm{Jac}(Y)$. Let σ be the endomorphism of $\mathrm{Jac}(Y)$ induced by the involution generating $\mathrm{Gal}(\pi)$.

The *Prym variety* P_π is the connected component containing 0 in the kernel of the map $\pi^* : \mathrm{Jac}(Y) \to \mathrm{Jac}(X)$. It is also the image of the map $1 - \sigma$ in $\mathrm{Jac}(Y)$. In other words,

$$P_\pi = \mathrm{Im}(1 - \sigma) = \mathrm{Ker}(1 + \sigma)^0.$$

The canonical principal polarization of $\mathrm{Jac}(Y)$ induces a principal polarization on P_π. Finally, $\mathrm{Jac}(Y)$ is isogenous to $\mathrm{Jac}(X) \oplus P_\pi$.

2.6 Moduli Spaces

In this paper, we consider:

- $\mathcal{M}_g$ the moduli space of curves of genus g over k;
- $\mathcal{A}_g$ the moduli space of principally polarized abelian varieties of dimension g over k;
- $\mathcal{R}_g$ the moduli space whose points represent unramified double covers $\pi : Y \to X$ over k, where X is a curve of genus g;
- $\mathcal{R}_g^{(f,f')}$ the stratum of $\mathcal{R}_g$ of points representing covers where X has p-rank f and P_π has p-rank f'.

3 Hasse-Witt Matrices of Genus 3 Curves and Their Prym Varieties

We continue to work over an algebraically closed field k of characteristic $p > 2$. Suppose $\pi : Y \to X$ is an unramified double cover of a non-hyperelliptic smooth curve of genus 3. In [4], Bruin describes the equations for X and P_π in terms of quadratic forms. We describe the Hasse-Witt matrices of X and P_π in terms of the quadratic forms, using results of Stöhr and Voloch in [17] and Yui in [22]. As an application, we answer Question 1 affirmatively when $3 \leq p \leq 19$ and $g = 3$ in Proposition 5.

3.1 The Prym Variety of an Unramified Double Cover of a Plane Quartic

A smooth curve X of genus 3 which is not hyperelliptic is isomorphic to a smooth plane quartic.

Lemma 1 ([4, Bruin]) *Suppose $\pi : Y \to X$ is an unramified double cover of a smooth plane quartic curve. Then there exist quadratic forms Q_1, Q_2, Q_3 in $k[u, v, w]$ such that $X \subset \mathbb{P}^2$ is given by the equation*

$$X : Q_1(u, v, w)Q_3(u, v, w) = Q_2(u, v, w)^2, \tag{1}$$

$Y \subset \mathbb{P}^4$ is given by the equations

$$Y : Q_1(u, v, w) = r^2, \ \ Q_2(u, v, w) = rs, \ \ Q_3(u, v, w) = s^2, \tag{2}$$

and the Prym variety P_π is isomorphic to Jac(Z) *for the smooth genus* 2 *curve Z with equation*

$$Z : z^2 = D(x) := -\det(M_1 + 2xM_2 + x^2M_3), \tag{3}$$

where M_i is the symmetric 3×3 matrix such that

$$(u, v, w)M_i(u, v, w)^T = Q_i(u, v, w).$$

Conversely, if $Q_1, Q_2, Q_3 \in k[u, v, w]$ are quadratic forms such that (1) *defines a smooth plane quartic X, then the equations above give an unramified double cover $\pi : Y \to X$ and a smooth genus* 2 *curve Z such that $P_\pi \simeq$* Jac(Z).

Proof This is proven in [4, Theorem 5.1(4)]. The fact that Z is smooth when X is smooth can be found in [4, Section 5, Case 4]. □

3.2 Hasse-Witt Matrices

Lemma 2 *Let $\pi : Y \to X$ be an unramified double cover of a smooth plane quartic curve, and suppose $P_\pi = \mathrm{Jac}(Z)$. Let $Q_1, Q_2, Q_3 \in k[u, v, w]$ be quadratic forms as in Lemma 1, and let $D(x) \in k[x]$ be defined as in Lemma 1* (3).

1. *Let*

$$q(u, v) = Q_2(u, v, 1)^2 - Q_1(u, v, 1)Q_3(u, v, 1).$$

Let $a_{i,j}$ be the values in k such that $q(u, v)^{p-1} = \sum_{i,j} a_{i,j}u^i v^j$. Then the Hasse-Witt matrix of X is

$$H_X = \begin{pmatrix} a_{p-1,p-1} & a_{2p-1,p-1} & a_{p-1,2p-1} \\ a_{p-2,p-1} & a_{2p-2,p-1} & a_{p-2,2p-1} \\ a_{p-1,p-2} & a_{2p-1,p-2} & a_{p-1,2p-2} \end{pmatrix}.$$

2. *Let $b_i \in k$ be the values in k such that $D(x)^{(p-1)/2} = \sum_i b_i x^i$. Then the Hasse-Witt matrix of Z is*

$$H_Z = \begin{pmatrix} b_{p-1} & b_{2p-1} \\ b_{p-2} & b_{2p-2} \end{pmatrix}.$$

Remark 5 In Lemma 2(1), the Hasse-Witt matrix is taken with respect to the basis of $\mathrm{H}^1(X, \mathcal{O}_X)$ given by the dual of the basis $\frac{du}{q_v}$, $u\frac{du}{q_v}$, $v\frac{du}{q_v}$ of $\mathrm{H}^0(X, \Omega^1_X)$. In Lemma 2(2), the Hasse-Witt matrix is taken with respect to the basis of $\mathrm{H}^1(Z, \mathcal{O}_Z)$ given by the dual of the basis $\frac{dx}{z}$, $x\frac{dx}{z}$ of $\mathrm{H}^0(Z, \Omega^1_Z)$.

Proof

1. Let $\omega_1, \dots, \omega_g$ be a basis for $\mathrm{H}^0(X, \Omega_X)$ and suppose that the action of the Cartier operator is given by

$$\mathscr{C}(\omega_i) = \sum_{j=1}^{g} c_{ij}\omega_j. \tag{4}$$

By Definition 1 and Remark 3, the Hasse-Witt matrix with respect to the dual basis is the matrix (c_{ij}^p).

The result [17, Theorem 1.1] of Stöhr and Voloch yields the following information in (5) and (6) about the action of the Cartier operator on the smooth plane curve X, with affine equation $q(u, v) = 0$. Consider the partial derivative operator $\nabla = \frac{\partial^{2p-2}}{\partial u^{p-1}\partial v^{p-1}}$. Then for any $h \in k(u, v)$,

$$\mathscr{C}\left(h\frac{du}{q_v}\right) = \left(\nabla(q^{p-1}h)\right)^{\frac{1}{p}} \frac{du}{q_v}. \tag{5}$$

Also, if $\alpha_{i,j} \in k$, then

$$\nabla\left(\sum_{i,j} \alpha_{i,j} u^i v^j\right) = \sum_{i,j} \alpha_{ip+p-1,jp+p-1} u^{ip} v^{jp}. \quad (6)$$

Write $\omega_i = h_i(u, v)\frac{du}{q_v}$. By (5) and (4),

$$\nabla(q^{p-1}h_i) = \sum_j c_{ij}^p h_j^p. \quad (7)$$

In this case, a basis for $H^0(X, \Omega_X^1)$ is $\omega_1 = \frac{du}{q_v}, \omega_2 = u\frac{du}{q_v}, \omega_3 = v\frac{du}{q_v}$. By definition, $q(u, v)^{p-1} = \sum_{i,j} a_{i,j} u^i v^j$. By (7) and (6), we have

$$\nabla(q^{p-1}) = \sum_{i,j} a_{ip+p-1,jp+p-1} u^{ip} v^{jp} = c_{11}^p + c_{12}^p u^p + c_{13}^p v^p$$

where c_{11}, c_{12}, c_{13} are the entries in the first row of the Hasse-Witt matrix. Note that $\deg(q) = 4$, so $\deg(q^{p-1}) = 4(p-1)$ and hence

$$\deg(\nabla(q^{p-1})) \leq 2(p-1).$$

Thus, the coefficient of $u^{ip}v^{jp}$ in $\nabla(q^{p-1})$ is zero unless $i + j \leq 1$. Equating nonzero coefficients gives $c_{11} = a_{p-1,p-1}, c_{12} = a_{2p-1,p-1}$ and $c_{13} = a_{p-1,2p-1}$.

Similarly, for the other two rows in the Hasse-Witt matrix,

$$\nabla(q^{p-1}u) = \sum_{i,j} a_{ip+p-1,jp+p-1} u^{ip+1} v^{jp} = c_{21}^p + c_{22}^p u^p + c_{23}^p v^p,$$

and

$$\nabla(q^{p-1}v) = \sum_{i,j} a_{ip+p-1,jp+p-1} u^{ip} v^{jp+1} = c_{31}^p + c_{32}^p u^p + c_{33}^p v^p.$$

2. Note that Z is smooth since X is smooth by Bruin [4, Section 5, Case 4]. The result follows from [3, Lemma 5.1]. Alternatively, the matrix H_Z is the transpose of the Cartier-Manin matrix for Z from [22, Proposition 1]. □

3.3 *The p-Ranks of X and Z*

By Proposition 4, the p-rank $f = f_X$ of X is the rank of $H_X H_X^{(p)} H_X^{(p^2)}$ and the p-rank $f' = f_Z$ of Z is the rank of $H_Z H_Z^{(p)}$.

Proposition 5 *Let* $3 \leq p \leq 19$*. For each pair* (f, f') *such that* $0 \leq f \leq 3$ *and* $0 \leq f' \leq 2$*, there exists an unramified double cover* $\pi : Y \to X$ *such that* X *is a smooth curve of genus* 3 *and* p*-rank* f *and* P_π *has* p*-rank* f'*; in other words,* $\mathscr{R}_3^{(f,f')}$ *is non empty when* $3 \leq p \leq 19$.

Proof The result holds (without any restriction on p) when $f' = 2$ or $f' = 1$ by Ozman and Pries [15, Proposition 6.4], as long as $(f, f') \neq (0, 1), (1, 1)$ when $p = 3$. To complete the proof, we provide an example below in each case when $f' = 0$ (and when $p = 3$ and $(f, f') = (0, 1), (1, 1)$). These examples were found with a computational search, using Lemma 2. □

In the examples below, we give the equations of the curves X, Z along with the coefficients of the quadratic forms that lead to these curves in the following format:

$$[q_{111}, q_{112}, q_{122}, q_{113}, q_{123}, q_{133}, q_{211}, q_{222}, q_{233}, q_{311}, q_{312}, q_{322}, q_{313}, q_{323}, q_{333}],$$

where:

- $Q_1 = q_{111}u^2 + q_{112}uv + q_{122}v^2 + q_{113}uw + q_{123}vw + q_{133}w^2$;
- $Q_2 = q_{211}u^2 + q_{222}v^2 + q_{233}w^2$;
- $Q_3 = q_{311}u^2 + q_{312}uv + q_{322}v^2 + q_{313}uw + q_{323}vw + q_{333}w^2$.

Example 1 $p = 3$

(f, f')	$X, Z, [q_{ijk}]$
$(3, 0)$	$X : 2u^4 + 2u^3v + u^3 + 2u^2v^2 + u^2v + 2u^2 + 2uv^3 + uv^2 + uv + 2u + v^3 + v^2 + 2v + 1$
	$Z : 2x^5 + x^4 + 2x^2 + x + z^2 + 1$
	$[q_{ijk}] = [2, 0, 2, 0, 0, 1, 1, 1, 1, 0, 1, 2, 2, 2, 2]$
$(2, 0)$	$X : 2u^4 + u^3v + 2u^3 + u^2v + 2uv^3 + uv^2 + 2v^3 + 2v^2 + 2$
	$Z : x^6 + 2x^5 + 2x^4 + x^2 + x + z^2$
	$[q_{ijk}] = [1, 0, 2, 0, 0, 0, 1, 1, 1, 0, 1, 2, 2, 1, 2]$
$(1, 0)$	$X : 2u^4 + 2u^3 + 2u^2v + 2u^2 + uv^2 + 2uv + x + 2v^4 + v^3 + v + 2$
	$Z : 2x^6 + 2x^5 + z^2 + 1$
	$[q_{ijk}] = [2, 0, 0, 1, 1, 1, 1, 1, 1, 0, 0, 1, 1, 1, 0]$
$(0, 0)$	$X : 2u^4 + 2u^3 + 2u^2v + 2u^2 + 2uv^2 + u + v^4 + 2v^3 + v + 1$
	$Z : 2x^6 + x + z^2 + 1$
	$[q_{ijk}] = [2, 0, 2, 0, 0, 1, 1, 1, 1, 0, 0, 1, 1, 1, 2]$
$(1, 1)$	$X : 2u^4 + u^2v^2 + u^2 + uv^3 + uv^2 + 2uv + 2u + 2v^4 + v^3 + 2v$
	$Z : 2x^6 + 2x^4 + x + y^2$
	$[q_{ijk}] = [0, 0, 1, 0, 0, 2, 1, 1, 1, 0, 1, 0, 1, 1, 2]$
$(0, 1)$	$X : 2u^4 + 2u^3v + 2u^3 + 2u^2 + uv^3 + uv^2 + 2uv + 2u + v^2 + 2$
	$Z : 2x^6 + 2x^3 + 2x^2 + x + y^2 + 1$
	$[q_{ijk}] = [2, 0, 1, 0, 0, 2, 1, 0, 0, 0, 1, 0, 1, 0, 1]$

Example 2 $p = 5$

(f, f')	$X, Z, [q_{ijk}]$
(3, 0)	$X : 4u^4 + 3u^3 + 4u^2v^2 + u^2v + 3uv^2 + 4u + v^3 + 3v^2 + 3v$
	$Z : 4x^6 + x^3 + 2x + z^2 + 3$
	$q_{ijk} = [1, 0, 1, 0, 0, 3, 1, 1, 0, 0, 0, 1, 3, 1, 0]$
(2, 0)	$X : 4u^4 + 3u^3 + 4u^2v^2 + u^2v + 3uv^2 + u + v^3 + 2v^2 + 2v$
	$Z : 4x^6 + 4x^3 + 3x + z^2 + 2$
	$[q_{ijk}] = [1, 0, 1, 0, 0, 2, 1, 1, 0, 0, 0, 1, 3, 1, 0]$
(1, 0)	$X : 4u^4 + 3u^2v^2 + 3u^2v + 2u^2 + 4uv^2 + uv + 2u + 4v^4 + 4v^3 + 4v^2 + 3$
	$Z : 2x^5 + x^3 + 2x^2 + 2x + z^2 + 2$
	$[q_{ijk}] = [3, 4, 4, 4, 4, 3, 1, 1, 1, 0, 0, 0, 0, 1, 3]$
(0, 0)	$X : 4u^4 + 3u^2v^2 + 3u^2v + 2u^2 + 4uv^2 + 3uv + 3v + 4v^4 + 4v^3 + v + 2$
	$Z : 2x^5 + 2x^2 + 2x + z^2 + 2$
	$[q_{ijk}] = [3, 4, 4, 1, 0, 1, 1, 1, 1, 0, 0, 0, 0, 1, 3]$

Example 3 $p = 7$

(f, f')	$X, Z, [q_{ijk}]$
(3, 0)	$X : 6u^4 + 5u^2v^2 + 3u^2v + 6u^2 + 6v^4 + v^3 + 3v^2 + v + 4$
	$Z : 6x^5 + 6x^3 + z^2 + 4$
	$[q_{ijk}] = [1, 0, 5, 0, 0, 5, 1, 1, 1, 0, 0, 0, 0, 3, 1]$
(2, 0)	$X : 6u^4 + 5u^2v^2 + 2u^2v + 2u^2 + 6v^4 + 4v^3 + 6v^2 + 6$
	$Z : 5x^5 + x^4 + 4x^3 + 6x^2 + 4x + z^2$
	$[q_{ijk}] = [1, 0, 2, 0, 0, 0, 1, 1, 1, 0, 0, 0, 0, 2, 4]$
(1, 0)	$X : 6u^4 + 5u^2v^2 + 2u^2v + u^2 + 6v^4 + 5v^2 + 4v + 5$
	$Z : 6x^5 + 6x^4 + x^2 + x + z^2$
	$[q_{ijk}] = [3, 0, 0, 0, 0, 6, 1, 1, 1, 0, 0, 0, 0, 3, 1]$
(0, 0)	$X : 6u^4 + u^2v^2 + 4u^2 + 3v^4 + 6v^2 + 6$
	$Z : 4x^5 + 4x^4 + 3x^2 + 3x + z^2$
	$[q_{ijk}] = [3, 0, 4, 0, 0, 0, 1, 1, 1, 0, 0, 1, 0, 0, 2]$

Example 4 $p = 11$

(f, f')	$X, Z, [q_{ijk}]$
$(3, 0)$	$X : 10u^4 + 9u^2v^2 + 5u^2v + 2u^2 + 10v^4 + 10v^3 + 4v^2 + 4v + 6$
	$Z : 9x^5 + 4x^4 + x^3 + 7x^2 + 8x + z^2 + 3$
	$[q_{ijk}] = [8, 0, 5, 0, 0, 2, 1, 1, 0, 0, 0, 0, 0, 2, 3]$
$(2, 0)$	$X : 10u^4 + 9u^2v^2 + 9u^2v + 9u^2 + 10v^4 + v^3 + v^2 + 7v + 7$
	$Z : 9x^5 + 9x^4 + 9x^3 + 2x^2 + 2x + z^2 + 1$
	$[q_{ijk}] = [10, 0, 6, 0, 0, 9, 1, 1, 0, 0, 0, 0, 0, 2, 2]$
$(1, 0)$	$X : 10u^4 + 9u^2v^2 + 9u^2v + 8u^2 + 10v^4 + 3v^3 + 10v^2 + 4v + 6$
	$Z : 9x^5 + 2x^4 + 3x^3 + 9x^2 + 2x + z^2 + 8$
	$[q_{ijk}] = [10, 0, 7, 0, 0, 2, 1, 1, 0, 0, 0, 0, 0, 2, 3]$
$(0, 0)$	$X : 10u^4 + 9u^2v^2 + 3u^2v + 5u^2 + 10v^4 + 9v^3 + 8v^2 + 4v + 1$
	$Z : 9x^5 + 8x^4 + 9x^3 + 3x^2 + 10x + z^2 + 8$
	$[q_{ijk}] = [7, 0, 10, 0, 0, 2, 1, 1, 1, 0, 0, 0, 0, 2, 1]$

Example 5 $p = 13$

(f, f')	$X, Z, [q_{ijk}]$
$(3, 0)$	$X : 12u^4 + 11u^2v^2 + 6u^2v + 11u^2 + 12v^4 + 12v^3 + 11v^2 + 3v + 12$
	$Z : 11x^5 + 10x^4 + 8x^3 + 3x^2 + 11x + z^2 + 1$
	$[q_{ijk}] = [3, 0, 6, 0, 0, 8, 1, 1, 1, 0, 0, 0, 0, 2, 0]$
$(2, 0)$	$X : 12u^4 + 11u^2v^2 + 2u^2v + 4u^2 + 12v^4 + 11v^3 + 5v^2 + 11v + 6$
	$Z : 11x^5 + 10x^4 + 8x^3 + 3x^2 + 11x + z^2 + 1$
	$[q_{ijk}] = [1, 0, 12, 0, 0, 12, 1, 1, 1, 0, 0, 0, 0, 2, 6]$
$(1, 0)$	$X : 12u^4 + 11u^2v^2 + 9u^2v + 9u^2 + 12v^4 + 7v^3 + 8v^2 + 4v + 1$
	$Z : 11x^5 + 7x^4 + 11x^3 + 6x^2 + 5x + z^2 + 1$
	$[q_{ijk}] = [2, 0, 3, 0, 0, 11, 1, 1, 1, 0, 0, 0, 0, 11, 12]$
$(0, 0)$	$X : 12u^4 + 11u^2v^2 + 9u^2v + 7u^2 + 12v^4 + 8v^3 + 6v^2 + 12v + 11$
	$Z : 6x^5 + 5x^4 + 3x^3 + 6x^2 + 6x + z^2 + 6$
	$[q_{ijk}] = [9, 0, 8, 0, 0, 12, 1, 1, 1, 0, 0, 0, 0, 1, 1]$

Example 6 $p = 17$

(f, f')	$X, Z, [q_{ijk}]$
(3, 0)	$X : 16u^4 + 15u^2v^2 + 15u^2 + 16v^4 + 5v^3 + 7v^2 + 6v + 3$
	$Z : 4x^5 + 8x^4 + 9x^3 + 9x^2 + x + z^2$
	$[q_{ijk}] = [0, 0, 13, 0, 0, 2, 1, 1, 1, 0, 0, 0, 0, 3, 2]$
(2, 0)	$X : 16u^4 + 15u^2v^2 + 10u^2v + u^2 + 16v^4 + 3v^3 + 4v^2 + 16$
	$Z : 4x^5 + 8x^4 + 9x^3 + 9x^2 + x + z^2$
	$[q_{ijk}] = [9, 0, 1, 0, 0, 0, 1, 1, 1, 0, 0, 0, 0, 3, 6]$
(1, 0)	$X : 16u^4 + 15u^2v^2 + 10u^2v + 3u^2 + 16v^4 + 4v^3 + 14v + 6$
	$Z : 4x^5 + 7x^4 + 5x^3 + 10x^2 + 9x + z^2 + 5$
	$[q_{ijk}] = [9, 0, 7, 0, 0, 16, 1, 1, 1, 0, 0, 0, 0, 3, 10]$
(0, 0)	$X : 16u^4 + 15u^2v^2 + 6u^2v + 15u^2 + 16v^4 + 9v^3 + 15v^2 + 15v + 16$
	$Z : 8x^5 + 7x^4 + 8x^3 + x^2 + 14x + z^2 + 11$
	$[q_{ijk}] = [6, 0, 9, 0, 0, 15, 1, 1, 1, 0, 0, 0, 0, 1, 0]$

Example 7 $p = 19$

(f, f')	$X, Z, [q_{ijk}]$
(3, 0)	$X : 18u^4 + 17u^2v^2 + 9u^2v + 3u^2 + 18v^4 + 5v^3 + 5v^2 + 18$
	$Z : 5x^5 + 11x^4 + 13x^3 + 8x^2 + 10x + z^2$
	$[q_{ijk}] = [3, 0, 8, 0, 0, 0, 1, 1, 1, 0, 0, 0, 0, 3, 8]$
(2, 0)	$X : 18u^4 + 17u^2v^2 + 12u^2v + 18u^2 + 18v^4 + 18v^3 + 9v^2 + 18$
	$Z : 5x^5 + 11x^4 + 13x^3 + 8x^2 + 10x + z^2$
	$[q_{ijk}] = [4, 0, 6, 0, 0, 0, 1, 1, 1, 0, 0, 0, 0, 3, 5]$
(1, 0)	$X : 18u^4 + 17u^2v^2 + 5u^2v + 18u^2 + 18v^4 + 6v^3 + 3v^2 + 6v + 4$
	$Z : 17x^5 + x^4 + x^3 + 18x^2 + 10x + z^2 + 13$
	$[q_{ijk}] = [12, 0, 3, 0, 0, 3, 1, 1, 1, 0, 0, 0, 0, 2, 8]$
(0, 0)	$X : 18u^4 + 17u^2v^2 + 17u^2v + 4u^2 + 18v^4 + v^3 + 14v^2 + 12v + 1$
	$Z : 16x^5 + 9x^4 + 14x^3 + 10x^2 + 8x + z^2 + 3$
	$[q_{ijk}] = [11, 0, 4, 0, 0, 10, 1, 1, 1, 0, 0, 0, 0, 5, 4]$

Remark 6 Since k is an algebraically closed field of odd characteristic p, it is possible to diagonalize the quadratic form Q_2 and take its coefficients to be 0 or 1. Even so, the complicated nature of the entries of H_X and H_Z makes it difficult to analyze the p-ranks algebraically.

The entries of H_X are quite complicated even in terms of the coefficients of $q(u, v) = Q_2(u, v, 1)^2 - Q_1(u, v, 1)Q_3(u, v, 1)$. For example, if $p = 3$ and $q(u, v) = \sum_{i,j} b_{ij}u^i v^j$, then the upper left entry of H_X is

$$2b_{00}b_{22} + 2b_{01}b_{21} + 2b_{02}b_{20} + 2b_{10}b_{12} + b_{11}^2.$$

Similarly, even the equation for $Z : z^2 = D(x)$ is rather complicated in terms of the coefficients of Q_1 and Q_3.

4 The Hasse-Witt Matrix of a Smooth Plane Quartic Defined as an Intersection in $\mathbb{P}^3$

In this section, we determine the Hasse-Witt matrix of a curve C of genus 3 defined as the intersection of a plane and degree 4 hypersurface in $\mathbb{P}^3$. We use this result in Sect. 5 to determine the Hasse-Witt matrix of each smooth plane quartic X which has an unramified double cover $\pi : Y \to X$ such that P_π is isomorphic to a fixed abelian surface.

As before, let k be an algebraically closed field of characteristic $p > 2$. Following [12], let C/k be a curve in $\mathbb{P}^3 = \text{Proj}(k[x, y, z, w])$ defined by $v = h = 0$ for homogeneous polynomials $v, h \in k[x, y, z, w]$ with $\gcd(v, h) = 1$. Let r and s denote the degrees of v and h, respectively. Let C^p denote the curve in $\mathbb{P}^3$ defined by $v^p = h^p = 0$. For $n \in \mathbb{Z}$, let $\mathcal{O}_{\mathbb{P}^3}(n)$ denote the nth tensor power of Serre's twisting sheaf.

Lemma 3 ([12, Lemma 3.1.3]) *The following diagram is commutative with exact rows, where the composite map* $\mathrm{H}^1(C, \mathcal{O}_C) \to \mathrm{H}^1(C, \mathcal{O}_C)$ *is induced by the Frobenius morphism on C and the map* F_1 *is the Frobenius morphism on* $\mathbb{P}^3$.

$$\begin{array}{ccccccc}
0 & \longrightarrow & \mathrm{H}^1(C,\mathcal{O}_C) & \longrightarrow & \mathrm{H}^3(\mathbb{P}^3,\mathcal{O}_{\mathbb{P}^3}(-r-s)) & \longrightarrow & \mathrm{H}^3(\mathbb{P}^3,\mathcal{O}_{\mathbb{P}^3}(-r))\oplus\mathrm{H}^3(\mathbb{P}^3,\mathcal{O}_{\mathbb{P}^3}(-s)) \\
 & & \downarrow & & \downarrow{\scriptstyle F_1^*} & & \downarrow \\
0 & \longrightarrow & \mathrm{H}^1(C^p,\mathcal{O}_{C^p}) & \longrightarrow & \mathrm{H}^3(\mathbb{P}^3,\mathcal{O}_{\mathbb{P}^3}((-r-s)p)) & \longrightarrow & \mathrm{H}^3(\mathbb{P}^3,\mathcal{O}_{\mathbb{P}^3}(-rp))\oplus\mathrm{H}^3(\mathbb{P}^3,\mathcal{O}_{\mathbb{P}^3}(-sp)) \\
 & & \downarrow & & \downarrow{\scriptstyle (vh)^{p-1}} & & \downarrow \\
0 & \longrightarrow & \mathrm{H}^1(C,\mathcal{O}_C) & \longrightarrow & \mathrm{H}^3(\mathbb{P}^3,\mathcal{O}_{\mathbb{P}^3}(-r-s)) & \longrightarrow & \mathrm{H}^3(\mathbb{P}^3,\mathcal{O}_{\mathbb{P}^3}(-r))\oplus\mathrm{H}^3(\mathbb{P}^3,\mathcal{O}_{\mathbb{P}^3}(-s))
\end{array}$$

Proof This is an excerpt from the diagram immediately preceding [12, Proposition 3.1.4]. □

For $t \in \mathbb{Z}_{>0}$, the k-vector space $\mathrm{H}^3(\mathbb{P}^3, \mathcal{O}_{\mathbb{P}^3}(-t))$ has basis

$$\{x^{k_1}y^{k_2}z^{k_3}w^{k_4} : (k_1, k_2, k_3, k_4) \in (\mathbb{Z}_{<0})^4, k_1 + k_2 + k_3 + k_4 = -t\}.$$

Lemma 4 *Suppose that $r \leq 3$. Then the following diagram is commutative with exact rows, where the map F is the Frobenius morphism on C and the map F_1 is the Frobenius morphism on $\mathbb{P}^3$.*

$$\begin{array}{ccccccc} 0 \longrightarrow & \mathrm{H}^1(C, \mathcal{O}_C) & \longrightarrow & \mathrm{H}^3(\mathbb{P}^3, \mathcal{O}_{\mathbb{P}^3}(-r-s)) & \xrightarrow{v} & \mathrm{H}^3(\mathbb{P}^3, \mathcal{O}_{\mathbb{P}^3}(-s)) \\ & \downarrow F^* & & \downarrow (vh)^{p-1}F_1^* & & \downarrow \\ 0 \longrightarrow & \mathrm{H}^1(C, \mathcal{O}_C) & \longrightarrow & \mathrm{H}^3(\mathbb{P}^3, \mathcal{O}_{\mathbb{P}^3}(-r-s)) & \xrightarrow{v} & \mathrm{H}^3(\mathbb{P}^3, \mathcal{O}_{\mathbb{P}^3}(-s)) \end{array} \tag{8}$$

Proof This follows from Lemma 3 and the fact that $\mathrm{H}^3(\mathbb{P}^3, \mathcal{O}_{\mathbb{P}^3}(-r)) = 0$ for $r \leq 3$. □

For $i, j \in \mathbb{Z}$, set $\delta_{ij} = \begin{cases} 1 & \text{if } i = j \\ 0 & \text{if } i \neq j \end{cases}$.

Set $t = 5$ and let

$$S_{-5} = \{(k_1, k_2, k_3, k_4) \in (\mathbb{Z}_{<0})^4, k_1 + k_2 + k_3 + k_4 = -5\}.$$

Write $S_{-5} = \{(k_1^{(i)}, k_2^{(i)}, k_3^{(i)}, k_4^{(i)})\}_{1 \leq i \leq 4}$, where $k_j^{(i)} = -1 - \delta_{ij}$ for $1 \leq i, j \leq 4$.

Proposition 6 *Let v and h be homogeneous polynomials in $k[x, y, z, w]$ with* $\gcd(v, h) = 1$. *Suppose that $r = \deg(v) = 1$ and $s = \deg(h) = 4$. Write $v = a_1x + a_2y + a_3z + a_4w$ and fix t, with $1 \leq t \leq 4$, such that $a_t \neq 0$. Let C/k be the curve in $\mathbb{P}^3 = \mathrm{Proj}(k[x, y, z, w])$ defined by $v = h = 0$.*

Write $(vh)^{p-1} = \sum c_{i_1,i_2,i_3,i_4}x^{i_1}y^{i_2}z^{i_3}w^{i_4}$. For $1 \leq i, j \leq 4$, write

$$\gamma_{i,j} = c_{p(1+\delta_{1j})-(1+\delta_{1i}),\ p(1+\delta_{2j})-(1+\delta_{2i}),\ p(1+\delta_{3j})-(1+\delta_{3i}),\ p(1+\delta_{4j})-(1+\delta_{4i})}.$$

Then the Hasse-Witt matrix of C is given by

$$\mathrm{HW}_C = (a_t^{p-1}\gamma_{i,j} - a_j^p a_t^{-1}\gamma_{i,t})_{1 \leq i,j \leq 4, i \neq t, j \neq t}.$$

Let $t = 4$ and $a_4 = 1$ and write $a = a_1$, $b = a_2$, and $c = a_3$. Then the matrix HW_C in Proposition 6 equals $\mathrm{HW}_C = (h_{i,j})$ where

$$h_{1,1} = c_{2p-2,p-1,p-1,p-1} - a^p c_{p-2,p-1,p-1,2p-1}$$
$$h_{1,2} = c_{p-2,2p-1,p-1,p-1} - b^p c_{p-2,p-1,p-1,2p-1}$$
$$h_{1,3} = c_{p-2,p-1,2p-1,p-1} - c^p c_{p-2,p-1,p-1,2p-1}$$
$$h_{2,1} = c_{2p-1,p-2,p-1,p-1} - a^p c_{p-1,p-2,p-1,2p-1}$$
$$h_{2,2} = c_{p-1,2p-2,p-1,p-1} - b^p c_{p-1,p-2,p-1,2p-1}$$

$$h_{2,3} = c_{p-1,p-2,2p-1,p-1} - c^p c_{p-1,p-2,p-1,2p-1}$$
$$h_{3,1} = c_{2p-1,p-1,p-2,p-1} - a^p c_{p-1,p-1,p-2,2p-1}$$
$$h_{3,2} = c_{p-1,2p-1,p-2,p-1} - b^p c_{p-1,p-1,p-2,2p-1}$$
$$h_{3,3} = c_{p-1,p-1,2p-2,p-1} - c^p c_{p-1,p-1,p-2,2p-1}.$$

Proof Consider the multiplication-by-v map

$$[\times v] : \mathrm{H}^3(\mathbb{P}^3, \mathcal{O}_{\mathbb{P}^3}(-5)) \to \mathrm{H}^3(\mathbb{P}^3, \mathcal{O}_{\mathbb{P}^3}(-4)).$$

By Lemma 4, computing the matrix of F^* is equivalent to computing the matrix of $(vh)^{p-1}F_1^*$ on the kernel of $[\times v]$.

First, we compute the matrix of $(vh)^{p-1}F_1^*$ on all of $\mathrm{H}^3(\mathbb{P}^3, \mathcal{O}_{\mathbb{P}^3}(-5))$. The k-vector space $\mathrm{H}^3(\mathbb{P}^3, \mathcal{O}_{\mathbb{P}^3}(-5))$ is 4-dimensional with basis

$$\{x^{k_1}y^{k_2}z^{k_3}w^{k_4} : (k_1, k_2, k_3, k_4) \in S_{-5}\}.$$

Explicitly, a basis is given by

$$e_1 = x^{-2}y^{-1}z^{-1}w^{-1}, e_2 = x^{-1}y^{-2}z^{-1}w^{-1}, e_3 = x^{-1}y^{-1}z^{-2}w^{-1}, e_4 = x^{-1}y^{-1}z^{-1}w^{-2}.$$

As in the proof of [12, Proposition 3.1.4], for each $j \in \{1, \dots, 4\}$, then

$$\begin{aligned}
(vh)^{p-1}F_1^*(e_j) &= (vh)^{p-1}F_1^*(x^{k_1^{(j)}}y^{k_2^{(j)}}z^{k_3^{(j)}}w^{k_4^{(j)}}) \\
&= (vh)^{p-1}x^{pk_1^{(j)}}y^{pk_2^{(j)}}z^{pk_3^{(j)}}w^{pk_4^{(j)}} \\
&= \sum c_{i_1,i_2,i_3,i_4}x^{i_1+pk_1^{(j)}}y^{i_2+pk_2^{(j)}}z^{i_3+pk_3^{(j)}}w^{i_4+pk_4^{(j)}} \\
&= \sum_{i=1}^{4} c_{k_1^{(i)}-pk_1^{(j)},k_2^{(i)}-pk_2^{(j)},k_3^{(i)}-pk_3^{(j)},k_4^{(i)}-pk_4^{(j)}}x^{k_1^{(i)}}y^{k_2^{(i)}}z^{k_3^{(i)}}w^{k_4^{(i)}} \\
&= \sum_{i=1}^{4} \gamma_{i,j}e_i.
\end{aligned}$$

Explicitly, the following 4-by-4 matrix H_0 represents the map $(vh)^{p-1}F_1^*$ on $\mathrm{H}^3(\mathbb{P}^3, \mathcal{O}_{\mathbb{P}^3}(-5))$, with respect to the basis e_1, e_2, e_3, e_4:

$$H_0 = \begin{pmatrix} c_{2p-2,p-1,p-1,p-1} & c_{p-2,2p-1,p-1,p-1} & c_{p-2,p-1,2p-1,p-1} & c_{p-2,p-1,p-1,2p-1} \\ c_{2p-1,p-2,p-1,p-1} & c_{p-1,2p-2,p-1,p-1} & c_{p-1,p-2,2p-1,p-1} & c_{p-1,p-2,p-1,2p-1} \\ c_{2p-1,p-1,p-2,p-1} & c_{p-1,2p-1,p-2,p-1} & c_{p-1,p-1,2p-2,p-1} & c_{p-1,p-1,p-2,2p-1} \\ c_{2p-1,p-1,p-1,p-2} & c_{p-1,2p-1,p-1,p-2} & c_{p-1,p-1,2p-1,p-2} & c_{p-1,p-1,p-1,2p-2} \end{pmatrix}. \tag{9}$$

Now we calculate the 3-by-3 matrix representing the restriction of $(vh)^{p-1}F_1^*$ to the kernel of $[\times v]$ on $\mathrm{H}^3(\mathbb{P}^3, \mathcal{O}_{\mathbb{P}^3}(-5))$. First, note that if $\ell \in \mathrm{H}^3(\mathbb{P}^3, \mathcal{O}_{\mathbb{P}^3}(-5))$ is in $\mathrm{Ker}([\times v])$, then $(vh)^{p-1}F_1^*(\ell)$ is also in $\mathrm{Ker}([\times v])$, by the commutativity of (8).

The k-vector space $\mathrm{H}^3(\mathbb{P}^3, \mathcal{O}_{\mathbb{P}^3}(-4))$ is one-dimensional with basis element $\lambda = x^{-1}y^{-1}z^{-1}w^{-1}$. Note that $v \cdot e_i = a_i\lambda$. Thus

$$\mathrm{Ker}([\times v]) = \left\{ \sum_{i=1}^{4} c_i e_i \mid \sum_{i=1}^{4} a_i c_i = 0 \right\}.$$

For $1 \le i, j \le 4$, write $\beta_j^{(i)} = a_i e_j - a_j e_i$. If $a_t \ne 0$, then $\mathrm{Ker}([\times v])$ has basis $\{\beta_j^{(t)}\}_{1\le j\le 4, j\ne t}$. It follows that

$$\begin{aligned}
(vh)^{p-1}F_1^*(\beta_j^{(t)}) &= a_t^p (vh)^{p-1}F_1^*(e_j) - a_j^p (vh)^{p-1}F_1^*(e_t) \\
&= a_t^p \sum_{i=1}^{4} \gamma_{i,j} e_i - a_j^p \sum_{i=1}^{4} \gamma_{i,t} e_i \\
&= \sum_{i=1}^{4} (a_t^p \gamma_{i,j} - a_j^p \gamma_{i,t}) e_i. \qquad (10)
\end{aligned}$$

The commutativity of the diagram (8) shows that $(vh)^{p-1}F_1^*(\beta_j^{(t)})$ is in $\mathrm{Ker}([\times v])$. Therefore, there are coefficients $\lambda_{i,j} \in k$ such that for $j \ne t$,

$$\begin{aligned}
(vh)^{p-1}F_1^*(\beta_j^{(t)}) &= \sum_{1\le i\le 4, i\ne t} \lambda_{i,j} \beta_i^{(t)} \\
&= \sum_{1\le i\le 4, i\ne t} \lambda_{i,j} (a_t e_i - a_i e_t). \qquad (11)
\end{aligned}$$

Comparing the coefficients of e_i for $i \ne t$ in (10) and (11), we see that

$$\lambda_{i,j} = a_t^{-1}(a_t^p \gamma_{i,j} - a_j^p \gamma_{i,t}).$$

This completes the proof of Proposition 6. □

5 The Fiber of the Prym Map When $g = 3$

In Sect. 3, we used a description from [4] of an unramified double cover $\pi : Y \to X$ of a plane quartic curve X and its Prym variety P_π in terms of quadratic forms. We then calculated the Hasse-Witt matrices of X and P_π and produced examples where X and P_π have specified p-ranks for small primes p. However, since the entries of the Hasse-Witt matrices are complicated, it is not clear how to apply this method for arbitrarily large primes p.

In the current section, we describe an alternative method in which we start with a smooth curve Z of genus 2 over k and construct smooth plane quartic curves X having an unramified double cover $\pi : Y \to X$ such that $P_\pi \simeq \mathrm{Jac}(Z)$. The advantage of this alternative method is that it allows us to prove an existence result for infinitely many primes. In particular, in Proposition 7, we prove that if $p \equiv 5 \bmod 6$, then there exists a smooth curve X defined over $\overline{\mathbb{F}}_p$ with genus 3 and p-rank 3 having an unramified double cover $\pi : Y \longrightarrow X$ such that P_π has p-rank 0.

Here is an outline of the section. In Sect. 5.1, we review a result of Verra that describes the geometry of the fibers of the Prym map $\mathscr{R}_3 \to \mathscr{A}_2$. In Sect. 5.2, we work with an explicit construction of the curves represented by points in a three-dimensional component of the fiber above $\mathrm{Jac}(Z)$. These curves occur as the intersection in $\mathbb{P}^3$ of a plane and the Kummer surface of $\mathrm{Jac}(Z)$. They are smooth plane curves X having an unramified double cover $\pi : Y \to X$ such that $P_\pi \simeq \mathrm{Jac}(Z)$. In Sect. 5.3, we describe the determinant of the Hasse-Witt matrix of such a curve X. The main application when $p \equiv 5 \bmod 6$ is in Sect. 5.4.

In Sect. 5.5, we use commutative algebra to characterize when X is non-ordinary (under conditions on the p-rank of Z). In Sect. 5.6, we fix $p = 3$ and apply the results of the section to a one-dimensional family of genus 2 curves Z with 3-rank 0. This allows us to deduce information about the locus of planes V for which X is non-ordinary and the geometry of the corresponding moduli space $\mathscr{R}_3^{(2,0)}$.

5.1 Review of Work of Verra

The material in this subsection is not used directly in this paper, but we include it because it provides important context. Let A be a principally polarized abelian surface. For example, one could take $A = \mathrm{Jac}(Z)$ for a smooth curve Z of genus 2. Let $\mathscr{A}_2$ be the moduli space of principally polarized abelian surfaces. Let s be the point of $\mathscr{A}_2$ representing A. We would like to consider the fiber of the Prym map $Pr_3 : \mathscr{R}_3 \to \mathscr{A}_2$ over s. (More precisely, let $\tilde{\mathscr{A}_2}$ denote the smooth toroidal compactification of $\mathscr{A}_2$ and $\bar{\mathscr{R}}_3$ the compactification of $\mathscr{R}_3$ and consider the fiber of $\bar{Pr}_3 : \bar{\mathscr{R}}_3 \to \tilde{\mathscr{A}_2}$ over s.)

Following [19, Section 2], let $\Theta \subset A$ be a symmetric theta divisor. Suppose that $\mathrm{Aut}(\Theta) \simeq \mathbb{Z}/2$. Under this mild condition on s, Verra proves in [19, Corollary 4.1] that $\bar{Pr}_3^{-1}(s)$ is a blow-up of $\mathbb{P}^3$. Moreover, $\bar{Pr}_3^{-1}(s)$ has one irreducible component N_s of dimension 3 and three components of dimension 2. By Verra [19, (3.14)-(3.16), page 442], the latter represent unramified double covers of hyperelliptic or singular curves whose Prym is isomorphic to A. The generic point of N_s represents an unramified double cover $\pi : Y \to X$ where X is a smooth plane quartic and $P_\pi \simeq A$. We briefly review the results of Verra in more detail below.

The linear system $|2\Theta|$ has dimension 3 and is thus isomorphic to $\mathbb{P}^3$. Every element Y in the linear system is a curve of arithmetic genus 5 with an involution. The linear system is base point free and its generic element is a smooth irreducible curve.

For each $Y \in |2\Theta|$, there is a morphism $\phi : A \to (\mathbb{P}^3)^\wedge$, where the wedge in the superscript indicates taking the dual space. Let $K = \phi(A)$. If $A = \mathrm{Jac}(Z)$ for some smooth irreducible curve Z of genus 2, then $\deg(\phi) = 2$ and K is the Kummer quartic surface of A.

By Verra [19, page 438], this yields a map $\psi : \mathbb{P}^3 - T \to \bar{Pr}_3^{-1}(s)$. The set T is analyzed in [19, (2.1)]; it consists of the set of points for which Y is not stable.

5.2 *Explicit Version of the Fiber of the Prym Map*

Let Z be a smooth curve of genus 2. The results of [19] give a way to find all smooth plane quartic curves X having an unramified double cover $\pi : Y \to X$ such that $P_\pi \simeq \mathrm{Jac}(Z)$. This is discussed in [4, Section 7], where Bruin shows how to recover a model of the form $X : Q_1Q_3 = (Q_2)^2$ from a smooth plane section X of the Kummer surface K.

Consider the Kummer surface

$$K = \mathrm{Jac}(Z)/\langle -1 \rangle \subset \mathbb{P}^3$$

associated to Z, namely, the quotient of $\mathrm{Jac}(Z)$ by the Kummer involution. It is a quartic surface, with 16 singularities corresponding to $\mathrm{Jac}(Z)[2] \simeq (\mathbb{Z}/2\mathbb{Z})^{2g}$. Let $\phi : \mathrm{Jac}(Z) \longrightarrow K$ be the degree 2 quotient map.

For a sufficiently general plane $V \subseteq \mathbb{P}^3$, the intersection

$$X = V \cap K$$

is a smooth plane quartic curve. This implies that X does not contain any of the branch points of ϕ. Thus the restriction of ϕ to $Y = \phi^{-1}(X)$ is an unramified double cover $\pi : Y \to X$. The Prym variety P_π is isomorphic to $\mathrm{Jac}(Z)$, as seen on [2, page 616]. Conversely, by Verra [19], if $\mathrm{Jac}(Z)$ is isomorphic to the Prym variety of an unramified double cover $\pi : Y \to X$, with X a smooth plane quartic, then X is isomorphic to a planar section of K and Y is its preimage in $\mathrm{Jac}(Z)$.

5.2.1 The Kummer Surface

Suppose that Z is a smooth curve of genus 2 with affine equation

$$Z : z^2 = D(x) := \sum_{i=0}^{6} d_i x^i .$$

Consider the Kummer surface $K = \text{Jac}(Z)/\langle -1 \rangle$ associated to Z, which is a quartic surface in $\mathbb{P}^3$. In this section, we write down the equation of K as found in [5, Chapter 3, Section 1].

The degree 2 quotient map $\phi : \text{Jac}(Z) \to K$ is defined explicitly as follows. A generic divisor of degree 2 on Z has the form $(x_1, z_1) + (x_2, z_2)$. Let Z_∞ be the divisor above $x = \infty$. Then

$$\phi\colon \text{Jac}(Z) \to K$$
$$[(x_1, z_1) + (x_2, z_2) - Z_\infty] \mapsto [1 : x_1 + x_2 : x_1x_2 : \beta_0],$$

where $\beta_0 = (F_0(x_1, x_2) - 2z_1z_2)/(x_1 - x_2)^2$ and $F_0(x_1, x_2)$ equals

$$2d_0 + d_1(x_1 + x_2) + 2d_2x_1x_2 + d_3(x_1 + x_2)x_1x_2 + 2d_4(x_1x_2)^2 \\ + d_5(x_1 + x_2)(x_1x_2)^2 + 2d_6(x_1x_2)^3.$$

The map ϕ realizes $\text{Jac}(Z)$ as a double cover of K that ramifies precisely at $\text{Jac}(Z)[2]$. It maps the 16 points of order 2 of $\text{Jac}(Z)$ to the 16 singularities of K.

Let $X_1, \ldots, X_4$ denote the coordinates on $\mathbb{P}^3$. By Cassels and Flynn [5, (3.1.8)], a projective model of the Kummer surface K in $\mathbb{P}^3$ is the zero locus of the following equation:

$$\kappa(X_1, X_2, X_3, X_4) = K_2X_4^2 + K_1X_4 + K_0 \tag{12}$$

with

$$\begin{aligned}
K_2 &= X_2^2 - 4X_1X_3 \\
K_1 &= -2(2d_0X_1^3 + d_1X_1^2X_2 + 2d_2X_1^2X_3 + d_3X_1X_2X_3 + 2d_4X_1X_3^2 + d_5X_2X_3^2 + 2d_6X_3^3) \\
K_0 &= (d_1^2 - 4d_0d_2)X_1^4 - 4d_0d_3X_1^3X_2 - 2d_1d_3X_1^3X_3 - 4d_0d_4X_1^2X_2^2 \\
&\quad +4(d_0d_5 - d_1d_4)X_1^2X_2X_3 + (d_3^2 + 2d_1d_5 - 4d_2d_4 - 4d_0d_6)X_1^2X_3^2 - 4d_0d_5X_1X_2^3 \\
&\quad +4(2d_0d_6 - d_1d_5)X_1X_2^2X_3 + 4(d_1d_6 - d_2d_5)X_1X_2X_3^2 - 2d_3d_5X_1X_3^3 - 4d_0d_6X_2^4 \\
&\quad -4d_1d_6X_2^3X_3 - 4d_2d_6X_2^2X_3^2 - 4d_3d_6X_2X_3^3 + (d_5^2 - 4d_4d_6)X_3^4,
\end{aligned}$$

where $d_0, \ldots, d_6$ are the coefficients of $D(x)$ in the equation for Z.

This model of the Kummer surface K in $\mathbb{P}^3$ arises from a projective model of $\text{Jac}(Z)$ in $\mathbb{P}^{15}$; explicit calculations are thus more efficient on the Kummer surface. Alternatively, by combining a few Frobenius identities of theta characteristic functions, one can derive another projective model of K parametrized by four theta constants [13, Section 7, Chapter IIIa].

5.2.2 Plane Quartics as Planar Sections of the Kummer Surface

Let K be the Kummer surface from (12). For a plane $V \subset \mathbb{P}^3$, consider the curve

$$X = V \cap K.$$

Recall that if X is smooth, then it has genus 3, and the pullback of $\mathrm{Jac}(Z) \to K$ to X yields an unramified double cover $\pi : Y \to X$ such that the Prym variety P_π is isomorphic to $\mathrm{Jac}(Z)$.

Let $V = V_{a,b,c,d}$ be a plane defined over k by

$$V : v(X_1, X_2, X_3, X_4) = aX_1 + bX_2 + cX_3 + dX_4 = 0. \tag{13}$$

The point $(0 : 0 : 0 : 1)$ is a singular point of the Kummer surface. For planes V which avoid the singularities of K, it is no restriction to take $d = 1$.

5.3 *The Hasse-Witt Matrix of X*

In Sect. 4, we determined the Hasse-Witt matrix for a curve X given as the intersection of a plane and quartic surface in $\mathbb{P}^3$. Recall that $X = V \cap K \subset \mathbb{P}^3$ where $K : \kappa = 0$ and $V : v = 0$ are defined in (12) and (13). Recall that c_{i_1,i_2,i_3,i_4} is the coefficient of $X_1^{i_1}X_2^{i_2}X_3^{i_3}X_4^{i_4}$ in $(v\kappa)^{p-1}$. In other words,

$$(v\kappa)^{p-1} = \sum_{i_1+i_2+i_3+i_4=5(p-1)} c_{i_1,i_2,i_3,i_4} X_1^{i_1}X_2^{i_2}X_3^{i_3}X_4^{i_4}.$$

Proposition 6 describes the Hasse-Witt matrix H_X of X in terms of the coefficients c_{i_1,i_2,i_3,i_4} and a, b, c, d.

Lemma 5 *Setting $d = 1$, then the coefficients of H_X are each homogeneous of degree $2(p-1)$ in $a, b, c, d_0, \ldots, d_6$.*

Proof First, note that the equation κ in (12) for K is homogeneous of degree 2 in $d_0, \ldots, d_6, X_4$. This is because $K_2X_4^2$, K_1X_4, and K_0 are each homogeneous of degree 2 in $d_0, \ldots, d_6, X_4$. Also, the equation v for V is homogeneous of degree 1 in a, b, c, X_4. Thus $(\kappa v)^{p-1}$ is homogeneous of degree $3(p-1)$ in $a, b, c, d_0, \ldots, d_6, X_4$. The coefficients of the 4×4 matrix H_0 from (9) are coefficients of $(\kappa v)^{p-1}$.

We now determine information about the coefficients of the Hasse-Witt matrix H_X. Set $d = 1$. Let U be the 3×3 matrix obtained by removing the 4th row and 4th column of H_0. Let

$$C = [c_{p-2,p-1,p-1,2p-1}, c_{p-1,p-2,p-1,2p-1}, c_{p-1,p-1,p-2,2p-1}]^T.$$

By Proposition 6, $H_X = U - C[a^p, b^p, c^p]$. The coefficients of U are of the form c_{i_1,i_2,i_3,i_4} with $i_4 = p - 1$; thus they are each homogeneous of degree $2(p-1)$ in $a, b, c, d_0, \ldots, d_6$. The coefficients of C are of the form c_{i_1,i_2,i_3,i_4} with $i_4 = 2p - 1$; thus they are each homogeneous of degree $p - 2$ in $a, b, c, d_0, \ldots, d_6$. Thus each coefficient of $H_X = U - C[a^p, b^p, c^p]$ is homogeneous of degree $2(p-1)$ in $a, b, c, d_0, \ldots, d_6$. □

5.4 An Existence Result for Each $p \equiv 5 \bmod 6$

Proposition 7 *If $p \equiv 5 \bmod 6$, then there exists a smooth curve X defined over $\overline{\mathbb{F}}_p$ with genus 3 and p-rank 3 having an unramified double cover $\pi : Y \longrightarrow X$ such that P_π has p-rank 0. More generally, $R_3^{(3,0)}$ is nonempty of dimension 4 for each prime $p \equiv 5 \bmod 6$.*

Proof Consider the genus 2 curve $Z : z^2 = x^6 - 1$; it is superspecial and thus has p-rank 0, when $p \equiv 5 \bmod 6$ [8, Proposition 1.11]. The Kummer surface K in $\mathbb{P}^3$ is the zero locus of

$$\kappa(X_1, X_2, X_3, X_4) = K_2 X_4^2 + K_1 X_4 + K_0 \tag{14}$$

with

$$\begin{aligned} K_2 &= X_2^2 - 4X_1X_3 \\ K_1 &= 4X_1^3 - 4X_3^3 \\ K_0 &= 4X_1^2X_3^2 - 8X_1X_2^2X_3 + 4X_2^4. \end{aligned}$$

So

$$\kappa = X_2^2X_4^2 - 4X_1X_3X_4^2 + 4X_1^3X_4 - 4X_3^3X_4 + 4X_1^2X_3^2 - 8X_1X_2^2X_3 + 4X_2^4.$$

Let $v = aX_1 + bX_2 + cX_3 + X_4$ (so $d = 1$). Let c_{i_1,i_2,i_3,i_4} be the coefficient of $X_1^{i_1}X_2^{i_2}X_3^{i_3}X_4^{i_4}$ in $(\kappa v)^{p-1}$. The Hasse-Witt matrix H_X of $X = V \cap K$ can be found in Proposition 6.

By Lemma 6 (below), when $p \equiv 5 \bmod 6$, then the determinant of H_X has degree $4(p-1)$ when considered as a polynomial in b. In particular, $\det(H_X)$ is a nonzero polynomial in a, b, c. The condition that X is singular is a nonzero polynomial condition in a, b, c. Therefore, there exists a triple $(a, b, c) \in \overline{\mathbb{F}}_p^3$ such that X is smooth and $\det(H_X) \neq 0$. This implies that X is ordinary, with p-rank 3, and the unramified double cover $\pi : Y \to X$ has the property that $P_\pi \simeq \mathrm{Jac}(Z)$ has p-rank 0. Thus $R_3^{(3,0)}$ is nonempty. The dimension result follows from [15, Proposition 5.2]. □

We remark that a result similar to Proposition 7 may hold when $p \equiv 5, 7 \bmod 8$ with $Z : z^2 = x^5 - x$ or when $p \equiv 2, 3, 4 \bmod 5$ with $Z : z^2 = x^5 - 1$.

The next lemma provides the cornerstone of the proof of Proposition 7.

Lemma 6 *Let* $p \equiv 5 \bmod 6$ *and let* X *be as in the proof of Proposition 7. When considered as a polynomial in* b*, the determinant of* H_X *has degree* $4(p-1)$.

Proof When considered as a polynomial in b, the coefficient c_{i_1,i_2,i_3,i_4} of $X_1^{i_1}X_2^{i_2}X_3^{i_3}X_4^{i_4}$ in $(\kappa v)^{p-1}$ has degree at most $p-1$. Any occurrence of b comes from the term bX_2 in v, so c_{i_1,i_2,i_3,i_4} has degree at most i_2 in b.

Note that κ has degree 2 in X_4, so κ^{p-1} has degree $2p-2$ in X_4. Any monomial in κ^{p-1} not divisible by X_2 arises as the product

$$(-4X_1X_3X_4^2)^{m_1}(4X_1^3X_4)^{m_2}(-4X_3^3X_4)^{m_3}(4X_1^2X_3^2)^{m_4} \tag{15}$$

for some $m_1, m_2, m_3, m_4 \in \mathbb{Z}^{\geq 0}$ with $m_1 + m_2 + m_3 + m_4 = p - 1$.

Claim 1 When considered as a polynomial in b, any term of the form $c_{i_1,p-1,i_3,2p-1}$ has degree at most $p-2$.

Proof of Claim 1 Any occurrence of b^{p-1} in $(\kappa v)^{p-1}$ comes from $\kappa^{p-1}(bX_2)^{p-1}$. The coefficient of $X_1^{i_1}X_3^{i_3}X_4^{2p-1}$ in κ^{p-1} is zero because κ^{p-1} has degree $2p-2$ in X_4. □

By Claim 1, in the middle column of H_X, the top and bottom entries,

$$c_{p-2,2p-1,p-1,p-1} - b^p c_{p-2,p-1,p-1,2p-1} \text{ and } c_{p-1,2p-1,p-2,p-1} - b^p c_{p-1,p-1,p-2,2p-1},$$

have degrees at most $2p-2$ in b. We consider the six terms in the expansion of $\det(H_X)$. The four terms that do not contain the central coefficient of H_X have degrees at most $2p-2+p-1+p-2 = 4p-5$. It remains to consider the product of the diagonal coefficients, and the product of the antidiagonal coefficients. We show that the former has degree at most $4p-6$ and the latter has degree $4p-4$ as polynomials in b.

Claim 2 When considered as a polynomial in b, each of the two terms $c_{k(p-1),p-1,\ell(p-1),p-1}$ for $(k,\ell) \in \{(1,2),(2,1)\}$ has degree at most $p-2$.

Proof of Claim 2 We must show that the coefficient of $X_1^{k(p-1)}X_3^{\ell(p-1)}X_4^{p-1}$ in κ^{p-1} is zero. Since X_2 does not divide this monomial, it appears as a product as in (15). In order to attain the desired powers of X_1 and X_3, we must have

$$m_1 + 3m_2 + 2m_4 = k(p-1)$$

and

$$m_1 + 3m_3 + 2m_4 = \ell(p-1).$$

Subtracting the two equations gives $3(m_2 - m_3) = \pm(p-1)$. But $3 \nmid (p-1)$ since $p \equiv 5 \bmod 6$. So the coefficient of $X_1^{k(p-1)} X_3^{\ell(p-1)} X_4^{p-1}$ in κ^{p-1} is zero, as required. □

Combining Claims 1 and 2 shows that the product of the diagonal entries of H_X has degree at most $p-2+2p-2+p-2 = 4p-6$ in b. Finally, we show that the product of the antidiagonal entries has degree $4p-4$ when considered as a polynomial in b.

Claim 3 When considered as polynomials in b, the following terms have degree $p-1$:

$$(1)\ c_{p-2,p-1,2p-1,p-1} \text{ and } (2)\ c_{2p-1,p-1,p-2,p-1}.$$

Proof of Claim 3 For (1), any occurrence of b^{p-1} in $(\kappa v)^{p-1}$ comes from $\kappa^{p-1}(bX_2)^{p-1}$. Hence we must show that the coefficient of $X_1^{p-2} X_3^{2p-1} X_4^{p-1}$ in κ^{p-1} is nonzero. Since X_2 does not divide this monomial, it arises as a product as in (15). In order to attain the desired power of X_4, we must have

$$2m_1 + m_2 + m_3 = p-1,$$

whereby $m_1 = m_4$. In order to attain the desired powers of X_1 and X_3, we must have

$$m_1 + 3m_2 + 2m_4 = 3(m_1 + m_2) = p-2$$

and

$$m_1 + 3m_3 + 2m_4 = 3(m_1 + m_3) = 2p-1.$$

So m_1 determines m_2, m_3, m_4. Write $p = 6k+5$ for some integer k, so

$$m_2 = 2k + 1 - m_1$$

and

$$m_3 = 4k + 3 - m_1.$$

The coefficient of $b^{p-1} X_1^{p-2} X_2^{p-1} X_3^{2p-1} X_4^{p-1}$ in $c_{p-2,p-1,2p-1,p-1}$ is

$$\sum_{m_1+m_2+m_3+m_4=p-1} \frac{(p-1)!}{m_1!m_2!m_3!m_4!} (-4)^{m_1} 4^{2k+1-m_1} (-4)^{4k+3-m_1} 4^{m_1}$$

$$= (-1)^{4k+3} 4^{p-1} \sum_{m_1+m_2+m_3+m_4=p-1} \frac{(p-1)!}{m_1!m_2!m_3!m_4!}.$$

Note that

$$\sum_{m_1+m_2+m_3+m_4=p-1} \frac{(p-1)!}{m_1!m_2!m_3!m_4!} = 4^{p-1}.$$

Therefore, the coefficient of $b^{p-1}X_1^{p-2}X_2^{p-1}X_3^{2p-1}X_4^{p-1}$ is the nonzero number

$$(-1)^{4k+3}4^{p-1}4^{p-1} = -4^{2p-2}.$$

For (2), consider the coefficient of b^{p-1} in $c_{2p-1,p-1,p-2,p-1}$. By the same strategy as above:

$$\begin{aligned} m_1 &= m_4; \\ 3m_1 + 3m_2 &= 2p - 1 = 12k + 9; \\ 3m_1 + 3m_3 &= p - 2 = 6k + 3. \end{aligned}$$

So, the coefficient of $b^{p-1}X_1^{2p-1}X_2^{p-1}X_3^{p-2}X_4^{p-1}$ in $c_{2p-1,p-1,p-2,p-1}$ is the nonzero number

$$\begin{aligned} &\sum_{m_1+m_2+m_3+m_4=p-1} \frac{(p-1)!}{m_1!m_2!m_3!m_4!}(-4)^{m_1}4^{4k+3-m_1}(-4)^{2k+1-m_1}4^{m_1} \\ &= (-1)^{2k+1}4^{p-1} \sum_{m_1+m_2+m_3+m_4=p-1} \frac{(p-1)!}{m_1!m_2!m_3!m_4!} = -4^{2p-2}, \end{aligned}$$

which completes the proof. □

Claim 4 When considered as a polynomial in b, the term $c_{p-1,p-2,p-1,2p-1}$ has degree $p-2$.

Proof of Claim 4 Consider the coefficient of b^{p-2} in $c_{p-1,p-2,p-1,2p-1}$. Since κ^{p-1} has degree $2p-2$ in X_4, any occurrence of $b^{p-2}X_1^{p-1}X_2^{p-2}X_3^{p-1}X_4^{2p-1}$ in $(\kappa v)^{p-1}$ comes from choosing the monomial bX_2 in $p-2$ factors v of v^{p-1} and X_4 in the remaining one. There are $p-1$ ways of doing so.

Now we compute the coefficient of $X_1^{p-1}X_3^{p-1}X_4^{2p-2}$ in κ^{p-1}. A monomial divisible by X_2 cannot be chosen in a factor κ of κ^{p-1}. Therefore, to obtain the exponent $2p-2$ of X_4, we need to pick the monomial $-4X_1X_3X_4^2$ in each factor κ of κ^{p-1}. Hence, the coefficient of $b^{p-2}X_1^{p-1}X_2^{p-2}X_3^{p-1}X_4^{2p-1}$ in $c_{p-1,p-2,p-1,2p-1}$ is $(p-1)(-4)^{p-1}$, which is not zero. □

Now we complete the proof of Lemma 6. By the claims, the largest power of b in $\det(H_X)$ arises in the product of the antidiagonal entries. By Claims 3 and 4, when considered as polynomials in b, the antidiagonal entries have degrees $p-1$, $p+p-2$, and $p-1$, respectively. Therefore, their product has degree

$$p - 1 + p + p - 2 + p - 1 = 4p - 4,$$

when considered as a polynomial in b. □

5.5 The Condition that X Is Non-ordinary

Let Z be an arbitrary smooth curve of genus 2 and $X = K \cap V$ with Hasse-Witt matrix H_X as in Sect. 5.3.

5.5.1 The Condition that X Is Non-ordinary

The curve X is non-ordinary if and only if $\det(H_X)$ is zero.

Proposition 8 *Setting $d = 1$, then $\det(H_X)$ is nonzero and homogeneous of degree $6(p-1)$ in $a, b, c, d_0, \ldots, d_6$.*

Proof By Lemma 5, each coefficient of the 3×3 matrix H_X is homogeneous of degree $2(p-1)$ in these variables. Thus it suffices to prove that $\det(H_X)$ is nonzero. We expect that this can be proven algebraically, but to avoid long computations, we continue with the following theoretical argument. If $\det(H_X)$ is identically 0, then a generic point of $\mathscr{R}_3 = \Pi^{-1}(\mathscr{M}_3)$ would represent an unramified double cover $\pi : Y \to X$ such that X is non-ordinary; this is false. □

We apply a fractional linear transformation to the variable x in order to reduce the number of variables defining Z, while preserving the degree of $\det(H_X)$ and its homogeneous property. Without loss of generality, we can suppose that no Weierstrass point of Z lies over $x = \infty$ and that 3 of the Weierstrass points are rational and lie over $x = 0, 1, -1$ (over a non-algebraically closed field, this may only be true after a finite extension). Then,

$$Z : z^2 = D(x) := (x^3 - x)(A_0x^3 + Ax^2 + Bx + C) \tag{16}$$

$$= A_0x^6 + Ax^5 + (B - A_0)x^4 + (C - A)x^3 - Bx^2 - Cx. \tag{17}$$

Note that $A_0 \neq 0$ by the hypothesis at ∞ and $C \neq 0$ since Z is smooth.

5.5.2 The Condition that X Is Non-ordinary and Z Has p-Rank 1

As in (16), write $Z : z^2 = D(x) := (x^3 - x)(A_0x^3 + Ax^2 + Bx + C)$. Then Z is not ordinary if and only if $\det(H_Z) = 0$.

Lemma 7 *The nonzero homogeneous polynomial $\det(H_X)$ does not vanish identically under the polynomial condition $\det(H_Z) = 0$, which is homogeneous of degree $p-1$ in A_0, A, B, C.*

Proof Since $D(x)$ is linear in A_0, A, B, C, the entries of the Hasse-Witt matrix H_Z are homogeneous of degree $(p-1)/2$ so $\det(H_Z)$ is homogeneous of degree $p-1$.

For the nonvanishing claim, it suffices to show that X is typically ordinary when Z has p-rank 1. This follows from the fact that each component of $\mathscr{R}_3^{(3,1)}$ has dimension 5, while each component of $\mathscr{R}_3^{(2,1)}$ has dimension 4, which is a special case of [15, Theorem 7.1]. □

For example, when $p = 3$ and $(A_0, A, B, C, a, b, c) = (1, 0, 1, 2t, 0, 1, 1)$ with $t \in \mathbb{F}_9$ a root of $t^2 - t - 1$, then $f = 1$ and $f' = 1$ and the curve X is smooth.

5.5.3 The Condition that X Is Non-ordinary and Z Has p-Rank 0

Lemma 8 *The curve Z is not ordinary under a polynomial condition on A_0, A, B, C which is homogeneous of degree $p - 1$. The curve Z has p-rank 0 under four polynomial conditions on A_0, A, B, C which are each homogeneous of degree $(p + 1)(p - 1)/2$.*

Proof The curve Z has p-rank 0 if and only if $H_Z H_Z^{(p)} = [0]$. The entries of $H_Z^{(p)}$ are homogeneous of degree $p(p-1)/2$, so the entries of $H_Z H_Z^{(p)}$ are homogeneous of degree $(p + 1)(p - 1)/2$. □

Proposition 9 *Let Z be a genus 2 curve with equation*

$$z^2 = (x^3 - x)(A_0 x^3 + Ax^2 + Bx + C).$$

Let V be a plane with equation $aX_1 + bX_2 + cX_3 + X_4$. Let $X = V \cap K$, and consider the unramified double cover $\pi : Y \to X$ given by the restriction of $\phi : \mathrm{Jac}(Z) \to K$. Then the condition that X is non-ordinary and Z has p-rank 0 is given by the vanishing of four homogeneous polynomials of degree $(p+1)(p-1)/2$ in A_0, A, B, C and the vanishing of one homogeneous polynomial $\det(H_X)$ of degree $6(p - 1)$ in a, b, c, A_0, A, B, C.

Proof The curve X is non-ordinary if and only if $\det(H_X)$ vanishes. By Proposition 8, $\det(H_X)$ is homogeneous of degree $6(p - 1)$ in a, b, c and the coefficients of $D(x)$, which are linear in A_0, A, B, C. The curve Z has p-rank 0 under the conditions in Lemma 8. □

We expect that the answer to Question 1 will be yes for all odd p when $g = 3$, $f = 2, 3$, and $f' = 0$. We now rephrase the question in those cases to a question in commutative algebra.

Question 2 For all odd primes p, is there a plane V for which $\det(H_X)$ does not vanish identically under the four conditions when $H_Z H_Z^{(p)} = [0]$? Is there a plane V for which $\det(H_X)$ does vanish for some Z such that $H_Z H_Z^{(p)} = [0]$?

The difficulty in showing that $\det(H_X)$ does not vanish identically when Z has p-rank 0 is that we do not know much about the variety of the ideal generated by the four polynomial conditions when $H_Z H_Z^{(p)} = [0]$ or the behavior of the derivatives of $\det(H_X)$ with respect to the variables a, b, c.

5.5.4 Example: When $p = 3$

Write $Z : z^2 = D(x) = (x^3 - x)(A_0x^3 + Ax^2 + Bx + C)$. Then

$$H_Z = \begin{pmatrix} -B & A \\ -C & B - A_0 \end{pmatrix}.$$

The four entries of $H_Z H_Z^{(3)}$ are:

$$B^4 - AC^3,\ C(B^3 - C^2(B - A_0)),\ A((B - A_0)^3 - BA^2), -CA^3 + (B - A_0)^4.$$

Recall that $C \neq 0$ since Z is smooth. If Z has 3-rank 0, then if any of $A, B, B - A_0$ are zero, then all of them are zero, which implies $A_0 = 0$, which contradicts the hypothesis at ∞. Thus $AB(B - A_0) \neq 0$ when Z has 3-rank 0. One can check that $H_Z H_Z^{(3)} = [0]$ if and only if

$$B^4 - AC^3 = 0, B^3 - C^2(B - A_0) = 0.$$

Write

$$(\upsilon\kappa)^2 = \sum_{i_1+i_2+i_3+i_4=10} c_{i_1,i_2,i_3,i_4} X_1^{i_1} X_2^{i_2} X_3^{i_3} X_4^{i_4},$$

and assume that $d = 1$. Then the Hasse-Witt matrix H_X is given by

$$\begin{pmatrix} c_{4,2,2,2} - a^3c_{1,2,2,5} & c_{1,5,2,2} - b^3c_{1,2,2,5} & c_{1,2,5,2} - c^3c_{1,2,2,5} \\ c_{5,1,2,2} - a^3c_{2,1,2,5} & c_{2,4,2,2} - b^3c_{2,1,2,5} & c_{2,1,5,2} - c^3c_{2,1,2,5} \\ c_{5,2,1,2} - a^3c_{2,2,1,5} & c_{2,5,1,2} - b^3c_{2,2,1,5} & c_{2,2,4,2} - c^3c_{2,2,1,5} \end{pmatrix} \quad (18)$$

By Proposition 4, the 3-rank of X is the stable rank of H_X, which is the rank of $H_X H_X^{(3)} H_X^{(3^2)}$. The entries of H_X are homogeneous of degree 4, and $\det(H_X)$ is homogeneous of degree 12 in a, b, c, A_0, A, B, C.

5.5.5 Example: Genus 3 Curves Having Pryms of 3-Rank 1 When $p = 3$

In Example 1, we showed that $\mathscr{R}_3^{(1,1)}$ and $\mathscr{R}_3^{(0,1)}$ are nonempty when $p = 3$, by finding curves X of 3-rank 1 or 0 having an unramified double cover $\pi : Y \to X$ such that P_π has 3-rank 1. Here we give a second proof of this using the methods of this section.

For $t, u \in k$ with $t \neq u$, consider the genus 2 curve

$$Z_{t,u} : z^2 = D(x) := x^6+x^3+1+x(x^3+1)(tx+u)=x^6+tx^5+ux^4+x^3+tx^2+ux+1.$$

One can check that $Z_{t,u}$ is smooth if $t \neq u$ and $Z_{t,u}$ has 3-rank 1 for $t \neq \pm u$.

Example 8 Let $p = 3$. Consider the plane quartic $X = V \cap K$ where K is the Kummer surface of $\mathrm{Jac}(Z_{t,u})$ and $V \subset \mathbb{P}^3$ is a plane. Then X has an unramified double cover $\pi : Y \to X$ such that $P_\pi \simeq Z_{t,u}$ has p-rank 1.

1. If $V : -X_2 + X_3 + X_4 = 0$ and $t = 1$, $u = 0$, then X is smooth with 3-rank $f = 1$.
2. If $V : -X_1 - X_2 + X_4 = 0$ and $t = 0$, $u = 1$, then X is smooth with 3-rank $f = 0$.

5.6 *The Moduli Space of Genus 3 Curves Having Pryms of 3-Rank 0 When $p = 3$*

In this section, we fix $p = 3$. In Sect. 5.6.1, we parametrize $\mathscr{M}_2^0$ (the 3-rank 0 stratum of $\mathscr{M}_2$) by a 1-parameter family of curves of genus 2 and 3-rank 0. Let α be the name of this parameter. In Sect. 5.6.2, we then analyze $\det(H_X)$ as V and α vary. This allows us to prove some information about the locus of the parameter space where X is non-ordinary. This implicitly provides geometric information about $\mathscr{R}_3^{(2,0)}$.

5.6.1 A Family of Genus 2 Curves with 3-Rank 0 When $p = 3$

For $\alpha \in k - \{0, 1, -1\}$, define

$$Z_\alpha : z^2 = A(x)B(x),$$

where

$$A(x)=x^3-\alpha x^2+\alpha x+(\alpha+1), \text{ and } B(x) = (x-\alpha)(x-(\alpha+1))(\alpha x+(\alpha+1)).$$

The importance of the next lemma is that it shows that $\mathrm{Jac}(Z_\alpha)$ parametrizes the 3-rank 0 stratum of $\mathscr{A}_2$, which is irreducible of dimension 1 when $p = 3$.

Lemma 9 *When $p = 3$, then the family $\{Z_\alpha\}_\alpha$ is a non-isotrivial family of smooth curves of genus 2 and 3-rank 0.*

Proof For $\alpha \in k - \{0, 1, -1\}$, the polynomial $A(x)B(x)$ has no repeated roots; hence, the curve Z_α is smooth and of genus 2. If Z_α is isomorphic to Z_β for $\alpha, \beta \in k - \{0, 1, -1\}$, then they have the same absolute Igusa invariants j_1, j_2, j_3 [9]. Using SageMath [18], we find that the absolute Igusa invariants of Z_α are:

$$j_1 = -\frac{(\alpha-1)^6}{\alpha(\alpha+1)^2},$$
$$j_2 = -\frac{(\alpha-1)^6}{\alpha(\alpha+1)^2},$$
$$j_3 = \frac{(\alpha-1)^2(\alpha^2-\alpha-1)^2}{\alpha^2}.$$

In particular, the absolute Igusa invariants are nonconstant functions of α; hence, the family $\{Z_\alpha\}_\alpha$ is non-isotrivial.

By Proposition 1, the Cartier-Manin matrix M of Z_α is

$$\begin{pmatrix} (\alpha+1)^3 & -(\alpha+1)^4 \\ 1 & -(\alpha+1) \end{pmatrix}$$

so the matrix $M^{(3)}M$ is

$$\begin{pmatrix} (\alpha+1)^9 & -(\alpha+1)^{12} \\ 1 & -(\alpha+1)^3 \end{pmatrix}\begin{pmatrix} (\alpha+1)^3 & -(\alpha+1)^4 \\ 1 & -(\alpha+1) \end{pmatrix} = \begin{pmatrix} 0 & 0 \\ 0 & 0 \end{pmatrix},$$

so Z_α has 3-rank $f' = 0$ by Proposition 4. □

5.6.2 The Locus of the Parameter Space Where X Is Non-ordinary When $p = 3$

Now, we compute the equation of the Kummer surface K_α of Z_α and choose a plane $V = V_{a,b,c,d}$ to obtain the smooth plane quartic $X_V^\alpha = K_\alpha \cap V$. Our goal is to find information about the 3-rank $f = f_V^\alpha$ of X_V^α as V and α vary. To do this, we determine the Hasse-Witt matrix $H := H_{X_V^\alpha}$ of X_V^α as in (18).

Proposition 10 *For a generic choice of plane V, the curve X_V^α is ordinary for all but finitely many α and has 3-rank 2 for at least one and at most finitely many α.*

Proof When $d = 1$, the value of the 3-rank of X_V^α is determined by polynomial conditions in a, b, c and α. In particular, X_V^α has 3-rank 3 if and only if the determinant of H_V^α is nonzero. So the first statement can be proven by checking, for a fixed plane V and fixed parameter α, whether $\det(H_V^\alpha) \neq 0$. The second statement can be proven by checking, for a fixed plane V, whether $\det(H_V^\alpha) = 0$ under a polynomial condition on α and whether there exists one value of α satisfying that polynomial condition for which X_V^α is smooth and has 3-rank 2.

Thus both statements follow from the next claim.

Claim For the plane $V : -X_2+X_4 = 0$, the curve X_V^α is ordinary for all but finitely many $\alpha \in k - \{0, 1, -1\}$, and X_V^α has 3-rank 2 for a nonzero finite number of $\alpha \in k$.

Proof of Claim When $V : -X_2 + X_4 = 0$, the Hasse-Witt matrix H of X_V^α is

$$\begin{pmatrix} a_{11} & a_{12} & a_{13} \\ a_{21} & a_{22} & a_{23} \\ a_{31} & a_{32} & a_{33} \end{pmatrix}$$

where

$$\begin{aligned}
a_{11} &= \alpha^{13} - \alpha^{11} - \alpha^{10} + \alpha^9 + \alpha^7 + \alpha^6 - \alpha^3 - \alpha^2 - 1,\\
a_{12} &= -\alpha^7 - \alpha^6 + \alpha^5 + \alpha^4 + \alpha^2 + \alpha,\\
a_{13} &= \alpha^{10} + \alpha^9 + \alpha^7 - \alpha^6 + \alpha^5 - \alpha^4,\\
a_{21} &= -\alpha^{16} - \alpha^{13} + \alpha^{11} + \alpha^9 + \alpha^8 + \alpha^7 - \alpha^5 + \alpha^4 - \alpha^3 - \alpha^2,\\
a_{22} &= \alpha^{13} + \alpha^9 + \alpha^8 + \alpha^7 - \alpha^6 - \alpha^4 - \alpha^3 - \alpha^2 - \alpha - 1,\\
a_{23} &= -\alpha^{13} + \alpha^{10} + \alpha^9 - \alpha^6 + \alpha^5 + \alpha^4 + \alpha^3 - \alpha^2 - \alpha - 1,\\
a_{31} &= \alpha^{13} + \alpha^{12} - \alpha^9 + \alpha^8 + \alpha^7 + \alpha^6 - \alpha^5 + \alpha^2 + \alpha,\\
a_{32} &= \alpha^{10} - \alpha^8 + \alpha^7 + \alpha^5 - \alpha^4 - \alpha^3 - \alpha^2 - \alpha - 1,\\
a_{33} &= \alpha^{12} - \alpha^{10} + \alpha^6 + \alpha^5 - \alpha^4 - \alpha - 1.
\end{aligned}$$

The determinant Det_H of H is

$$\begin{aligned}
\mathrm{Det}_H = \alpha^3(\alpha+1)^4(\alpha-1)^5(\alpha^3+\alpha^2+\alpha-1)(\alpha^5+\alpha^4+\alpha^3+\alpha^2-\alpha+1)\\
(\alpha^5-\alpha^4+\alpha^3+\alpha^2+\alpha+1)(\alpha^6-\alpha^4+\alpha^3-\alpha+1)(\alpha^7+\alpha^5+\alpha-1).
\end{aligned}$$

For each $\alpha \in k - \{0, 1, -1\}$ which is not a root of Det_H, the Hasse-Witt matrix is invertible and so X_V^α is ordinary. For $\alpha \in k$ satisfying $\alpha^3 + \alpha^2 + \alpha - 1 = 0$, a computation in SageMath [18] shows that X_V^α is smooth and the matrix $HH^{(3)}H^{(3^2)}$ has rank 2. Therefore, for these values of α, X_V^α has 3-rank 2 by Proposition 4. □

□

Proposition 10 does not eliminate the possibility that there exists a plane V such that $\det(H_V^\alpha) = 0$ for all α.

Proposition 11 *For a generic choice of $\alpha \in k$, the curve X_V^α is ordinary for a generic choice of V and has* 3*-rank* 2 *under a codimension* 1 *condition on V.*

Proof The first statement already follows from Proposition 10. The second statement can be proven by checking, for fixed $\alpha \in k$, whether $\det(H_V^\alpha) = 0$ under a polynomial condition on a, b, c and whether there exists one possibility for a, b, c satisfying that polynomial condition for which X_V^α is smooth and has 3-rank 2.

Let $\alpha \in \mathbb{F}_9$ be fixed to be a root of the polynomial $t^2 + 2t + 2$. If $d = 1$, the Hasse-Witt matrix $H = (a_{ij})_{i,j}$ of X_V^α, for arbitrary a, b, c, is given by

$$\begin{aligned}
a_{11} &= a^3c + b^2 + ac + (\alpha+1)(a^3 - bc + a) + (\alpha-1)(ab - b) - \alpha c^2\\
a_{12} &= b^3c + (\alpha+1)b^3\\
a_{13} &= c^4 - ac + (\alpha+1)c^2(c-1) + (-\alpha+1)b^2 + -\alpha(ab + bc)
\end{aligned}$$

$$a_{21} = a^3b - ab + (-\alpha - 1)(a^3 + ac + c) + (-\alpha + 1)(a^2 + c^2 + a - bc)$$
$$a_{22} = b^4 + (-\alpha - 1)b^3 - \alpha c^2 + \alpha b$$
$$a_{23} = bc^3 + (-\alpha - 1)c^3 + \alpha(a^2 - ac + bc + c^2) + (\alpha - 1)ab$$
$$a_{31} = a^4 - a^2 + (-\alpha + 1) + (\alpha + 1)(a^3 - ab - b) + (-\alpha + 1)(b^2 - bc - c) + \alpha ac$$
$$a_{32} = ab^3 + (\alpha + 1)b^3 + \alpha bc$$
$$a_{33} = ac^3 + a^2 + (\alpha + 1)(c^3 + ac - c) - \alpha(b^2 + bc + ab).$$

Then one can check that $\det(H)$ is nonvanishing in a, b, c. Also, when (a, b, c) equals $(2, 0, 2)$, then one can check that the curve X_V^α is smooth with 3-rank 2. □

Proposition 11 does not eliminate the possibility that there exists $\alpha \in k$ such that $\det(H_V^\alpha) = 0$ for all planes V.

6 Points on the Kummer Surface

Suppose that Z is a supersingular curve of genus 2 defined over a finite field $\mathbb{F}_q$ of characteristic p. This section contains a result about the number of $\mathbb{F}_q$-points on the Kummer surface K of $\text{Jac}(Z)$. The material in this section is probably well known to experts, but we could not find it in the literature. The connection between this section and the rest of the paper is found in Question 3 below: we remark that if $X = V \cap K$ for some plane V and if p divides $\#X(\mathbb{F}_q)$, then the p-rank of X is at least 1.

Let Z be a genus 2 curve over $\mathbb{F}_q$. Suppose that Z has equation $z^2 = D(x)$, and define a quadratic twist W of Z by $\lambda z^2 = D(x)$ for $\lambda \in \mathbb{F}_q^\times \setminus (\mathbb{F}_q^\times)^2$. The isomorphism class of W does not depend on the choice of non-square element λ.

Lemma 10 *Let Z be a genus 2 curve over $\mathbb{F}_q$, and let W be its quadratic twist. Let $K = \text{Jac}(Z)/\langle -1 \rangle$. Then*

$$|K(\mathbb{F}_q)| = (|\text{Jac}(Z)(\mathbb{F}_q)| + |\text{Jac}(W)(\mathbb{F}_q)|)/2.$$

Proof The degree 2 cover $\phi : \text{Jac}(Z) \to K$ is defined over $\mathbb{F}_q$. Now let $\psi : \text{Jac}(Z) \to \text{Jac}(W)$ be the isomorphism of abelian varieties over $\mathbb{F}_{q^2}$ induced by the isomorphism of the underlying curves given by $(x, z) \mapsto (x, \sqrt{\lambda} z)$. Let τ be a generator for $\text{Gal}(\mathbb{F}_{q^2}/\mathbb{F}_q)$. Then $\tau\psi\tau^{-1} = -\psi$.

Let $P \in K(\mathbb{F}_q)$. Write $\phi^{-1}(P) = \{Q, -Q\}$. Since $P \in K(\mathbb{F}_q)$ and ϕ is defined over $\mathbb{F}_q$, then $\{\sigma(Q), -\sigma(Q)\} = \{Q, -Q\}$ for all $\sigma \in \text{Gal}(\overline{\mathbb{F}}_q/\mathbb{F}_q)$. Therefore, Q is defined over $\mathbb{F}_{q^2}$ and either $\tau(Q) = Q$, whereby $Q \in \text{Jac}(Z)(\mathbb{F}_q)$, or else $\tau(Q) = -Q$, whereby $\psi(Q) \in \text{Jac}(W)(\mathbb{F}_q)$. The points $P \in K(\mathbb{F}_q)$ for which $\phi^{-1}(P) = \{Q, -Q\}$ with $Q = -Q$ are precisely those for which $Q \in \text{Jac}(Z)(\mathbb{F}_q)$ and $\psi(Q) \in \text{Jac}(W)(\mathbb{F}_q)$. Therefore, we see that every point in $K(\mathbb{F}_q)$ is counted twice in $|\text{Jac}(Z)(\mathbb{F}_q)| + |\text{Jac}(W)(\mathbb{F}_q)|$. □

The zeta function of a genus 2 curve $Z/\mathbb{F}_q$ is

$$\mathscr{Z}(T) = \exp\left(\sum_{k\geq 1} \frac{|Z(\mathbb{F}_{q^k})|}{k} T^k\right) = \frac{L_{Z/\mathbb{F}_q}(T)}{(1-T)(1-qT)},$$

where $L_{Z/\mathbb{F}_q}(T) = 1 + a_1 T + a_2 T^2 + q a_1 T^3 + q^2 T^4 = \prod_{i=1}^{4} (1 - \alpha_i T)$ with $\alpha_1\alpha_3 = \alpha_2\alpha_4 = q$.

Lemma 11 *Let Z be a genus 2 curve over $\mathbb{F}_q$, $A = \mathrm{Jac}(Z)$, and $K = A/\langle -1\rangle$. Then*

$$|\mathrm{Jac}(Z)(\mathbb{F}_q)| = 1 + a_1 + a_2 + a_1 q + q^2$$

and

$$|K(\mathbb{F}_q)| = 1 + a_2 + q^2$$

where the a_i are the coefficients of $L_{Z/\mathbb{F}_q}(T)$ as defined above.

Proof The second statement follows from the first, using Lemma 10 and the fact that if $W/\mathbb{F}_q$ is the quadratic twist of Z, then

$$L_{W/\mathbb{F}_q}(T) = L_{Z/\mathbb{F}_q}(-T) = 1 - a_1 T + a_2 T^2 - q a_1 T^3 + q^2 T^4.$$

For the first statement, note that

$$|\mathrm{Jac}(Z)(\mathbb{F}_q)| = \frac{|Z(\mathbb{F}_q)|^2 + |Z(\mathbb{F}_{q^2})|}{2} - q. \tag{19}$$

Equating the coefficients of T and T^2 in

$$\frac{L_{Z/\mathbb{F}_q}(T)}{(1-T)(1-qT)} = \exp\left(\sum_{k\geq 1} \frac{|Z(\mathbb{F}_{q^k})|}{k} T^k\right)$$

gives $a_1 = |Z(\mathbb{F}_q)| - (q+1)$ and $a_2 = \frac{1}{2}|Z(\mathbb{F}_q)|^2 + \frac{1}{2}|Z(\mathbb{F}_{q^2})| - (q+1)|Z(\mathbb{F}_q)| + q$. The result now follows from (19). □

Corollary 3 *Let Z be a supersingular genus 2 curve over $\mathbb{F}_q$, let $A = \mathrm{Jac}(Z)$, and let $K = A/\langle -1\rangle$. Then*

$$|K(\mathbb{F}_q)| \equiv 1 \bmod q.$$

Proof If Z is supersingular, then $q \mid a_2$. The result now follows. □

Question 3 Suppose Z is supersingular and K is the Kummer surface of Jac(Z). Does there exist a plane $V \subset \mathbb{P}^3$ defined over $\mathbb{F}_q$ such that p divides $\#X(\mathbb{F}_q)$ where $X = V \cap K$? If so, then the p-rank of X is at least 1.

7 Results for Arbitrary g

In this section, when $3 \leq p \leq 19$, we use the results from Sect. 3 about genus 3 curves in characteristic p to verify the existence of smooth curves X of arbitrary genus $g \geq 3$ having an unramified double cover whose Prym variety has small p-rank. Specifically, we work inductively to study the dimension of certain moduli strata $\mathscr{R}_g^{(f,f')}$ for $g \geq 3$ in characteristic p with $3 \leq p \leq 19$. The reader is strongly advised to read [15] before reading this section.

A highlight of this approach is that X is smooth and we can control its p-rank f. Indeed, the original proof of [21], found in [2, Section 2], shows that $\bar{\mathscr{R}}_g^{(f'+1,f')}$ is nonempty for each $g \geq 2$ and $0 \leq f' \leq g-1$; in other words, there is a *possibly singular* curve of genus g and p-rank $f'+1$ with an unramified double cover π such that P_π has p-rank f'. We omit the details of this argument.

In this section, the word *component* means irreducible component. Although the phrasing is slightly redundant, we emphasize that a component of a given dimension is *nonempty* because this property of the component is the most difficult to prove and is sufficient to yield the existence applications.

7.1 Increasing the p-Rank of the Prym Variety

The next result allows us to use geometric information about $\mathscr{R}_g^{(f,f')}$ to deduce geometric information about $\mathscr{R}_g^{(f,F')}$ when $f' \leq F' \leq g-1$.

Proposition 12 ([15, Proposition 5.2]) *Let $g \geq 3$. Suppose that $\mathscr{R}_g^{(f,f')}$ is nonempty and has a component of dimension $g-2+f+f'$ in characteristic p. Then $\mathscr{R}_g^{(f,F')}$ is nonempty and has a component of dimension $g-2+f+F'$ in characteristic p for each F' such that $f' \leq F' \leq g-1$.*

7.2 Background on Boundary of $\mathscr{R}_g$

The strategy used in [15] is to use unramified double covers of *singular* curves of given genus and p-rank to produce unramified double covers of *smooth* curves of the same genus and p-rank, with control over the p-rank of the Prym variety. This strategy must be implemented very precisely because, in general, the p-rank of both the curve and the Prym will increase when deforming X away from the boundary. In fact, there are situations where this is guaranteed to happen.

This section contains background about p-ranks of unramified double covers of singular curves. Let $\bar{\mathscr{R}}_g$ be the compactification of $\mathscr{R}_g$ as defined and analyzed in [6, Section 1.4]. The points of $\bar{\mathscr{R}}_g \backslash \mathscr{R}_g$ represent unramified double covers of singular stable curves of genus g.

Let $\bar{\mathscr{R}}_{g;1} = \bar{\mathscr{R}}_g \times_{\bar{\mathscr{M}}_g} \bar{\mathscr{M}}_{g;1}$ be the moduli space whose points represent unramified double covers $\pi : Y \to X$ together with marked points $y \in Y$ and $x \in X$ such that $\pi(y) = x$, as in [15, Section 2.3]. Adding a marking increases the dimension of the moduli space by 1. By Ozman and Pries [15, Lemma 2.1], there is a surjective morphism $\psi_R : \bar{\mathscr{R}}_{g;1} \to \bar{\mathscr{R}}_g$ whose fibers are irreducible.

Suppose that $g = g_1 + g_2$, with $g_i \geq 1$. We recall some material about the boundary divisor $\Delta_{g_1:g_2}[\bar{\mathscr{R}}_g]$ from [6, Section 1.4]. This boundary divisor is the image of the clutching map

$$\kappa_{g_1:g_2} : \bar{\mathscr{R}}_{g_1;1} \times \bar{\mathscr{R}}_{g_2;1} \to \bar{\mathscr{R}}_g,$$

defined on a generic point as follows: Let τ_1 be a point of $\bar{\mathscr{R}}_{g_1;1}$ representing $(\pi_1 : C_1' \to C_1, x_1' \mapsto x_1)$ and let τ_2 be a point of $\bar{\mathscr{R}}_{g_2;1}$ representing $(\pi_2 : C_2' \to C_2, x_2' \mapsto x_2)$. Let X be the curve with components C_1 and C_2, formed by identifying x_1 and x_2 in an ordinary double point. Let Y be the curve with components C_1' and C_2', formed by identifying x_1' and x_2' (resp. $x_1'' = \sigma(x_1')$ and $x_2'' = \sigma(x_2')$) in an ordinary double point. Then $\kappa_{g_1:g_2}(\tau_1, \tau_2)$ is the point representing the unramified double cover $Y \to X$. This is illustrated in Fig. 1.

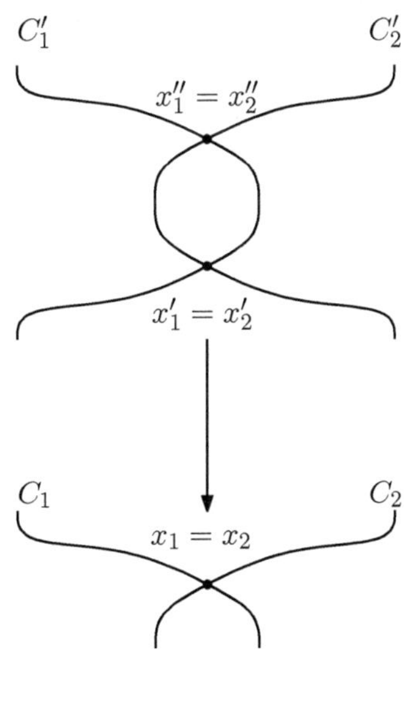

Fig. 1 $\Delta_{g_1:g_2}$

In [15, Section 3.4.1], the authors analyze the p-rank stratification of this boundary divisor. By Ozman and Pries [15, Lemmas 3.6-3.7], the clutching morphism restricts to the following:

$$\kappa_{g_1:g_2} : \mathscr{R}_{g_1;1}^{(f_1,f_1')} \times \mathscr{R}_{g_2;1}^{(f_2,f_2')} \to \Delta_{g_1:g_2}[\bar{\mathscr{R}}_g^{(f_1+f_2,f_1'+f_2'+1)}]. \tag{20}$$

The following lemma is useful in the inductive arguments in Sect. 7.5.

Lemma 12 *Suppose that $S_i \subset \mathscr{R}_{g_i}^{(f_i,f_i')}$ has dimension d_i for $i = 1, 2$. Then it follows that $d_1 + d_2 + 2$ is the dimension of*

$$\mathscr{K} = \kappa_{g_1:g_2}(\psi_R^{-1}(S_1) \times \psi_R^{-1}(S_2)).$$

Furthermore, if S_i is a component of $\mathscr{R}_{g_i}^{(f_i,f_i')}$ for $i = 1, 2$, then $\mathscr{K}$ is contained in a component of $\bar{\mathscr{R}}_g^{(f_1+f_2,f_1'+f_2'+1)}$ whose dimension is at most $d_1 + d_2 + 3$.

Proof The first statement follows from the facts that the fibers of ψ_R have dimension 1 and the clutching maps are finite. The second statement follows from (20) and [20, page 614]; see [15, Lemma 3.1]. □

7.3 Some Extra Results When $p = 3$

The $p = 3$ case is guaranteed to be more difficult, because $\mathscr{R}_2^{(0,0)}$ and $\mathscr{R}_2^{(1,0)}$ are empty in that case [7, Section 7.1]. In other words, when $p = 3$, if $\pi : Y \to X$ is an unramified double cover of a genus 2 curve such that P_π is non-ordinary, then X is ordinary. This is the key reason why there are some extra hypotheses when $p = 3$ in [15, Propositions 6.1, 6.4, Theorem 7.2].

We now have the extra information that all pairs (f, f') occur when $p = 3$ and $g = 3$. In this section, we use this to confirm that the extra hypotheses when $p = 3$ can be removed from most of the results of [15, Sections 6 and 7]. This will allow us to work more uniformly for odd p in the next section.

Let $\tilde{\mathscr{A}}_{g-1}$ denote the toroidal compactification of $\mathscr{A}_{g-1}$. Let $\tilde{\mathscr{A}}_{g-1}^{f'}$ denote the p-rank f' stratum of $\tilde{\mathscr{A}}_{g-1}$. Let $V_g^{f'} = \bar{Pr}_g^{-1}(\tilde{\mathscr{A}}_{g-1}^{f'}) \cap \mathscr{R}_g$.

The next result about the p-rank stratification of $R_3^{(f,1)}$ is true for all $p \geq 5$ by Ozman and Pries [15, Proposition 6.4].

Lemma 13 *The result [15, Proposition 6.4] does not require the hypothesis that $f \neq 0, 1$ when $p = 3$. In other words, if $p = 3$ and $0 \leq f \leq 3$, then $\Pi^{-1}(\mathscr{M}_3^f)$ is irreducible and $\mathscr{R}_3^{(f,1)} = \Pi^{-1}(\mathscr{M}_3^f) \cap V_3^1$ is nonempty with dimension $2 + f$.*

Proof The only place that this hypothesis was used was to verify that $\mathscr{R}_3^{(f,1)}$ is nonempty, which we have now verified in Proposition 5. □

The next result about non-ordinary Prym varieties is true for all $p \geq 5$ by Ozman and Pries [15, Theorem 7.1]. We say that P_π is almost ordinary if its p-rank satisfies $f' = g - 2$.

Lemma 14 *The result [15, Theorem 7.1] does not require the hypothesis $f \geq 2$ when $p = 3$ and $g \geq 3$. In other words, if $p = 3$, $g \geq 3$, and $0 \leq f \leq g$, then $\mathscr{R}_g^{(f,g-2)}$ is nonempty, and each of its components has dimension $2g - 4 + f$. More generally, let S be a component of $\mathscr{M}_g^f$, then the locus of points of $\Pi^{-1}(S)$ representing unramified double covers for which the Prym variety P_π is almost ordinary is nonempty and codimension 1 in $\Pi^{-1}(S)$.*

Proof The only place this hypothesis was used was to apply [15, Proposition 6.4] in the "base cases" and "nonempty" paragraphs. So the result follows by Lemma 13. □

The hypothesis $p \geq 5$ also appears in [15, Corollary 7.3], because a key point of the proof is that $\mathscr{R}_2^{(1,0)}$ and $\mathscr{R}_2^{(0,0)}$ are nonempty, which is false when $p = 3$. We generalize [15, Corollary 7.3] to include the case $p = 3$ in Sect. 7.5.

7.4 A Dimension Result

The following result is also needed in Sect. 7.5.

Proposition 13 *Let $3 \leq p \leq 19$. Then $\mathscr{R}_3^{(3,0)}$ contains a component of dimension 4, and $\mathscr{R}_3^{(2,0)}$ contains a component of dimension 3.*

Proof Let $3 \leq p \leq 19$ and $f = 2, 3$. By Proposition 5, there exists an unramified double cover $\pi_f : Y_f \to X_f$ such that X_f is a smooth plane quartic with p-rank f and P_{π_f} has p-rank 0. This shows that $\mathscr{R}_3^{(f,0)}$ is nonempty. Let S_f denote a component of $\mathscr{R}_3^{(f,0)}$ containing the point representing π_f. By purity, we have $\dim(S_f) \geq 1 + f$. To finish the proof, it suffices to show that $\dim(S_f) = 1 + f$.

Consider the morphism $Pr_3 : \mathscr{R}_3 \to \mathscr{A}_2$. Let $\mathscr{X}$ be an irreducible component of the p-rank 0 stratum $\mathscr{A}_2^0$ of $\mathscr{A}_2$. By Koblitz [11, Theorem 7, page 163], $\mathscr{X}$ has dimension 1 (in fact, it is rational [14, page 117]). As in Sect. 5.1, $Pr_3^{-1}(\mathscr{X})$ has one irreducible component $N_{\mathscr{X}}$ of relative dimension 3 and three components of relative dimension 2. Since $\dim(\mathscr{X}) = 1$, it follows that $\dim(N_{\mathscr{X}}) = 4$.

Let $f = 3$. Since X_3 is a smooth plane quartic, then π_3 is represented by a point of $N_{\mathscr{X}}$ for some component $\mathscr{X}$ of $\mathscr{A}_2^0$. Thus $S_3 \subset N_{\mathscr{X}}$. This implies that $\dim(S_3) \leq 4$, which completes the proof when $f = 3$.

In fact, for any component $\mathscr{X}$ of $\mathscr{A}_2^0$ containing the point representing P_{π_3}, the fact that the smooth plane quartic X_3 has p-rank 3 implies that the generic point of $N_{\mathscr{X}}$ is in $\mathscr{R}_3^{(3,0)}$. This is relevant when P_{π_3} is superspecial, in which case the point that represents it is in multiple components $\mathscr{X}$.

Let $f = 2$. Since X_2 is a smooth plane quartic, then π_2 is represented by a point of $N_{\mathscr{Z}'}$ for some component $\mathscr{Z}'$ of $\mathscr{A}_2^0$. Thus $S_2 \subset N_{\mathscr{Z}'}$. We claim that the generic point of $N_{\mathscr{Z}'}$ represents an unramified double cover where X has p-rank 3, not 2. This is clear when $3 \le p \le 11$, because $\mathscr{A}_2^0$ is irreducible for those primes by Katsura and Oort [10, Theorem 5.8] so there is only one component of $\mathscr{A}_2^0$, so $\mathscr{Z}' = \mathscr{Z}$. For $13 \le p \le 19$, we note in the tables given in Sect. 3.3 that the same curve Z is used in the cases (3, 0) and (2, 0). This means that the Prym varieties P_{π_2} and P_{π_3} are isomorphic. So $\mathscr{Z}'$ is one of the components $\mathscr{Z}$ of $\mathscr{A}_2^0$ containing the point representing P_{π_3}, and the claim is true by the previous paragraph. Since $N_{\mathscr{Z}'}$ has dimension 4 and its generic point represents a cover where X has p-rank 3, it follows that $\dim(S_2) \le 3$. □

7.5 Final Result

In this section, in characteristic $3 \le p \le 19$, we verify the existence of unramified double covers $\pi : Y \to X$ where X has genus g and p-rank f and P_π has p-rank f', for arbitrary g as long as f is bigger than approximately $2g/3$ and f' is bigger than approximately $g/3$. This is most interesting when either $\frac{g}{3} \le f' < \frac{g}{2} - 1$ or $p = 3$ because [15, Corollary 7.2] resolves the case when $\frac{g}{2} - 1 \le f' \le g - 1$ with no conditions on g and f when $p \ge 5$.

We first include an inductive result which holds for any odd prime p. This strengthens [15, Theorem 7.2].

Theorem 1 *Let f_0 be such that $\mathscr{R}_3^{(f_1,0)}$ has a (nonempty) component of dimension $1 + f_1$ in characteristic p for each f_1 such that $f_0 \le f_1 \le 3$.*

Let $g \ge 2$ and write $g = 3r + 2s$ for integers $r, s \ge 0$. Suppose that $rf_0 \le f \le g$ (with $f \ge rf_0 + 2s$ when $p = 3$) and $r + s - 1 \le f' \le g - 1$.

Then $\mathscr{R}_g^{(f,f')}$ has a (nonempty) component of dimension $g - 2 + f + f'$ in characteristic p.

Proof In light of Proposition 12, it suffices to prove the result when $f' = r + s - 1$. The proof is by induction on $r + s$. In the base case $(r, s) = (0, 1)$, then $g = 2$, and the result is true by Ozman and Pries [15, Proposition 6.1]. In the base case $(r, s) = (1, 0)$, then $g = 3$, and the result is true by hypothesis. Now suppose that $r + s \ge 2$. As an inductive hypothesis, suppose that the result is true for all pairs (r', s') such that $1 \le r' + s' < r + s$.

Case 1: suppose that $r \ge 1$. This implies $g \ge 5$. Let $g_1 = 3$ and $g_2 = g - 3$. By the hypotheses on f, it is possible to choose f_1, f_2 such that $f_1 + f_2 = f$ and $f_0 \le f_1 \le 3$ and $f_0(r - 1) \le f_2 \le g_2$ (with $f_2 \ge f_0(r - 1) + 2s$ if $p = 3$). Let $f_1' = 0$ and $f_2' = r + s - 2$. By hypothesis, $\mathscr{R}_3^{(f_1,0)}$ is nonempty and has a component S_1 of dimension $d_1 = 1 + f_1$. By the inductive hypothesis applied to $(r - 1, s)$, it follows that $\mathscr{R}_{g_2}^{(f_2,r+s-2)}$ is nonempty and has a component S_2 of dimension $d_2 = g_2 - 2 + f_2 + (r + s - 2)$.

By Lemma 12, the image $\mathscr{K}$ of the clutching morphism κ has dimension equal to $d_1 + d_2 + 2$. Furthermore, Lemma 12 shows that $\mathscr{K}$ is contained in a component $\mathscr{W}$ of $\bar{\mathscr{R}}_g^{(f_1+f_2, f_1'+f_2'+1)} = \bar{\mathscr{R}}_g^{(f, r+s-1)}$ whose dimension is at most

$$d_1 + d_2 + 3 = (g_2 + 3) - 2 + (f_1 + f_2) + (r + s - 1) = g - 2 + f + f'.$$

Also $\dim(\mathscr{W}) \geq g - 2 + f + f'$ by purity. Thus $\dim(\mathscr{W}) = g - 2 + f + f'$ and the generic point of $\mathscr{W}$ is not contained in $\mathscr{K}$. The generic points of S_1 and S_2 represent unramified double covers of smooth curves by hypothesis. Thus the generic point of $\mathscr{W}$ is not contained in any other boundary component of $\bar{\mathscr{R}}_g$. It is thus contained in $\mathscr{R}_g$, and so it represents an unramified double cover of a smooth curve.

Case 2: suppose that $s \geq 1$. This implies $g \geq 4$. Let $g_1 = 2$ and $g_2 = g - 2$. By the hypotheses on f, when $p \geq 5$, it is possible to choose f_1, f_2 such that $f_1 + f_2 = f$ and $0 \leq f_1 \leq 2$ and $f_0 r \leq f_2 \leq g_2$. When $p = 3$, let $f_1 = 2$ and $f_2 = f_0 r + 2(s - 1)$. Let $f_1' = 0$ and $f_2' = r + s - 2$. By Ozman and Pries [15, Proposition 6.1], $\mathscr{R}_2^{(f_1,0)}$ is nonempty and has a component S_1 of dimension $d_1 = f_1$. By the inductive hypothesis applied to $(r, s-1)$, it follows that $\mathscr{R}_{g_2}^{(f_2, r+s-2)}$ is nonempty and has a component S_2 of dimension $d_2 = g_2 - 2 + f_2 + (r + s - 2)$. The proof then follows the approach of Case (1). □

Corollary 4 *Let $f_0 = 2$ and $3 \leq p \leq 19$. Let $g \geq 2$ and write $g = 3r + 2s$ for integers $r, s \geq 0$. Suppose that $2r \leq f \leq g$ (with $f \geq 2r + 2s$ when $p = 3$) and $r + s - 1 \leq f' \leq g - 1$.*

Then $\mathscr{R}_g^{(f,f')}$ has a (nonempty) component of dimension $g - 2 + f + f'$ in characteristic p.

In particular, this holds in the following situations:

1. *If $g = 3r$ and (f, f') is such that $2r \leq f \leq g$ and $r - 1 \leq f' \leq g - 1$;*
2. *If $g = 3r + 2$ and (f, f') is such that $2r \leq f \leq g$ and $r \leq f' \leq g - 1$, (with $f \geq 2r + 2$ when $p = 3$);*
3. *If $g = 3r + 4$ and (f, f') is such that $2r \leq f \leq g$ (with $f \geq 2r + 4$ when $p = 3$) and $r + 1 \leq f' \leq g - 1$.*

Proof By Proposition 13, the hypothesis in Theorem 1 holds for $f_0 = 2$. The result is immediate from Theorem 1. □

Acknowledgements This project began at the *Women in Numbers Europe 2* workshop in the Lorentz Center, Leiden. We are very grateful to the Lorentz Center for their hospitality and support. Elias benefited from a Leibniz fellowship at the Oberwolfach Research Institute during part of this work. Ozman was partially supported by Bogazici University Research Fund Grant Number 10842 and by the BAGEP Award of the Science Academy with funding supplied by Mehveş Demiren in memory of Selim Demiren. Pries was partially supported by NSF grant DMS-15-02227. We would like to thank Bouw and Bruin for helpful conversations.

References

1. J.D. Achter, R. Pries, Monodromy of the p-rank strata of the moduli space of curves. Int. Math. Res. Not. (15):Art. ID rnn053, 25 (2008)
2. A. Beauville, Prym varieties: a survey, in *Theta Functions—Bowdoin 1987, Part 1 (Brunswick, ME, 1987)*. Proceedings of Symposia in Pure Mathematics, vol. 49 (American Mathematical Society, Providence, RI, 1989), pp. 607–620
3. I. Bouw, The p-rank of ramified covers of curves. Compos. Math. **126**(3), 295–322 (2001)
4. N. Bruin, The arithmetic of Prym varieties in genus 3. Compos. Math. **144**(2), 317–338 (2008)
5. J.W.S. Cassels, E.V. Flynn, *Prolegomena to a Middlebrow Arithmetic of Curves of Genus 2*. London Mathematical Society Lecture Note Series, vol. 230 (Cambridge University Press, New York, 1996)
6. A. Chiodo, D. Eisenbud, G. Farkas, F Schreyer. Syzygies of torsion bundles and the geometry of the level ℓ modular variety over $\overline{\mathcal{M}}_g$. Invent. Math. **194**(1), 73–118 (2013)
7. C. Faber, G. van der Geer, Complete subvarieties of moduli spaces and the Prym map. J. Reine Angew. Math. **573**, 117–137 (2004)
8. T. Ibukiyama, T. Katsura, F. Oort, Supersingular curves of genus two and class numbers. Compos. Math. **57**(2), 127–152 (1986)
9. J.-I. Igusa, Arithmetic variety of moduli for genus two. Ann. Math. (2) **72**, 612–649 (1960)
10. T. Katsura, F. Oort, Families of supersingular abelian surfaces. Compos. Math. **62**(2), 107–167 (1987)
11. N. Koblitz, p-adic variation of the zeta-function over families of varieties defined over finite fields. Compos. Math. **31**(2), 119–218 (1975)
12. M. Kudo, S. Harashita, Superspecial curves of genus 4 in small characteristic. Finite Fields Appl. **45**, 131–169 (2017)
13. D. Mumford. *Tata Lectures on Theta II*. Modern Birkhäuser Classics, reprint of the 1984 edition (2007)
14. F. Oort, Subvarieties of moduli spaces. Invent. Math. **24**, 95–119 (1974)
15. E. Ozman, R. Pries, Ordinary and almost ordinary Prym varieties. Asian J. Math. (to appear)
16. J.-P. Serre, Sur la topologie des variétés algébriques en caractéristique p, in *Symposium Internacional de topología algebraica International Symposium on Algebraic Topology* (Universidad Nacional Autónoma de México and UNESCO, Mexico City, 1958), pp. 24–53
17. K.-O. Stöhr, J.F. Voloch, A formula for the Cartier operator on plane algebraic curves. J. Reine Angew. Math. **377**, 49–64 (1987)
18. The Sage Developers, SageMath, the sage mathematics software system (version x.y.z) (2017). http://www.sagemath.org
19. A. Verra, The fibre of the Prym map in genus three. Math. Ann. **276**(3), 433–448 (1987)
20. A. Vistoli, Intersection theory on algebraic stacks and on their moduli spaces. Invent. Math. **97**(3), 613–670 (1989)
21. W. Wirtinger, Untersuchungen über Thetafunktionen (B. G. Teubner, Leipzig, 1895)
22. N. Yui, On the Jacobian varieties of hyperelliptic curves over fields of characteristic $p > 2$. J. Algebra **52**(2), 378–410 (1978)

Elliptic Fibrations on Covers of the Elliptic Modular Surface of Level 5

Francesca Balestrieri, Julie Desjardins, Alice Garbagnati, Céline Maistret, Cecília Salgado, and Isabel Vogt

Abstract We consider the K3 surfaces that arise as double covers of the elliptic modular surface of level 5, $R_{5,5}$. Such surfaces have a natural elliptic fibration induced by the fibration on $R_{5,5}$. Moreover, they admit several other elliptic fibrations. We describe such fibrations in terms of linear systems of curves on $R_{5,5}$. This has a major advantage over other methods of classification of elliptic fibrations, namely, a simple algorithm that has as input equations of linear systems of curves in the projective plane yields a Weierstrass equation for each elliptic fibration. We deal in detail with the cases for which the double cover is branched over the two reducible fibers of type I_5 and for which it is branched over two smooth fibers, giving a complete list of elliptic fibrations for these two scenarios.

F. Balestrieri
Mathematical Institute, University of Oxford, Oxford, UK
e-mail: balestrieri@maths.ox.ac.uk

J. Desjardins
Université Grenoble Alpes, Institut Fourier, CNRS UMR, St Martin d'Hères, France
e-mail: julie.desjardins@imj-prg.fr

A. Garbagnati
Dipartimento di Matematica, Università Statale degli Studi di Milano, Milano, Italy
e-mail: alice.garbagnati@unimi.it

C. Maistret
University of Bristol, Bristol, UK
e-mail: cm16281@bristol.ac.uk

C. Salgado (✉)
Instituto de Matemática, Cidade Universitária, Universidade Federal do Rio de Janeiro (UFRJ), Rio de Janeiro, Brazil
e-mail: salgado@im.ufrj.br

I. Vogt
Department of Mathematics, Massachusetts Institute of Technology, Cambridge, MA, USA
e-mail: ivogt@mit.edu

I. I. Bouw et al. (eds.), *Women in Numbers Europe II*, Association for Women in Mathematics Series 11, https://doi.org/10.1007/978-3-319-74998-3_9

1 Introduction

Let $S/\mathbb{C}$ be a smooth projective surface and $B/\mathbb{C}$ be a smooth projective curve. We say that a proper flat map $\mathscr{E}: S \to B$ is an elliptic fibration if the generic fiber S_b is a smooth genus 1 curve and a section $O: B \to S$ is given. Given a section, we regard the generic fiber of $\mathscr{E}$ as an elliptic curve over the function field $k(B)$, and so we can work with a Weierstrass form of $\mathscr{E}$. We will say that an elliptic fibration is relatively minimal if there are no contractible curves contained in its fibers. For the remainder of this paper, all elliptic fibrations will be assumed to be relatively minimal and not of product type.

Not all surfaces S admit elliptic fibrations, and if S admits an elliptic fibration, a lot is known about the base curve and about the maximal number of elliptic fibrations on it. More precisely if S is of general type, then S admits no elliptic fibrations; if the Kodaira dimension of S is nonpositive, then the curve B is rational; if a surface S admits more than one elliptic fibration as above, then it is a K3 surface (a surface with trivial canonical bundle and trivial irregularity). In particular if S is either a K3 surface or a rational surface, then $B \simeq \mathbb{P}^1$. Every relatively minimal rational elliptic surface is the total space of a pencil of plane cubics; such surfaces admit only the obvious elliptic fibration. We refer to [6] and to [11] for more on the theory of elliptic fibrations on surfaces.

1.1 K3 Surfaces Arising from Rational Elliptic Surfaces and Their Elliptic Fibrations

In this paper we will consider K3 surfaces S that are double covers of rational elliptic surfaces R, branched over two fibers of the elliptic fibration $\mathscr{E}_R : R \to \mathbb{P}^1$. More precisely, such K3 surfaces are the minimal resolution of the fiber product $\bar{S}$ of a rational elliptic surface $\mathscr{E}_R : R \to \mathbb{P}^1$ and a degree 2 map $\mathbb{P}^1 \to \mathbb{P}^1$, which is necessarily branched over two points.

We recall that an involution on a K3 surface is non-symplectic if it acts by -1 on $H^0(S, K_S) \simeq \mathbb{C}$. Call ι the involution on S agreeing with the cover involution of the $2:1$ map $\bar{S} \to R$; then this involution is non-symplectic and fixes the inverse image of two fibers of the fibration $\mathscr{E}_R : R \to \mathbb{P}^1$, which are the branch curves. The quotient surface S/ι is rational and is either R or a blowup of R (denoted by $\widetilde{R}$ in what follows). It was proved by Zhang (see [15]) that every K3 surface S admitting a non-symplectic involution ι whose fixed locus contains curves of genus at most 1 arises by a base change of order two from a rational elliptic fibration $\mathscr{E}_R : R \to \mathbb{P}^1$ as described above.

The K3 surface S obtained as above is naturally equipped with one elliptic fibration $\mathscr{E}_S : S \to \mathbb{P}^1$, induced via pullback from $\mathscr{E}_R$. In [4] the relationship between elliptic fibrations on S and linear systems on the rational surface R (or $\widetilde{R}$) is studied.

The elliptic fibrations on S fall into one of the three categories below according to the action of ι on the fibers. Note that $\mathscr{E}_S$ belongs to the second one:

- if ι preserves each fiber of the fibration, then it acts on the fibers as the elliptic involution. The elliptic fibration on S is therefore induced by fibrations in rational curves on $\widetilde{R}$. We will call these pencils "conic bundles" if they are rational fibrations on R and "generalized conic bundles" if they are rational fibrations on $\widetilde{R}$ (but not on R);
- if ι preserves the fibration, but not each fiber of the fibration, this implies that ι acts on the base of the fibration (with two fixed points). In this case the elliptic fibration on S is induced by a pencil of genus 1 curves on R, whose members split in the double cover. We call these pencils "splitting genus 1 pencils";
- if ι does not preserve the elliptic fibration, we call the fibration of type 3. A fibration is of type 3 if and only if the class of the fiber of the fibration, in the Néron-Severi group of S, is not preserved by ι.

As a result of this classification and of the technique introduced in [4], in good cases one may classify the singular fibers of all elliptic fibrations $\mathscr{E}\colon S \to \mathbb{P}^1$ in terms of the singular fibers of more tractable linear series on S/ι. Our focus here is on even finer information: obtaining explicit Weierstrass equations of elliptic fibrations on such K3 surfaces. Using Tate's algorithm, one may then read off the singular fibers from the order of vanishing of the coefficients and discriminant of the Weierstrass equation (see, e.g., the table on page 41 of [6]). In Sects. 5.2 and 6.1.2, we give methods and algorithms for determining Weierstrass equations coming from conic bundles and splitting genus 1 pencils on rational elliptic surfaces, under some assumptions. This is our main result.

Theorem 1 *Let S be a K3 surface arising from the rational elliptic surface $\mathscr{E}_R\colon R \to \mathbb{P}^1$ as described above. Let $\mathscr{E}$ be an elliptic fibration on S that is not of type 3. Then, under certain conditions, one obtains a Weierstrass equation for $\mathscr{E}$ by applying:*

- *the algorithm of Sect. 5.2 if $\mathscr{E}$ is induced by a (generalized) conic bundle with prescribed properties (see Sect. 5.2 for details);*
- *the algorithm of Sect. 6.1.2 if $\mathscr{E}$ is induced by a splitting genus 1 pencil.*

1.2 Outline of the Paper

We focus particularly on K3 surfaces arising as double covers of $R_{5,5}$, the elliptic modular surface of level 5. This is the universal elliptic curve over the modular curve $X_1(5)$, and the evident map $\mathscr{E}_{R_{5,5}}\colon R_{5,5} \to X_1(5) \simeq \mathbb{P}^1$ is the unique elliptic fibration. The fibers of $\mathscr{E}_{R_{5,5}}$ are smooth except for two nodal rational curves (type I_1) and two 5-gons (type I_5); this property also determines $R_{5,5}$ and implies that the Mordell–Weil group $\mathrm{MW}(\mathscr{E}_R) = \mathbb{Z}/5\mathbb{Z}$. The geometry of this surface as the total space of a pencil of plane cubics is described in Sect. 2.

In Sect. 3 we classify the conic bundles on $R_{5,5}$ up to automorphisms. The main result of this section is Proposition 1.

In the remainder of the paper, we study elliptic fibrations on different K3 surfaces arising from $R_{5,5}$ by choosing different base changes. In Sect. 4 we describe such K3 surfaces, writing their equations as double covers of $\mathbb{P}^2$, as well as giving the Weierstrass form of the elliptic fibrations induced by $\mathscr{E}_R$.

In Sect. 5 we consider the elliptic fibrations induced on K3 surfaces by the conic bundles classified in Sect. 3. We observe directly that the same conic bundle induces elliptic fibrations with very different properties on K3 surfaces according to the choice of the branch curves.

In Sects. 6 and 7, we restrict our attention to two K3 surfaces obtained by choosing maximally different branch fibers: in Sect. 6 we consider the very special case where the branch fibers are $2I_5$, and in Sect. 7 we consider the generic case where the branch fibers are $2I_0$.

When the double cover is branched over the two 5-gons, the K3 surface is called $S_{5,5}$, and the involution ι fixes the union of ten rational curves. There is a unique such K3 surface possessing such a non-symplectic involution. This K3 surface admits 13 types of elliptic fibrations, classified by Nishiyama in [9]. In this special case we are able to determine equations for all elliptic fibrations on $S_{5,5}$ using our techniques and algorithms. We observe that in this case there are no fibrations of type 3. The main result of this Section is Proposition 2.

When the double cover is branched on two smooth fibers, the K3 surface moves in the two-dimensional family of the K3 surfaces admitting an elliptic fibration with a 5-torsion section. We call a very general member of this family $X_{5,5}$, and using the lattice-theoretic technique of [9], we list all the admissible configurations of fibers of an elliptic fibration on $X_{5,5}$ in Table (7.1) proving Proposition 3. In this case the elliptic fibrations cannot be induced by splitting genus 1 pencils or by generalized conic bundles. On the other hand, there are plenty of elliptic fibrations of type 3, and we describe one of them in detail.

2 The Surface $R_{5,5}$

Let $R_{5,5}$ be the elliptic modular surface of level 5. We know from [1, Tableau] (with coordinates $X = x_0$, $Y = x_0 - x_1$, $Z = x_2$) that the surface $R_{5,5}$ is the blowup of $\mathbb{P}^2$ in the basepoints of the pencil $\mathscr{P}$ of cubics:

$$\lambda x_1 x_2(x_0 - x_1) + \mu x_0(x_0 - x_1 - x_2)(x_0 - x_2) = 0. \tag{1}$$

We will denote this blowup morphism by $\beta\colon R_{5,5} \to \mathbb{P}^2$. The cubic corresponding to $\lambda = 0$ is reducible and consists of the three lines $m_1 : x_0 = 0$, $m_2 : x_0 - x_1 - x_2 = 0$, $m_3 : x_0 - x_2 = 0$; the cubic corresponding to $\mu = 0$ is reducible and consists of the three lines $\ell_1 : x_0 - x_1 = 0$, $\ell_2 : x_1 = 0$, $\ell_3 : x_2 = 0$ (Fig. 1).

Fig. 1 The reducible cubics and basepoints of the pencil $\mathscr{P}$

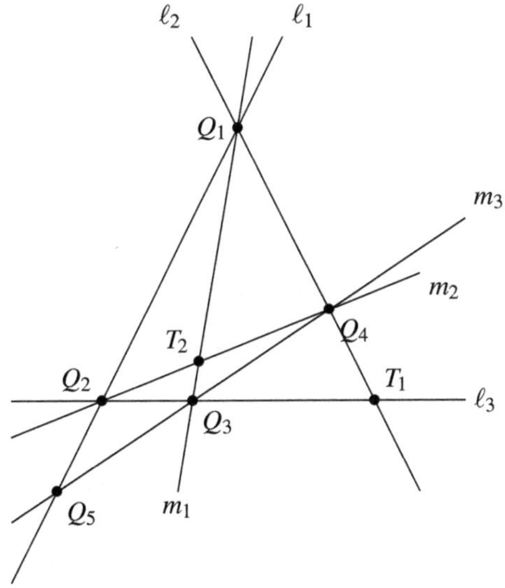

The nine basepoints of this pencil of cubics are as follows: $Q_1 := (0 : 0 : 1)$, the point Q_1' infinitely near to Q_1 and corresponding to the tangent direction m_1. $Q_2 := (1 : 1 : 0)$, the point Q_2' infinitely near to Q_2 and corresponding to the tangent direction m_2, $Q_3 := (0 : 1 : 0)$, the point Q_3' infinitely near to Q_3 and corresponding to the tangent direction ℓ_3, $Q_4 := (1 : 0 : 1)$, the point Q_4' infinitely near to Q_4 and corresponding to the tangent direction ℓ_2, and $Q_5 := (1 : 1 : 1)$.

In the following we will denote by T_1 the point $(1 : 0 : 0)$, which is not a basepoint of the pencil and corresponds to the intersection of the lines ℓ_2 and ℓ_3, and by T_2 the point $(0 : 1 : -1)$, which is not a basepoint of the pencil and corresponds to the intersection of the lines m_1 and m_2.

Let h denote the preimage of the class of a line; then $\mathrm{NS}(R_{5,5})$ is spanned by h and the components of the exceptional divisors of the blowup $\beta : R_{5,5} \to \mathbb{P}^2$. We will denote the (irreducible) exceptional divisor corresponding to Q_i (resp. Q_i') by E_i (resp. F_i) for $i = 1, 2, 3, 4$. At Q_5 there is only E_5. Note that $F_i^2 = -1$, $E_i^2 = -2$ for $i = 1, 2, 3, 4$, $E_5^2 = -1$, $E_i E_j = 0$ if $i \neq j$, $F_i F_j = 0$ if $i \neq j$, $E_i F_i = 1$. By slight abuse of notation, let ℓ_1, ℓ_2, ℓ_3 and m_1, m_2, m_3 denote the proper transforms on $R_{5,5}$ of the corresponding lines in $\mathbb{P}^2$. We have the following relations in $\mathrm{NS}(R_{5,5})$:

$$
\begin{aligned}
\ell_1 &= h - E_1 - F_1 - E_2 - F_2 - E_5 & m_1 &= h - E_1 - 2F_1 - E_3 - F_3 \\
\ell_2 &= h - E_1 - F_1 - E_4 - 2F_4 & m_2 &= h - E_2 - 2F_2 - E_4 - F_4 \\
\ell_3 &= h - E_2 - F_2 - E_3 - 2F_3 & m_3 &= h - E_3 - F_3 - E_4 - F_4 - E_5.
\end{aligned}
\tag{2}
$$

The Weierstrass equation of the elliptic fibration of $R_{5,5}$ is obtained by (1) choosing $x_2 = 1$ and applying standard transformations. It is

$$y^2 = x^3 + A(\lambda : \mu)x + B(\lambda : \mu), \quad \text{where} \tag{3}$$

$$A(\mu) := \frac{-1}{48}\mu^4 - \frac{1}{4}\mu^3\lambda - \frac{7}{24}\mu^2\lambda^2 + \frac{1}{4}\mu\lambda^3 - \frac{1}{48}\lambda^4, \text{ and}$$

$$B(\mu) := \frac{1}{864}\mu^6 + \frac{1}{48}\mu^5\lambda + \frac{25}{288}\mu^4\lambda^2 + \frac{25}{288}\mu^2\lambda^4 - \frac{1}{48}\mu\lambda^5 + \frac{1}{864}\lambda^6.$$

The discriminant is $\frac{1}{16}\mu^5\lambda^5(-\lambda^2 + 11\mu\lambda + \mu^2)$, so there are two fibers of type I_5 over $(\lambda : \mu) = (0 : 1)$ and $(\lambda : \mu) = (1 : 0)$. Moreover there are two fibers of type I_1 over $(\lambda : \mu) = (1 : -\frac{11}{2} \pm \frac{5}{2}\sqrt{5})$.

Now the function

$$(\lambda : \mu) \mapsto (x(\lambda : \mu); y(\lambda : \mu)) = \left(\frac{\mu^2 - 6\mu\lambda + \lambda^2}{12}; \frac{\mu^2\lambda}{2}\right)$$

is a 5-torsion section of this fibration. It is known (see, e.g., [11, Section 9.5]) that the elliptic fibration on $R_{5,5}$ has Mordell–Weil group equal to $\mathbb{Z}/5\mathbb{Z}$.

The negative curves on $R_{5,5}$ are:

1. the ten components of the two fibers of type I_5 denoted by $\Theta_0^{(1)}$, $\Theta_1^{(1)}$, $\Theta_2^{(1)}$, $\Theta_3^{(1)}$, $\Theta_4^{(1)}$ on the first fiber and $\Theta_0^{(2)}$, $\Theta_1^{(2)}$, $\Theta_2^{(2)}$, $\Theta_3^{(2)}$, $\Theta_4^{(2)}$ on the second fiber (these are all (-2)-curves);
2. the five sections P_0, P_1, P_2, P_3, and P_4, where P_0 meets the components $\Theta_0^{(1)}$ and $\Theta_0^{(2)}$, P_1 meets the components $\Theta_1^{(1)}$ and $\Theta_2^{(2)}$, P_2 meets the components $\Theta_2^{(1)}$ and $\Theta_4^{(2)}$, P_3 meets the components $\Theta_3^{(1)}$ and $\Theta_1^{(2)}$, and P_4 meets the components $\Theta_4^{(1)}$ and $\Theta_3^{(2)}$ (these sections are all (-1)-curves).

The dual graph of this configuration is given in Fig. 2. We observe that Fig. 2 is a generalization of the Petersen graph (it is exactly the Petersen graph if one does not consider the empty edges), and we point out that this graph represents several intersecting objects in algebraic geometry and in tropical geometry (see, e.g., [10]).

We may then make the following choice of identifications:

$$\begin{array}{lllll} \Theta_0^{(1)} = m_1 & \Theta_1^{(1)} = E_3 & \Theta_2^{(1)} = m_3 & \Theta_3^{(1)} = E_4 & \Theta_4^{(1)} = m_2 \\ \Theta_0^{(2)} = E_1 & \Theta_1^{(2)} = \ell_2 & \Theta_2^{(2)} = \ell_3 & \Theta_3^{(2)} = E_2 & \Theta_4^{(2)} = \ell_1 \\ P_0 = F_1 & P_1 = F_3 & P_2 = E_5 & P_3 = F_4 & P_4 = F_2. \end{array} \tag{4}$$

We observe that there is an automorphism σ_5 on $R_{5,5}$ of order 5, which is the translation by the section P_1. It acts on the negative curves as follows: $\sigma_5(\Theta_i^{(1)}) = \Theta_{i+1}^{(1)}$ and $\sigma_5(\Theta_i^{(2)}) = \Theta_{i+2}^{(2)}$, where $i+1$ and $i+2$ are considered modulo 5; $\sigma_5(P_k) = P_{k+1}$, where $k+1$ is considered modulo 5.

There is also an automorphism σ_2 of order 2 on $R_{5,5}$ which is the elliptic involution on the elliptic curve (3) over the function field $k(\mu)$. Note that σ_2 restricts to the elliptic involution on each smooth fiber of the fibration (3). It acts on the

Fig. 2 Dual graph of negative curves on $R_{5,5}$. The symbol filled circle denotes a (-2)-curve, and open circle denotes a (-1)-curve

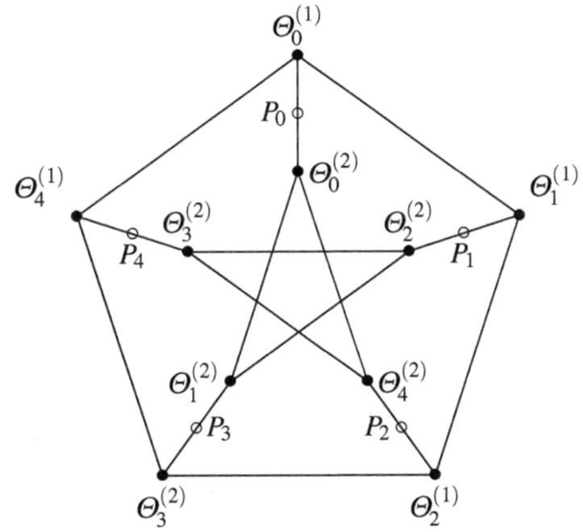

negative curves as follows: $\sigma_2(\Theta_i^{(j)}) = \Theta_{-i}^{(j)}$, where $i \in \mathbb{Z}/5\mathbb{Z}$, $j = 1, 2$ and $\sigma_2(P_k) = P_{-k}$, where $k \in \mathbb{Z}/5\mathbb{Z}$. There is also an automorphism α of $R_{5,5}$ lying above the involution of $\mathbb{P}^1$ $\alpha\colon (\lambda : \mu) \mapsto (-\mu : \lambda)$. In terms of the Weierstrass equation (3), we have

$$\alpha\colon (x, y, \mu) \mapsto (x/\mu^2, -y/\mu^3, -1/\mu).$$

Note that the automorphism $\sigma_5^2\alpha\sigma_5^3$ is induced by the element $\begin{pmatrix} 0 & 1 & 0 \\ -1 & 1 & 1 \\ 1 & 0 & 0 \end{pmatrix}$ in $\mathrm{PGL}_3(\mathbb{C})$. From this description, the action on $\mathrm{NS}(R_{5,5})$ is apparent $\alpha(\Theta_0^{(1)}) = \Theta_0^{(2)}$ and $\alpha(\Theta_0^{(2)}) = \Theta_0^{(1)}$; $\alpha(\Theta_i^{(1)}) = \Theta_i^{(2)}$ and $\alpha(\Theta_i^{(2)}) = \Theta_{-i}^{(1)}$; and finally $\Theta(P_0) = P_0$, while the remaining sections are permuted so as to preserve intersections.

3 Conic Bundles on $R_{5,5}$

In this section we classify the conic bundles on $R_{5,5}$ by considering their reducible fibers proving Proposition 1.

The key result we use is that on a rational elliptic surface, every conic bundle has at least one reducible fiber. Further, any reducible fiber must be of type A_n or D_m, as shown in Fig. 3 (see, e.g., [4]).

Proposition 1 *There are exactly three conic bundles on $R_{5,5}$ up to automorphisms, i.e., the conic bundles B_1, B_2, and B_3 induced by the pencils of plane rational curves with Eqs.* (5), (6), *and* (7), *respectively.*

Proof

Step 1: Classification of the reducible fibers. Every conic bundle has at least one reducible fiber, so in order to classify the conic bundles, it suffices to find all

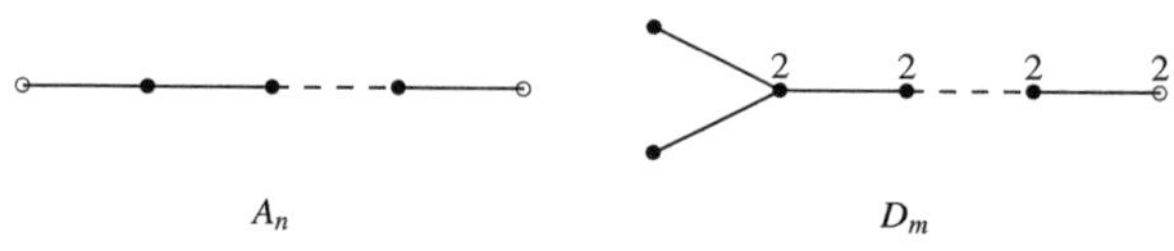

Fig. 3 Possible reducible fibers of conic bundles on (minimal) rational elliptic surfaces. The numbers n and m refer each to the number of components. The multiplicity of a component is indicated above the corresponding vertex if it is not 1. The symbol filled circle denotes a (-2)-curve, and open circle denotes a (-1)-curve

the possible reducible fibers. The components of the reducible fibers are negative curves. As $R_{5,5}$ (and in fact any extremal rational elliptic surface; see [6, VIII.1.2]) has only finitely many curves of negative self-intersection, one simply must find all possible A_n and D_m (for $m \geq 3$) configurations among the curves with negative self-intersection.

Every reducible fiber contains at least a (-1)-curve, as shown in Fig. 3. Since we are looking for a classification up to automorphisms, and the automorphism σ_5 permutes the (-1)-curves on $R_{5,5}$, we can always assume that one of the reducible fibers of the conic bundle contains the (-1)-curve P_0. Moreover we recall that the action of σ_2 switches $\Theta_i^{(j)}$ with $\Theta_{5-i}^{(j)}$ and the action of α switches the two I_5-fibers. Up to the action of σ_2 and α, the reducible fibers of type A_n that contain P_0 as a component are necessarily of one of the following type:

- $P_0 + \sum_{i=0}^{k} \Theta_i^{(1)} + P_k$, $k = 1, 2, 3$, and in this case, we have a fiber of type A_{k+3}.

Up to the action of σ_2, the reducible fibers of type D_m that contain P_0 as a component are necessarily of the following types:

- $2P_0 + \Theta_0^{(1)} + \Theta_0^{(2)}$, and in this case, we have a fiber of type D_3;
- $2P_0 + 2\Theta_0^{(1)} + \Theta_1^{(1)} + \Theta_4^{(1)}$, and in this case, we have a fiber of type D_4.

Step 2: At most three conic bundles. We consider the conic bundles associated to the possible singular fibers described above and show that some of them are equivalent up to the action of σ_5 and σ_2. Moreover we describe the conic bundles that we find and we give equations for them:

(1) $|B_1| = |P_0 + \Theta_0^{(1)} + \Theta_1^{(1)} + P_1|$. The components of one reducible fiber are P_0, $\Theta_0^{(1)}$, $\Theta_1^{(1)}$, P_1. This fiber is of type A_4. The curves $\Theta_2^{(1)}$, $\Theta_4^{(1)}$, $\Theta_0^{(2)}$, and $\Theta_2^{(2)}$ are sections of the bundle. There is another reducible fiber of type A_4 which is formed by the curves P_2, $\Theta_4^{(2)}$, $\Theta_3^{(2)}$, P_4 and one of type D_3 which is formed by P_3 (with multiplicity 2), $\Theta_3^{(1)}$, $\Theta_1^{(2)}$.

Using the identifications made earlier, this class can also be written as:

$$B_1 = F_1 + m_1 + E_3 + F_3 = F_1 + (h - E_1 - 2F_1 - E_3 - F_3) + E_3 + F_3$$
$$= h - E_1 - F_1.$$

Unwinding what this means geometrically, $|B_1|$ comes via proper transform from the pencil of lines through Q_1 in $\mathbb{P}^2$ which has equation

$$x_1 = \tau x_0. \tag{5}$$

Under this description we can also understand the singular fibers: they correspond to the special lines m_1, ℓ_1, and ℓ_2. For example, the line ℓ_2 corresponds to the reducible fiber $\beta^*(\ell_2) - E_1 - F_1 = \ell_2 + E_4 + 2F_4 = \Theta_1^{(2)} + \Theta_3^{(1)} + 2P_3$.

The conic bundle $|B_1|$ is sent to other conic bundles by σ_5, by σ_2, and by their powers. Each of these has exactly three reducible fibers of types A_4, A_4, and D_3.

We observe that the fiber of type D_3 of the conic bundle $|B_1|$ is sent to $2P_0 + \Theta_0^{(1)} + \Theta_2^{(2)}$ by the automorphism σ_5^2, so the conic bundle with reducible fiber $2P_0 + \Theta_0^{(1)} + \Theta_0^{(2)}$ is equivalent to $|B_1|$ up to automorphisms.

Similarly the A_4-fiber $P_2 + \Theta_4^{(2)} + \Theta_3^{(2)} + P_4$ is sent to $P_3 + \Theta_1^{(2)} + \Theta_0^{(2)} + P_0$ by σ_5, so also the conic bundle with reducible fiber $P_0 + \Theta_0^{(2)} + \Theta_1^{(2)} + P_3$ is equivalent to $|B_1|$ up to automorphisms.

(2) $|B_2| = |P_0 + \Theta_0^{(1)} + \Theta_1^{(1)} + \Theta_2^{(1)} + P_2|$. The components of one reducible fiber are P_0, $\Theta_0^{(1)}$, $\Theta_1^{(1)}$, $\Theta_2^{(1)}$, P_2. This fiber is of type A_5. The curves $\Theta_3^{(1)}$, $\Theta_4^{(1)}$, $\Theta_0^{(2)}$, $\Theta_4^{(2)}$, and P_1 are sections of the bundle. There is another reducible fiber of type A_5 which is formed by the curves P_3, $\Theta_1^{(2)}$, $\Theta_2^{(2)}$, $\Theta_3^{(2)}$, P_4.

Similarly here we can write:

$$B_2 = F_1 + m_1 + E_3 + m_3 + E_5 = 2h - E_1 - F_1 - E_3 - 2F_3 - E_4 - F_4.$$

Hence B_2 corresponds to the pencil of conics through Q_1, Q_3, Q_3', and Q_4. More explicitly this is given by conics passing through Q_1 and Q_4, and tangent to ℓ_3 at Q_3 and in $\mathbb{P}^2$ this pencil is given by the equation

$$x_1 x_2 = \tau(x_0 x_2 - x_0^2). \tag{6}$$

The two reducible fibers correspond to the reducible conics $m_1 \cup m_3$ and $\ell_2 \cup \ell_3$.

The A_5-fiber of $|B_2|$ whose components are P_3, $\Theta_1^{(2)}$, $\Theta_2^{(2)}$, $\Theta_3^{(2)}$, P_4 is sent to the reducible fiber $P_0 + \sum_{i=0}^{2} \Theta_i^{(2)} + P_1$ by σ_5^2.

(3) $|B_3| = |P_0 + \Theta_0^{(1)} + \Theta_1^{(1)} + \Theta_2^{(1)} + \Theta_3^{(1)} + P_3|$. The components of one reducible fiber are P_0, $\Theta_0^{(1)}$, $\Theta_1^{(1)}$, $\Theta_2^{(1)}$, $\Theta_3^{(1)}$, P_3. This fiber is of type A_6. The curves $\Theta_0^{(2)}$, $\Theta_1^{(2)}$, P_1, and P_2 are sections of the bundle. The curve $\Theta_4^{(1)}$ is a multisection of degree 2. There is another reducible fiber of type D_4 which is formed by the curves P_4, $\Theta_3^{(2)}$, $\Theta_4^{(2)}$, $\Theta_2^{(2)}$.

We can also describe this using:

$$B_3 = F_1 + m_1 + E_3 + m_3 + E_4 + F_4 = 2h - E_1 - F_1 - E_3 - 2F_3 - E_5.$$

Therefore B_3 comes from the pencil of conics in $\mathbb{P}^2$ through Q_1, Q_3, Q_3', and Q_5; that is conics through Q_1 and Q_5, tangent to ℓ_3 at Q_3, and in $\mathbb{P}^2$ this pencil is given by the equation:

$$x_1x_2 = (\tau + 1)x_0x_2 - \tau x_0^2. \tag{7}$$

We can again understand the reducible fibers as coming from reducible conics $\ell_1 \cup \ell_3$ and $m_1 \cup m_3$.

The D_4-fiber is sent to the fiber $2P_0 + 2\Theta_0^{(2)} + \Theta_1^{(2)} + \Theta_4^{(2)}$ by σ_5.

Step 3: Exactly three conic bundles. It remains only to prove that the conic bundles B_i for $i = 1, 2, 3$ are all inequivalent up to automorphisms. Since the reducible fibers of $|B_1|$ are $(2A_4, D_3)$, the reducible fibers of $|B_2|$ are $(2A_5)$, and the reducible fibers of $|B_3|$ are (A_6, D_4), they cannot be equivalent up to automorphisms. Hence there are three conic bundles on $R_{5,5}$ up to automorphisms.

4 K3 Surfaces Obtained by $R_{5,5}$

Now we consider K3 surfaces obtained from $R_{5,5}$ by a base change of order 2 branched on two fibers. Of course the K3 surface obtained depends on the branch fibers. Let us explicitly give the description of the K3 surfaces that we can obtain in this way. They will be both described as elliptic fibrations (induced by the one of $R_{5,5}$) and as double covers of $\mathbb{P}^2$.

4.1 *The Branch Fibers Are $2I_5$: The K3 Surface $S_{5,5}$*

Let us consider the K3 surface $S_{5,5}$ obtained by a base change of order 2 of $R_{5,5}$ whose branch locus corresponds to the two fibers of type I_5. This means that all the components of the fibers of type I_5 are in the branch locus of the double cover $S_{5,5} \dashrightarrow R_{5,5}$.

4.1.1 The Surface $\widetilde{R}$

The double cover of $R_{5,5}$ branched over the two fibers of type I_5 has ordinary double point singularities at the ten points over the nodes in the branch fibers. In order to obtain a K3 surface, one can blow up these ten points on the double cover, introducing ten exceptional divisors. Equivalently one may first blow up the ten nodes of the branch fibers to obtain a non-minimal rational elliptic surface $\widetilde{R}$ and then consider the double cover of this surface branched over the strict transforms of the branch fibers. Note that in the preimage of the branch fibers, all the 10 exceptional curves occur with multiplicity 2, and so they are not in the branch locus.

We will make use of this non-minimal rational elliptic surface $\widetilde{R}$; it is simply the blowup of $\mathbb{P}^2$ in $9 + 10$ points, some of which are infinitely near to each other. The ten additional points are T_1, T_2 and the two points on each of the exceptional divisors E_i for $i = 1, 2, 3, 4$ corresponding to the tangent directions at Q_i specified by, respectively, ℓ_1 and ℓ_2, ℓ_1 and ℓ_3, m_1 and m_3, and finally m_2 and m_3.

We will denote by E_{Q_5}, E_{T_1}, and E_{T_2} the exceptional divisors over Q_5, T_1, and T_2, respectively. These three divisors are (-1)-curves.

We will denote by E_i, F_i, G_i, and H_i the four exceptional divisors over Q_i for $i = 1, 2, 3, 4$. For each i, the divisor E_i is a (-4)-divisor intersecting F_i, G_i, and H_i, which are orthogonal (-1)-curves. We make the identification that F_i is the tangent direction corresponding to the basepoint of the pencil of cubics, and H_i and G_i correspond to the tangent directions specified in the following table:

i	1	2	3	4
G_i	ℓ_1	ℓ_3	m_1	m_3
H_i	ℓ_2	ℓ_1	m_3	m_2.

Note that the E_i and F_j are the strict transforms of the curves of the same name on $R_{5,5}$.

The strict transforms of the lines ℓ_j and m_j, $j = 1, 2, 3$ on $\widetilde{R}$ are the following:

$$\begin{aligned}
\ell_1 &:= h - E_1 - F_1 - 2G_1 - H_1 - E_2 - F_2 - G_2 - 2H_2 - E_{Q_5};\\
\ell_2 &:= h - E_1 - F_1 - G_1 - 2H_1 - E_4 - 2F_4 - G_4 - H_4 - E_{T_1};\\
\ell_3 &:= h - E_2 - F_2 - 2G_2 - H_2 - E_3 - 2F_3 - G_3 - H_3 - E_{T_1};\\
m_1 &:= h - E_1 - 2F_1 - G_1 - H_1 - E_3 - F_3 - 2G_3 - 2H_3 - E_{T_2};\\
m_2 &:= h - E_2 - 2F_2 - G_2 - H_2 - E_4 - F_4 - G_4 - 2H_4 - E_{T_2};\\
m_3 &:= h - E_3 - F_3 - G_3 - 2H_3 - E_4 - F_4 - 2G_4 - H_4 - E_{Q_5}.
\end{aligned}$$

The sections of the non-relatively minimal fibration on $\widetilde{R}$ are F_j, $j = 1, 2, 3, 4$ and E_{Q_5} (i.e., the strict transform of the sections of the fibration on $R_{5,5}$).

4.1.2 Geometric Description of $S_{5,5}$ and Its Néron–Severi Group

The surface $S_{5,5}$ admits a non-symplectic involution ι which is the cover involution of the double cover $S_{5,5} \to \widetilde{R}$. This involution fixes ten rational curves (the curves m_i, ℓ_i,$i = 1, 2, 3$ and E_j with $j = 1, 2, 3, 4$), and it acts trivially on the Néron–Severi group.

The elliptic fibration $\mathcal{E}_{S_{5,5}}\colon S_{5,5} \to \mathbb{P}^1$ induced by $\mathcal{E}_{R_{5,5}}\colon R_{5,5} \to \mathbb{P}^1$ has two fibers of type I_{10} (induced by the fibers of type I_5 on $R_{5,5}$) and four other singular fibers, all of type I_1. The trivial lattice of the fibration (generated by the class of the generic fiber, the class of the zero section, and the classes of the nontrivial components of the reducible fibers) has rank 20. The trivial lattice is a sublattice of the Néron–Severi group, and since the Néron–Severi group of a K3 surface has rank at most 20, we conclude that it is exactly 20. By the Shioda–Tate formula, there are no sections of infinite order for the fibration $\mathscr{E}_{S_{5,5}} : S_{5,5} \to \mathbb{P}^1$. The 5-torsion sections of the fibration on $R_{5,5}$ induce 5-torsion sections of $\mathscr{E}_{S_{5,5}}$. Hence, $\mathrm{MW}(\mathscr{E}_{S_{5,5}}) \supseteq \mathbb{Z}/5\mathbb{Z}$. The possible torsion parts of the Mordell–Weil group of an elliptic fibration on a K3 surface are $\mathbb{Z}/n\mathbb{Z}$ for $2 \le n \le 8$, $\mathbb{Z}/2\mathbb{Z} \times \mathbb{Z}/m\mathbb{Z}$ for $m = 2, 4, 6$, and $(\mathbb{Z}/k\mathbb{Z})^2$ for $k = 3, 4$ (see, e.g., [12, Thm 7.1]). So we conclude that $\mathrm{MW}(\mathscr{E}_{S_{5,5}}) = \mathbb{Z}/5\mathbb{Z}$.

The curves $\Theta_i^{(j)}$ are in the branch locus, and we denote by $\Omega_i^{(j)}$ the rational curve on $S_{5,5}$ which maps $1 : 1$ to $\Theta_i^{(j)}$. Moreover, we have ten other rational curves on $S_{5,5}$: the curves $\Omega_{i_1,i_1-1}^{(j)}$, for $i_1, i_1 - 1 \in \mathbb{Z}/5\mathbb{Z}$, $j = 1, 2$, which are the curves resolving the singularities of the intersection point between $\Theta_{i_1}^{(j)}$ and $\Theta_{i_1-1}^{(j)}$ and are the double cover of the ten exceptional curves of the blowup $\widetilde{R} \to R_{5,5}$. The curves P_i are not in the branch locus and we denote by Q_i their $2 : 1$ cover in $S_{5,5}$. The dual graph of this configuration is given in Fig. 4. We observe that this gives exactly the diagram given in [14, Figure 1] as dual graph of certain rational curves on the K3 surface whose transcendental lattice is $\langle 2 \rangle^2$, which is a different way to describe the surface $S_{5,5}$.

Let $\pi : S_{5,5} \to \widetilde{R}$ denote the double cover. As can be deduced from (4) and the above identifications, pushing forward curve classes has the following effect:

Fig. 4 Dual graph of relevant negative curves on $S_{5,5}$

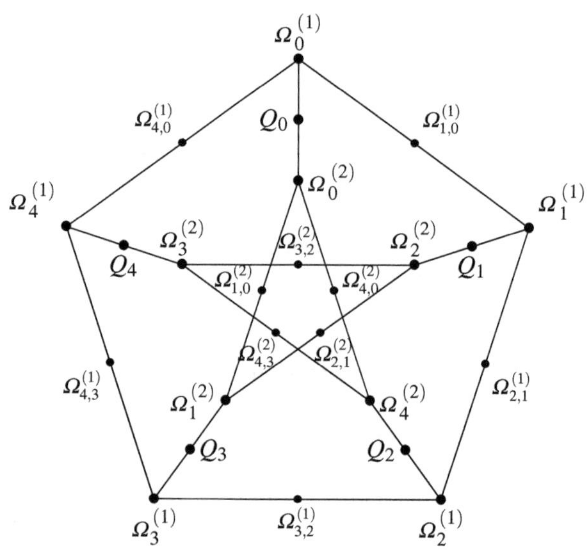

$$\begin{array}{lllll}
\pi_*\Omega_0^{(1)} = m_1 & \pi_*\Omega_{3,2}^{(1)} = 2G_4 & \pi_*\Omega_0^{(2)} = E_1 & \pi_*\Omega_{3,2}^{(2)} = 2G_2 & \pi_*Q_0 = 2F_1 \\
\pi_*\Omega_{1,0}^{(1)} = 2G_3 & \pi_*\Omega_3^{(1)} = E_4 & \pi_*\Omega_{1,0}^{(2)} = 2H_1 & \pi_*\Omega_3^{(2)} = E_2 & \pi_*Q_3 = 2F_4 \\
\pi_*\Omega_1^{(1)} = E_3 & \pi_*\Omega_{4,3}^{(1)} = 2H_4 & \pi_*\Omega_1^{(2)} = \ell_2 & \pi_*\Omega_{4,3}^{(2)} = 2H_2 & \pi_*Q_1 = 2F_3 \\
\pi_*\Omega_{2,1}^{(1)} = 2H_3 & \pi_*\Omega_4^{(1)} = m_2 & \pi_*\Omega_{1,0}^{(2)} = 2E_{T_1} & \pi_*\Omega_4^{(2)} = \ell_1 & \pi_*Q_4 = 2F_2 \\
\pi_*\Omega_2^{(1)} = m_3 & \pi_*\Omega_{4,0}^{(1)} = 2E_{T_2} & \pi_*\Omega_2^{(2)} = \ell_3 & \pi_*\Omega_{4,0}^{(2)} = 2G_1 & \pi_*Q_2 = 2E_5.
\end{array} \tag{8}$$

4.1.3 Weierstrass Equation of $S_{5,5}$

By the Weierstrass equation (3), the fibers of $\mathscr{E}_{R_{5,5}}$ of type I_5 are the fibers over $\mu = 0$ and $\mu = \infty$. So the base change branched on these fibers is given by $\mu \to \mu^2$, and the elliptic fibration on $S_{5,5}$ induced by the one on $R_{5,5}$ is

$$y^2 = x^3 + A(\mu)x + B(\mu), \quad \text{where} \tag{9}$$

$$A(\mu) := \frac{-1}{48}\mu^8 - \frac{1}{4}\mu^6 - \frac{7}{24}\mu^4 + \frac{1}{4}\mu^2 - \frac{1}{48}, \text{ and}$$

$$B(\mu) := -\frac{1}{864}\mu^{12} - \frac{1}{48}\mu^{10} - \frac{25}{288}\mu^8 - \frac{25}{288}\mu^4 + \frac{1}{48}\mu^2 - \frac{1}{864}.$$

The discriminant is $\frac{1}{16}\mu^{10}(-1 + 11\mu^2 + \mu^4)$, so there are, as expected, two fibers of type I_{10} over $\mu = 0$ and $\mu = \infty$. Moreover there are four fibers of type I_1 over $\mu = \pm\sqrt{-\frac{11}{2} \pm \frac{5}{2}\sqrt{5}}$.

4.1.4 Double Cover of $\mathbb{P}^2$

On the other hand, $\widetilde{R}$ and $R_{5,5}$ are blowups of $\mathbb{P}^2$, and the branch fibers of $\pi : S_{5,5} \dashrightarrow R_{5,5}$ correspond to the cubics $f_3 := x_1x_2(x_0 - x_1) = 0$ and $g_3 := x_0(x_0 - x_1 - x_2)(x_0 - x_2) = 0$. This exhibits $S_{5,5}$ as a double cover of $\mathbb{P}^2$ branched along the union of these two cubics. So we obtain a different equation for $S_{5,5}$, as a double cover of $\mathbb{P}^2$, i.e.,

$$w^2 = x_1x_2(x_0 - x_1)x_0(x_0 - x_1 - x_2)(x_0 - x_2). \tag{10}$$

We observe that $S_{5,5}$ is rigid (both in the moduli space of the elliptic K3 surfaces with prescribed reducible fibers and in the moduli space of the K3 surfaces with a non-symplectic involution with a prescribed fixed locus), since both $R_{5,5}$ and the choice of the branch fibers are.

4.2 The Branch Fibers Are $2I_0$: The K3 Surface $X_{5,5}$

Let us consider the K3 surface $X_{5,5}$ obtained by a base change of order 2 of $R_{5,5}$ whose branch locus corresponds to two fibers of type I_0. Let us assume it is very general among the K3 obtained in this way. This K3 surface lies in a two-dimensional family of K3 surfaces (see [5]), whose parameters depend on the choice of the two branch fibers (see (12)).

4.2.1 Geometric Description of $X_{5,5}$ and Its Néron–Severi Group

The surface $X_{5,5}$ admits a non-symplectic involution ι which is the cover involution of the double cover $X_{5,5} \to R_{5,5}$ and which fixes two elliptic curves.

The elliptic fibration $\mathcal{E}_{X_{5,5}} : X_{5,5} \to \mathbb{P}^1$ induced by $\mathcal{E}_{R_{5,5}} : R_{5,5} \to \mathbb{P}^1$ has four fibers of type I_5 and four fibers of type I_1. Moreover it has a 5-torsion section, induced by one of the sections of the elliptic fibration on $\mathcal{E}_{R_{5,5}}$. The Néron–Severi group and the transcendental lattice of this K3 surface are computed in [5], and a set of generators of the Néron–Severi group is given by the class of the fiber of the fibration, the zero section, one section of order 5, and the irreducible components of the reducible fibers of the fibration.

The curves $\Theta_i^{(j)}$ are not in the branch locus, and we denote by $\Omega_i^{(j,k)}$ for $k = 1, 2$ the two disjoint rational curves which are mapped to $\Theta_i^{(j)}$ by the quotient map $X_{5,5} \to R_{5,5}$. The curves P_i are not in the branch locus and we denote by Q_i their $2:1$ cover in $X_{5,5}$. The dual graph of this configuration is shown in Fig. 5. This diagram is also a tropical surface, similar to the one given in [10, Figure 1], as pointed out by B. Sturmfels.

4.2.2 Weierstrass Equation of $X_{5,5}$

Let us denote by μ_1 and μ_2 two arbitrary points of $\mathbb{P}^1_\mu$ such that the fibers of (3) over μ_1 and μ_2 are smooth. Let $X_{5,5}$ be the surface obtained from $R_{5,5}$ by a base change of order 2 branched in μ_1 and μ_2. We already observed that the surface $X_{5,5}$ lives in a two-dimensional family of K3 surfaces and its equation depends on the two parameters μ_1 and μ_2.

Let us consider the base change $\mathbb{P}^1_{(\alpha:\beta)} \to \mathbb{P}^1_{(\mu:\lambda)}$ branched over $(\mu : \lambda) = (\mu_1 : 1)$ and $(\mu_2 : 1)$, i.e., the base change given by

$$\mu = \mu_1\alpha^2 + \beta^2, \qquad \lambda = \alpha^2 + \beta^2/\mu_2. \tag{11}$$

It induces on $X_{5,5}$ the elliptic fibration:

$$y^2 = x^3 + A(\alpha : \beta)x + B(\alpha : \beta) \tag{12}$$

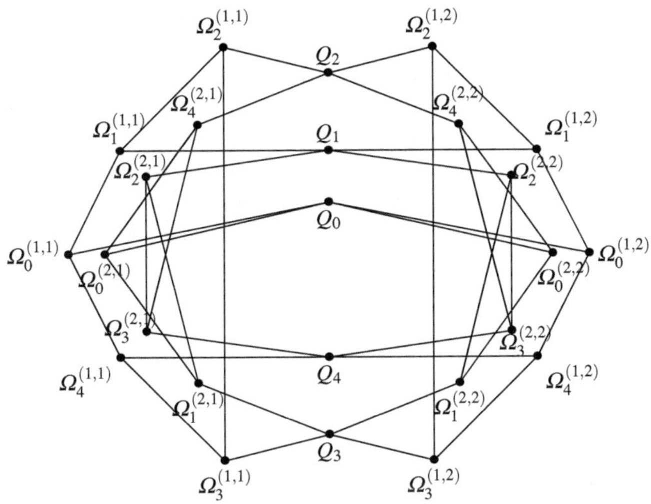

Fig. 5 Dual graph of relevant negative curves on $X_{5,5}$

whose discriminant is

$$\Big(\left(\alpha^2\mu_2+\beta^2\right)^5 \left(\mu_1\alpha^2+\beta^2\right)^5 \left(\alpha^4\mu_1^2\mu_2^2+11\alpha^4\mu_1\mu_2^2-\alpha^4\mu_2^2+11\alpha^2\mu_1\mu_2\beta^2+ \right.$$
$$\left. +2\mu_1\alpha^2\mu_2^2\beta^2+11\alpha^2\mu_2^2\beta^2-2\alpha^2\mu_2\beta^2+\mu_2^2\beta^4+11\beta^4\mu_2-\beta^4\right) \Big) / \left(16\mu_2^7\right) \tag{13}$$

For generic values of μ_1 and μ_2, the elliptic fibration has $4I_5+4I_1$ as singular fibers.

4.2.3 Double Cover of $\mathbb{P}^2$

On the other hand, $X_{5,5}$ is the double cover of $\mathbb{P}^2$ branched on the union of the two cubics $x_1x_2(x_0-x_1)+\mu_1x_0(x_0-x_1-x_2)(x_0-x_2)=0$ and $x_1x_2(x_0-x_1)+\mu_2x_0(x_0-x_1-x_2)(x_0-x_2)=0$. So $X_{5,5}$ can be described by the equation

$$w^2=\big(x_1x_2(x_0-x_1)+\mu_1x_0(x_0-x_1-x_2)(x_0-x_2)\big)$$
$$\big(x_1x_2(x_0-x_1)+\mu_2x_0(x_0-x_1-x_2)(x_0-x_2)\big). \tag{14}$$

4.3 *Branch Fibers I_5 and I_1*

If one uses as branch fibers a fiber of type I_5 and one of type I_1, one obtains a rigid K3 surface (in the moduli space of the elliptic K3 surfaces with prescribed reducible fibers), whose singular fibers are $I_{10}+2I_5+I_2+2I_1$, and theoretically one has four different admissible choices to do that. Indeed one can choose the fiber of type I_5 which is the branch fiber to be the fiber either over $\mu_1 = 0$ or over $\mu_1 = \infty$, and similarly one can choose the fiber of type I_1 which is the branch fiber to be the fiber either over $\mu_2 = (-11+5\sqrt{5})/2$ or over $\mu_2 = (-11-5\sqrt{5})/2$. In order to obtain the Weierstrass equation of these elliptic fibrations, it suffices to apply the base change (11) with the chosen values for μ_1 and μ_2.

We assume that $\mu_2 = (-11+5\sqrt{5})/2$, and we obtain the following two Weierstrass equations. If $\mu_1 = 0$, the base change (11) applied to $\mu_1 = 0$ and $\mu_2 = (-11+5\sqrt{5})/2$ gives an elliptic fibration:

$$y^2 = x^3 + A(\alpha:\beta)x + B(\alpha:\beta)$$

whose discriminant is

$$-\frac{1}{512}\left(-\alpha^2+5\beta^2\sqrt{5}\right)\left(-2\alpha^2-11\beta^2+5\beta^2\sqrt{5}\right)^5\alpha^2\beta^{10}.$$

In order to choose $\mu_1 = \infty$, one has to slightly change the equation of the base change (11), which now is $\mu = \alpha^2$ and $\lambda = \alpha^2/\mu_2 + \beta^2$, and one obtains

$$y^2 = x^3 + A(\alpha:\beta)x + B(\alpha:\beta)$$

whose discriminant is

$$\frac{1}{5120}\left(-375125+167761\sqrt{5}\right)\left(-25\alpha^2+\beta^2\sqrt{5}\right)\left(-2\alpha^2+11\beta^2+5\beta^2\sqrt{5}\right)^5\beta^2\alpha^{10}.$$

We observe that in the first case the K3 surface obtained is described as a double cover of $\mathbb{P}^2$ by the equation

$$w^2 = x_1x_2(x_0-x_1)\left(x_0x_1x_2(13+5\sqrt{5}) - x_1^2x_2(11+5\sqrt{5}) + 2x_0^3 - 4x_0^2x_2 - 2x_1x_0^2 + 2x_0x_2^2\right),$$

in the second by the equation

$$w^2 = x_0(x_0-x_1-x_2)(x_0-x_2)\left(x_0x_1x_2(13+5\sqrt{5}) - x_1^2x_2(11+5\sqrt{5}) + 2x_0^3 - 4x_0^2x_2 - 2x_1x_0^2 + 2x_0x_2^2\right).$$

4.4 Branch Fibers I_5 and I_0

If one chooses as branch fibers one fiber of type I_5 and one of type I_0, one obtains a one-dimensional family of K3 surfaces, whose singular fibers are $I_{10}+2I_5+4I_1$, and theoretically one has two different admissible ways to do that. Indeed one can choose that the fiber of type I_5 which is the branch fiber is the fiber either over $\mu_1 = 0$ or over $\mu_1 = \infty$, while μ_2 is the parameter of the family. In order to obtain the equations of these elliptic fibrations, one has to apply the base changes $(\mu = \beta^2, \lambda = \alpha^2 + \beta^2/\mu_2)$ or $(\mu = \alpha^2, \lambda = \alpha^2/\mu_2 + \beta^2)$ to Eq. (3), exactly as in the previous sections.

Similarly one can describe these K3 surfaces as a double cover of $\mathbb{P}^2$ substituting in (14) the appropriate values of μ_1 and μ_2.

4.5 Branch Fibers I_1 and I_0

If one chooses as branch fibers one fiber of type I_1 and one of type I_0, one obtains a one-dimensional family of K3 surfaces, whose singular fibers are $4I_5 + I_2 + 2I_1$, and theoretically one has two different admissible ways to do that. Indeed one can choose that the fiber of type I_1 which is the branch fiber is the fiber either over $\mu_1 = (-11 - 5\sqrt{5})/2$ or over $\mu_1 = (-11 + 5\sqrt{5})/2$, while μ_2 is the parameter of the family. In order to obtain the equations of these elliptic fibrations, one has to apply the base change (11) with the chosen μ_1 to Eq. (3), and to obtain an equation of this surface as a double cover of $\mathbb{P}^2$, one has to substitute the chosen μ_1 in (14).

4.6 Branch Fibers $2I_1$

If one chooses as branch fibers the two fibers of type I_1, one obtains a rigid K3 surface, whose reducible fibers are $4I_5 + 2I_2$. In order to obtain the equation of this elliptic fibration, one has to apply the base change (11) with the chosen $\mu_1 = (-11 - 5\sqrt{5})/2$ and $\mu_2 = (-11 + 5\sqrt{5})/2$ to Eq. (3). Similarly to obtain an equation of this surface as a double cover of $\mathbb{P}^2$, one has to substitute $\mu_1 = (-11 - 5\sqrt{5})/2$ and $\mu_2 = (-11 + 5\sqrt{5})/2$ in (14).

5 Elliptic Fibrations on K3 Surfaces Induced by the Conic Bundles

The aim of this section is to describe both geometrically and by the Weierstrass equations the elliptic fibrations induced by the conic bundles B_i (described in Sect. 3) on the K3 surfaces described in Sect. 4. We also provided a general method to find these Weierstrass equations, under some assumption on the conic bundles (see Sect. 5.2).

5.1 An Example

Let us consider the K3 surface $S_{5,5}$, whose equation as a double cover of $\mathbb{P}^2$ is given by (10). Let us consider also the conic bundle $|B_1|$ from Sect. 3. By Garbagnati and Salgado [4, Theorem 5.3], the conic bundle $|B_1|$ induces an elliptic fibration on $S_{5,5}$ with three reducible fibers of type I_2^*: one whose components are Q_0, $\Omega_{4,0}^{(1)}$, $\Omega_0^{(1)}$, $\Omega_{1,0}^{(1)}$, $\Omega_1^{(1)}$, Q_1, and $\Omega_{2,1}^{(1)}$; one whose components are Q_2, $\Omega_{4,0}^{(2)}$, $\Omega_4^{(2)}$, $\Omega_{4,3}^{(2)}$, $\Omega_3^{(2)}$, Q_4, and $\Omega_{3,2}^{(2)}$; and one whose components are $\Omega_{4,3}^{(1)}$, $\Omega_{3,2}^{(1)}$, $\Omega_3^{(1)}$, Q_3, $\Omega_1^{(2)}$, $\Omega_{2,1}^{(2)}$, and $\Omega_{1,0}^{(2)}$.

5.1.1 Equation of the Elliptic Fibration on $S_{5,5}$ Induced by $|B_1|$

Let us consider the conic bundle B_1 on $R_{5,5}$ associated to the pencil of lines $x_1 = \tau x_0 \subset \mathbb{P}^2$. It induces an elliptic fibration on $S_{5,5}$. To find the equation of this elliptic fibration, we use the equation of $S_{5,5}$ as double cover of $\mathbb{P}^2$, i.e., Eq. (10), and we substitute in $x_1 = \tau x_0$ in (10).

This gives

$$w^2 = (\tau x_0)x_2(x_0 - \tau x_0)x_0(x_0 - \tau x_0 - x_2)(x_0 - x_2).$$

We put $x_2 = 1$ and we obtain

$$w^2 = \tau(1-\tau)x_0^3(x_0 - \tau x_0 - 1)(x_0 - 1).$$

Let us consider the change of coordinates $w \mapsto wx_0$ and divide both the members by x_0^2. We obtain

$$w^2 = \tau(1-\tau)x_0(x_0 - \tau x_0 - 1)(x_0 - 1).$$

This is the equation of an elliptic fibration over $\mathbb{P}^1_\tau$.

Moreover one can explicitly compute the Weierstrass form: first one uses the change of coordinates $w \mapsto \tau^2(1-\tau)w$ and $x_0 \mapsto \tau x_0$ obtaining

$$w^2\tau^4(1-\tau)^2 = \tau^2(1-\tau)x_0(\tau x_0(1-\tau) - 1)(\tau x_0 - x_2)$$

and so

$$w^2 = x_0\left(x_0 - \frac{1}{\tau(1-\tau)}\right)\left(x_0 - \frac{1}{\tau}\right)$$

Second, one considers the change of coordinates given by $w \mapsto w/\tau^3(1-\tau)^3$ and $x_0 \mapsto x_0/\tau^2(1-\tau)^2$ and multiplies all the equation by $\tau^6(1-\tau)^6$. So one obtains

$$w^2 = x_0(x_0 - \tau(1-\tau))(x_0 - \tau(1-\tau)^2). \tag{15}$$

The discriminant is $\tau^8(1-\tau)^8$, and so, by Tate's algorithm, there are three fibers of type I_2^* over $\tau = 0$, $\tau = 1$, $\tau = \infty$.

5.2 An Algorithm to Compute Weierstrass Equations

The aim of this section is to formalize systematically what we did above.

Setup Let V be a K3 surface obtained by a base change of order 2 from a rational elliptic surface R. Therefore, V can be described as double cover of $\mathbb{P}^2$ branched on the union of two (possibly reducible) plane cubics from the pencil determining R. It has an equation of the form

$$w^2 = f_3(x_0 : x_1 : x_2)g_3(x_0 : x_1 : x_2). \tag{16}$$

Let B be a conic bundle on R, e.g., a basepoint-free linear system of rational curves giving $R \to \mathbb{P}^1_\tau$. Pushing forward to $\mathbb{P}^2$, B is given by a pencil of plane rational curves with equation $h(x_0 : x_1 : x_2, \tau)$. The polynomial $h(x_0 : x_1 : x_2, \tau)$ is homogeneous in x_0, x_1, x_2, say of degree $e \geq 1$ and linear in τ.

As the anticanonical series on R coincides with the elliptic fibration, the adjunction formula implies that every curve with equation $h(x_0 : x_1 : x_2, \tau)$ meets both of the branch curves (the proper transforms on R of) $f_3 = 0$ and $g_3 = 0$ in two additional points. It therefore meets (the proper transform of) their union $f_3g_3 = 0$ in four points. (Note that there may be additional points of intersection on $\mathbb{P}^2$ which are separated in the blowup R.) Therefore the preimage in V is the double cover of a rational curve branched over four points, e.g., the standard presentation of an elliptic curve. For general τ, we must find an isomorphism of the curve $h(x_0 : x_1 : x_2, \tau) = 0$ with $\mathbb{P}^1$ and extract the images of the four intersection points with $f_3g_3 = 0$.

When $e \leq 3$, an isomorphism with $\mathbb{P}^1$ is provided by projection from a point of order $e - 1$ on the curve (e.g., any point in $\mathbb{P}^2$ if $e = 1$, a point on the conic if $e = 2$, and a double point of the cubic if $e = 3$). Such a point necessarily exists (in the case $e = 3$ the singularity must be a basepoint of the pencil) and is also necessarily a basepoint of the original pencil of cubics giving $\mathscr{E}_R$. Up to acting by $\mathrm{PGL}_3(\mathbb{C})$, we may assume that this point is $(0 : 1 : 0)$.

Algorithm When $e \leq 3$

1. Compute the resultant of the polynomials $f_3(x_0 : x_1 : x_2)g_3(x_0 : x_1 : x_2)$ and $h(x_0 : x_1 : x_2, \tau)$ with respect to the variable x_1. The result is a polynomial $r(x_0 : x_2, \tau)$ which is homogeneous in x_0 and x_2, corresponding to the images of *all* of the intersection points $\{f_3g_3 = 0\} \cap \{h_\tau = 0\}$ after projection from $(0 : 1 : 0)$.
2. Since B is a conic bundle, $r(x_0 : x_2, \tau)$ will be of the form

$$a(x_0 : x_2, \tau)^2 b(x_0 : x_2, \tau)c(\tau),$$

where a and b are homogeneous in x_0 and x_2, the degree of a depends upon e, and the degree of b in x_0 and x_2 is 4.

3. The equation of V is now given by $w^2 = r(x_0 : x_2, \tau)$, which is birationally equivalent to
$$w^2 = c(\tau)b(x_0 : x_2, \tau), \tag{17}$$
by the change of coordinates $w \mapsto wa(x_0 : x_2, \tau)$. Since for almost every τ, Eq. (17) is the equation of a $2 : 1$ cover of $\mathbb{P}^1_{(x_0:x_2)}$ branched in four points, (17) is the equation of the genus 1 fibration on the K3 surface V induced by the conic bundle B.
4. If there is a section of fibration (17), then it is possible to obtain the Weierstrass form by standard transformations.

Remark 2 The algorithm can be applied exactly in the same way to the generalized conic bundles, and not only to the conic bundles.

When $e \geq 4$, projection from a point may suffice, for example, if all curves have a basepoint of degree $e - 1$. However there are several conic bundles whose general member cannot be parametrized by lines.

We now consider a parametrization which can be done by conics. Let $\mathscr{P}$ be a pencil of rational curves of degree e passing through the (possibly infinitely near) basepoints $T_1, \dots T_r$ with certain multiplicities. Let $\mathscr{C}$ be a pencil of conics whose basepoints are among $T_1, \dots T_r$ and such that $2e - 1$ intersection points between a generic curve in $\mathscr{P}$ and a generic curve in $\mathscr{C}$ are in $\{T_1, \dots T_r\}$. So there is exactly one extra intersection between a generic curve in $\mathscr{P}$ and a generic curve in $\mathscr{C}$. This allows to parametrize the curves in $\mathscr{P}$ by the curves in $\mathscr{C}$. If moreover the basepoints of $\mathscr{C}$ are three distinct points and one infinitely near point, we will say that the condition ($\star$) is satisfied. So ($\star$) is satisfied if the curves in $\mathscr{P}$ can be parametrized by a pencil of conics passing through four points, exactly two of which are infinitely near.

If the condition ($\star$) is satisfied, up to changing coordinates by some matrix $M \in \mathrm{PGL}_3(\mathbb{C})$, we may assume that the basepoints of $\mathscr{C}$ are $p_1 = (0 : 0 : 1)$, $p_2 = (0 : 1 : 0)$, $p_3 = (1 : 0 : 0)$ and the infinitely near point p_1' corresponds to the line $x_0 = x_1$. The pencil of degree 2 maps $\mathbb{P}^1_z \to \mathbb{P}^2$ sending
$$1 \mapsto p_1, \qquad \infty \mapsto p_2, \qquad 0 \mapsto p_3$$
with derivative at $z = 1$ in the direction of the line $x_0 = x_1$ is given by
$$(x_0 : x_1 : x_2) = (z - 1 : z(z - 1) : p \cdot z), \qquad p \in \mathbb{P}^1.$$

In the following we are interested in pencils of quartics which satisfy the condition ($\star$), so we study the effect of this condition on quartic curves. We say that a pencil of quartics satisfies (†) if, up to a change of coordinates, it is of one of the following types: (1) each quartic is double at p_2 and has a tacnode at p_1 with principal tangent specified by p_1'; (2) each quartic is double at p_2 and p_3 and has a cusp at p_1 with principal tangent specified by p_1'.

We recall that, by the construction of the conic bundles, all the basepoints of $\mathscr{P}$ (and thus also of $\mathscr{C}$) are also singular points for the sextic $f_3 g_3 = 0$.

Algorithm Assuming (†)

1. Factor $h(z-1 : z(z-1) : pz, \tau) = c(\tau)z^a(z-1)^b r(z, p, \tau)$ where $r(z, p, \tau)$ is linear in z and $a+b$ is 5 or 6 depending on the multiplicity of h at p_2. The solution z_0 of $r(z, p, \tau) = 0$ gives the rational parameterization
$$h(z_0 - 1 : z_0(z_0 - 1) : pz_0, \tau) = 0$$
with parameter $p \in \mathbb{P}^1$.
2. A birational equation of the K3 surface is given by
$$w^2 = (f_3 g_3)(z_0 - 1 : z_0(z_0 - 1) : pz_0).$$
3. If there is a section of fibration (17), then it is possible to obtain the Weierstrass form by standard transformations.

This algorithm can be generalized to pencil of curves satisfying ($\star$).

5.3 *The Elliptic Fibrations Induced by Conic Bundles*

Here we can describe and compute the equations of all the elliptic fibrations induced by the conic bundles on the different K3 surfaces introduced. For this purpose, we apply the algorithm, described in the previous section, to the equations of the conic bundles given in Sect. 3 and to the Weierstrass equations given in Sect. 4.

5.3.1 The K3 Surface $S_{5,5}$

The conic bundle $|B_2|$ induces an elliptic fibration on $S_{5,5}$ with two reducible fibers of type I_4^*: one whose components are Q_0, $\Omega_{0,4}^{(1)}$, $\Omega_0^{(1)}$, $\Omega_{1,0}^{(1)}$, $\Omega_1^{(1)}$, $\Omega_{2,1}^{(1)}$, $\Omega_2^{(1)}$, Q_2, and $\Omega_{3,2}^{(1)}$ and one whose components are Q_3, $\Omega_{1,0}^{(2)}$, $\Omega_1^{(2)}$, $\Omega_{2,1}^{(2)}$, $\Omega_2^{(2)}$, $\Omega_{3,2}^{(2)}$, $\Omega_3^{(2)}$, Q_4, and $\Omega_{4,3}^{(2)}$.

The Weierstrass equation is computed applying the algorithm to
$$f_3 g_3 = x_0 x_1 x_2 (x_0 - x_1)(x_0 - x_2)(x_0 - x_1 - x_2),$$
by (10), and $h(x_0 : x_1 : x_2, \tau) = x_1 x_2 - \tau(x_0 x_2 - x_0^2)$. One obtains the Weierstrass equation:
$$y^2 = x^3 - \frac{\tau^2(-\tau^2 + \tau^4 + 1)}{3}x - (1/27)\frac{\tau^3(-2+\tau^2)(-1+2\tau^2)(\tau^2+1)}{27}. \tag{18}$$

So the discriminant $\Delta(\tau)$ is $-\tau^{10}(-1+\tau)^2(\tau+1)^2$. Hence, by Tate's algorithm, this fibration has two fibers of type I_4^* over $\tau = 0, \infty$ and two fibers of type I_2 over $\tau = \pm 1$.

The conic bundle $|B_3|$ induces an elliptic fibration on $S_{5,5}$ with two reducible fibers: one of type I_6^* whose components are Q_0, $\Omega_{0,4}^{(1)}$, $\Omega_0^{(1)}$, $\Omega_{1,0}^{(1)}$, $\Omega_1^{(1)}$, $\Omega_{2,1}^{(1)}$,

$\Omega_2^{(1)}$, $\Omega_{3,2}^{(1)}$, $\Omega_3^{(1)}$, Q_3, and $\Omega_{4,3}^{(1)}$ and one of type III^* whose components are Q_4, $\Omega_3^{(2)}$, $\Omega_{3,2}^{(2)}$, $\Omega_{4,3}^{(2)}$, $\Omega_2^{(2)}$, $\Omega_4^{(2)}$, $\Omega_{2,1}^{(2)}$, and $\Omega_{0,4}^{(2)}$.

The Weierstrass equation is

$$y^2 = x^3 - \frac{\tau^3(\tau^3 + 6\tau^2 + 9\tau + 3)}{3}x - \frac{\tau^5(\tau + 3)(2\tau^3 + 12\tau^2 + 18\tau + 9)}{27}. \tag{19}$$

So the discriminant $\Delta(\tau)$ is $-\tau^9(\tau + 4)(\tau + 1)^2$. Hence, by Tate's algorithm, this fibration has one fiber of type I_6^* over $\tau = \infty$, one fiber of type III^* over $\tau = 0$, one fiber of type I_2 over $\tau = -1$, and one fiber of type I_1 over $\tau = -4$.

5.3.2 The K3 Surface $X_{5,5}$

Here we consider the elliptic fibrations induced by B_i on $X_{5,5}$. We recall that in this case one has to apply the algorithm to (14). We summarize the results in the following table, where r denotes the rank of the Mordell–Weil group of the fibration.

	Δ	singular fibers	r
B_1	$-\tau^6(\tau-1)^6(\mu_1-\mu_2)^4$ $(\tau^3-2\tau^2-2\tau^2\mu_2+6\tau\mu_2+\tau+\tau\mu_2^2-4\mu_2)$ $(\tau^3-2\tau^2-2\tau^2\mu_1+6\tau\mu_1+\tau+\tau\mu_1^2-4\mu_1)$	$1I_0^*$ $\tau=0$ $2I_6$ $\tau=\infty, 1$ $6I_1$	2
B_2	$-\tau^8(-\tau^3+2\tau^2+\tau^2\mu_2+2\tau\mu_2-\tau+\mu_2)(-\tau+\mu_2)$ $(-\tau^3+2\tau^2+\tau^2\mu_1+2\tau\mu_1-\tau+\mu_1)(-\tau+\mu_1)(\mu_1-\mu_2)^4$	$2I_8$ $\tau=0,\infty$ $8I_1$	2
B_3	$-\tau^8(\tau+2\tau^2+\tau^3-6\mu_2\tau-2\mu_2\tau^2+4\mu_2^2+\tau\mu_2^2)$ $(\tau^3-6\mu_1\tau+2\tau^2+\tau+\mu_1^2\tau+4\mu_1^2-2\mu_1\tau^2)(\mu_1-\mu_2)^4$	I_{10} $\tau=\infty$ I_2^* $\tau=0$ $6I_1$	1

(20)

5.3.3 Other K3 Surfaces

As we saw above, all the equations for K3 surfaces that are double covers of $R_{5,5}$ branched over some special fibers can be obtained from the general equation for $X_{5,5}$ by substituting particular values of μ_1 and μ_2.

In particular in order to find the Weierstrass equations of the elliptic fibrations induced by the conic bundles B_i on a specific K3 surface, it suffices to substitute in (5.3.2) the appropriate values of μ_1 and μ_2. As an example here, we construct a table analogous to (5.3.2) if the K3 surface is obtained by $R_{5,5}$ branching along one fiber of type I_5 and one smooth fiber. We already noticed that one has two different choices for the I_5 branch fiber. Once one chooses the branch fiber I_5, the construction is not symmetric in I_5. For $i = 1, 2$ the reducible fibers of the conic bundle $|B_i|$ are symmetric up to switching the I_5-fibers, so the conic bundle $|B_i|$

is associated to elliptic fibrations with the same property choosing differently the I_5 branch fiber. For the conic bundles $|B_1|$ and $|B_2|$, we will choose the I_5 branch fiber to be the one over $\mu_1 = 0$. The conic bundle $|B_3|$ has a unique reducible fiber, supported over one specific I_5, so the elliptic fibrations induced by this conic bundle have not necessarily the same properties if one changes the I_5 branch fibers. So we give the equations of the elliptic fibrations if one chose both $\mu_1 = 0$ and $\mu_1 = \infty$.

For all the conic bundles, μ_2 is the parameter of the one-dimensional family of K3 surfaces we are considering. So we obtain the following (where $r = \mathrm{rank}(MW)$, and both the reducible fibers and r are given for generic choice of μ_2):

	Δ	singular fibers	r
B_1	$-\tau^7\mu_2^4(-1+\tau)^8(\tau^3-2\tau^2-2\tau^2\mu_2$ $+\tau+6\tau\mu_2+\tau\mu_2^2-4\mu_2)$	$I_1^*\ \tau=0,\ I_2^*\ \tau=1$ $I_6\ \tau=\infty,\ 3I_1$	1
B_2	$-\tau^{10}\mu_2^4(-1+\tau)^2(\mu_2+\tau^2\mu_2+2\tau\mu_2$ $-\tau^3-\tau+2\tau^2)(\mu_2-\tau)$	$I_4^*\ \tau=0,\ I_2\ \tau=1$ $I_8,\ \tau=\infty,\ 4I_1$	1
B_3, $\mu_1=0$	$-\mu_2^4\tau^9(\tau+1)^2(\tau+2\tau^2+\tau^3-6\mu_2\tau$ $-2\mu_2\tau^2+4\mu_2^2+\tau\mu_2^2)$	$III^*\ \tau=\infty,\ I_2\ \tau=-1$ $I_{10}\ \tau=0,\ 3I_1$	1
B_3, $\mu_1=\infty$	$-\tau^8(4+\tau)(\tau+2\tau^2+\tau^3-6\mu_2\tau$ $-2\mu_2\tau^2+4\mu_2^2+\tau\mu_2^2)$	$I_2^*\ \ \tau=0$ $I_6^*\ \tau=\infty,\ 4I_1$	1

(21)

6 The K3 Surface $S_{5,5}$ and Its Elliptic Fibrations

The aim of this section is to prove the following results, computing the equations of all the elliptic fibrations on $S_{5,5}$.

Proposition 2 *The K3 surface $S_{5,5}$ admits 13 different elliptic fibrations. One of them is induced by $\mathcal{E}_R$, three by conic bundles, three by splitting genus one pencils, and six by generalized conic bundles. The equations of these elliptic fibrations are given in* (9) *(the one induced by $\mathcal{E}_R$); in* (15), (18), *and* (19) *(the ones induced by the conic bunldes); in* (6.2.1) *(the ones induced by splitting genus 1 pencil); and in* (6.2.2), (30) *(the ones induced by generalized conic bundles).*

To prove this, we deeply analyze the elliptic fibrations on $S_{5,5}$ induced by splitting genus 1 pencils and generalized conic bundles, in particular giving an algorithm to find the Weierstrass equation of any elliptic fibration induced by a splitting genus 1 pencil.

The K3 surface $S_{5,5}$ can be also described as the (unique!) K3 surface which admits a non-symplectic involution fixing ten rational curves. Indeed, by our construction it is clear that ι fixes ten rational curves (the inverse image of the components of the two I_5 fibers). The fact that this K3 surface is unique is classically known, and due to Nikulin, see [8]. The transcendental lattice of this K3 surface is $T_S \simeq \langle 2\rangle \oplus \langle 2\rangle$. The elliptic fibrations on this K3 surface are classified by Nishiyama (see [9, Table 1.2]), who used a lattice-theoretic method that we will apply later to a different K3 surface, in Sect. 7. The complete list of the elliptic fibrations is the following:

n^o	singular fibers	MW
1	$2II^* + 2I_2$	$\{1\}$
2	$II^* + I_6^* + 2I_1$	$\{1\}$
3	$I_{12}^* + 2I_2 + 2I_1$	$\mathbb{Z}/2\mathbb{Z}$
4	$2III^* + I_0^*$	$\mathbb{Z}/2\mathbb{Z}$
5	$III^* + I_6^* + I_2 + I_1$	$\mathbb{Z}/2\mathbb{Z}$
6	$I_{18} + I_2 + 4I_1$	$\mathbb{Z}/3\mathbb{Z}$

n^o	singular fibers	MW
7	$I_{14}^* + 4I_1$	$\{1\}$
8	$I_8^* + I_2^* + 2I_1$	$\mathbb{Z}/2\mathbb{Z}$
9	$2I_4^* + 2I_2$	$(\mathbb{Z}/2\mathbb{Z})^2$
10	$I_{16} + I_4 + 4I_1$	$\mathbb{Z}/4\mathbb{Z}$
11	$IV^* + I_{12} + 4I_1$	$\mathbb{Z} \times \mathbb{Z}/3\mathbb{Z}$
12	$3I_2^*$	$(\mathbb{Z}/2\mathbb{Z})^2$
13	$2I_{10} + 4I_1$	$\mathbb{Z}/5\mathbb{Z}$

(22)

Since the involution ι acts as the identity on the Néron–Severi group of $S_{5,5}$, there are no fibrations of type 3 on this K3 surface.

The fibration 13 in Table (22) is the one induced by the fibration $\mathscr{E}_{R_{5,5}}$, and it has Eq. (9). By Sect. 5, the fibrations 5, 9, and 12 are induced by conic bundles, and their equations are (19), (18), and (15), respectively.

The other fibrations are induced either by generalized conic bundles or by splitting genus 1 pencils.

An elliptic fibration induced by a splitting genus 1 pencil corresponds to a fibration of genus 1 curves on a non-relatively minimal rational elliptic surface (i.e., this fibration admits (-1)-curves as components of some fibers). The original relatively minimal rational elliptic surface $R_{5,5}$ can be recovered from this non-minimal surface by blowing down some divisors. A different choice of divisors to blow-down, namely, the (-1)-curves which are components of the fibers of the splitting genus 1 pencil, gives us another rational elliptic surface on which the splitting genus 1 pencil above corresponds to a relatively minimal elliptic fibration. Hence each fibration given by a splitting genus 1 pencil is indeed induced by a rational elliptic surface (different from $R_{5,5}$) by a base change of order 2 whose branch locus consists of ten curves. Considering the list of elliptic fibrations on $S_{5,5}$ given in Table (22), one observes immediately that the fibrations 6, 10, and 11 could be of this type, i.e., they could be induced by splitting genus 1 pencils (this is in fact proved in [4]). Indeed they present some fibers which appear in pairs (each pair is good candidate to be the inverse image of a unique fiber on the rational elliptic surface) and at most two other fibers, which do not appear an even number of time, but which are either of type I_{2n} or of type IV^*. These fibers are the one obtained by base change of order 2 branched over I_n or IV-fibers, so they could be the ramification fibers of the base change. The other fibrations in Table (22) have not the same properties and, thus, cannot be induced by rational elliptic surface by a base change of order 2. We already observed that the fibration 13 is induced by the elliptic fibration on $R_{5,5}$ and that the fibrations 5, 9, and 12 are induced by conic bundles, so the fibrations 1, 2, 3, 4, 7, and 8 are induced by generalized conic bundles.

6.1 Splitting Genus 1 Fibrations

6.1.1 An Example, the Fibration 6

We give an example of splitting genus 1 pencil of curves: we are looking for a fiber of type I_{18}, and it is given by

$$W_X := Q_0+\Omega_0^{(2)}+\Omega_{4,0}^{(2)}+\Omega_4^{(2)}+\Omega_{4,3}^{(2)}+\Omega_3^{(2)}+\Omega_{3,2}^{(2)}+\Omega_2^{(2)}+\Omega_{2,1}^{(2)}+\Omega_1^{(2)}+ \\ +Q_3+\Omega_3^{(1)}+\Omega_{3,2}^{(1)}+\Omega_2^{(1)}+\Omega_{2,1}^{(1)}+\Omega_1^{(1)}+\Omega_{1,0}^{(1)}+\Omega_0^{(1)}.$$

Using (8), the class $W_R := \pi_* W_X$ is

$$\begin{aligned} &2F_1+E_1+2G_1+\ell_1+2H_2+E_2+2G_2+\ell_3+2E_{T_1}+\ell_2+2F_4+E_4+2G_4+m_3+2H_3 \\ &+E_3+2G_3+m_1=5h-2E_1-2F_1-2G_1-4H_1-E_2-2F_2-G_2-H_2-2E_3-4F_3 \\ &\qquad -2G_3-2H_3-E_4-F_4-G_4-2H_4-2E_{Q_5}-E_{T_2}. \end{aligned} \tag{23}$$

Since we know that the curves in the linear system described split in the double cover, we can assume that the class W_R is both the push-down and the geometric image of a fiber of our fibration (i.e., $\pi_*(W_X)=\pi(W_X)=W_R$).

This class is the strict transform on $\widetilde{R}$ of quintics in $\mathbb{P}^2$ with the following properties: they have a tacnode in Q_1 with principal tangent ℓ_2; they have a tacnode in Q_3 with principal tangent ℓ_3; they have a node in Q_5; the tangent in Q_2 is m_2; the tangent in Q_4 is m_2; they pass through T_2.

This gives the following families of quintics

$$\begin{aligned} &-bx_0^5+bx_0^4x_1+bx_0^4x_2-ex_0^3x_1x_2+ex_0^2x_1^2x_3+ex_0^2x_1x_2^2+ \\ &\qquad (-3b-e)x_0x_1^2x_2^2+bx_1^3x_2^2+bx_1^2x_2^3=0. \end{aligned} \tag{24}$$

This is the equation of the splitting genus 1 pencil that we are looking for. It indeed corresponds to a pencil of curves of genus 1 parametrized by $(b:e)$.

We now have to intersect the branch sextic given in Eq. (10) with (24). The resultant of the polynomials (10) and (24) with respect to the variable x_0 is

$$-x_1^{12}x_2^{12}b^3(x_1-x_2)^4(x_2+x_1)^2(e+2b).$$

We observe that all the solutions in (x_1,x_2) have even multiplicities, as is necessarily if the double cover splits into two curves, isomorphic to the base curve (see [4, Section 3]).

More precisely, the curve splits after the base change of order two branched in $(b:e)=(0:1)$ and $(b:e)=(1:-2)$, cf. Lemma 1 below. In order to write down explicitly the Weierstrass form of the elliptic fibration on $X_{5,5}$, one first finds the rational elliptic fibration given by the splitting pencil of genus 1 curves (24), and then one performs the base change of order two.

This can be computed by any computer algebra system, and it is

$$y^2 = x^3 + A(b:e)x + B(b:e),$$

where

$$A(b:e) := \frac{23}{48}b^4 - \frac{5}{12}b^3 e - \frac{1}{8}b^2 e^2 + \frac{1}{12}be^3 - \frac{1}{48}e^4$$

and

$$B(b:e) := -\frac{181}{864}b^6 - \frac{17}{144}b^5 e + \frac{31}{288}b^4 e^2 - \frac{1}{54}b^3 e^3 - \frac{5}{288}b^2 e^4 + \frac{1}{144}be^5 - \frac{1}{864}e^6.$$

The discriminant of this rational elliptic fibration is

$$\frac{1}{16}b^9(e + 2b)(e^2 - 5be + 13b^2)$$

and the fibration has one fiber of type I_9 and three other singular fibers, all of type I_1.

Now we consider the base change of order 2 branched in $(b : e) = (0 : 1)$ and $(b : e) = (1 : -2)$. It can be directly written as a map $\mathbb{P}^1_{(\beta:\epsilon)} \to \mathbb{P}^1_{(b:e)}$, where $b = \beta^2$ and $e = -\beta^2 + 2\beta\epsilon + \epsilon^2$.

So we obtain a new elliptic fibration on $S_{5,5}$ whose equation is

$$y^2 = x^3 + A'(\beta:\epsilon)x + B'(\beta:\epsilon), \tag{25}$$

$$A'(\beta:\epsilon) := \frac{23}{48}\beta^8 - \frac{5}{12}\beta^6(-\beta^2 + 2\beta\epsilon + \epsilon^2) - \frac{1}{8}\beta^4(-\beta^2 + 2\beta\epsilon + \epsilon^2)^2 +$$

$$+\frac{1}{12}\beta^2(-\beta^2 + 2\beta\epsilon + \epsilon^2)^3 - (1/48)(-\beta^2 + 2\beta\epsilon + \epsilon^2)^4$$

and

$$\begin{aligned} B'(\beta:\epsilon) := &-(1/108)\beta^{12} - (5/9)\beta^{11}\epsilon - (7/18)\beta^{10}\epsilon^2 + (28/27)\beta^9\epsilon^3 + (7/12)\beta^8\epsilon^4 \\ &-(11/12)\beta^7\epsilon^5 + -(35/72)\beta^6\epsilon^6 + (1/3)\beta^5\epsilon^7 + (5/24)\beta^4\epsilon^8 - (5/108)\beta^3\epsilon^9 \\ &-(1/18)\beta^2\epsilon^{10} - (1/72)\epsilon^{11}\beta - (1/864)\epsilon^{12}. \end{aligned}$$

The discriminant is

$$(1/16)\beta^{18}(19\beta^4 - 14\beta^3\epsilon - 3\beta^2\epsilon^2 + 4\beta\epsilon^3 + \epsilon^4)(\epsilon + \beta)^2.$$

This fibration has a fiber of type I_{18}, as expected, one fiber of type I_2 (in $(\beta : \epsilon) = (1 : -1)$), and four fibers of type I_1. By construction this elliptic fibration exhibits $S_{5,5}$ as the double cover of a rational elliptic surface with one fiber of type I_9 and three fibers of type I_1.

-down of the class M_2 is

$$-E_1 - F_1 - G_1 - 2H_1 - E_2 - F_2 - 2G_2 - H_2 - E_3 - 2F_3$$
$$-G_3 - H_3 - E_4 - 2F_4 - G_4 - H_4 - 2E_{T_2} - 2E_5$$

sponds to a pencil of quartics with bitangent lines l_2 (in Q_2 and Q_4) $_1$ and Q_3) and having two nodes in E_{T_2} and E_5. The equation of this

$$x_0^3x_1 + bx_0^2x_1^2 - 2bx_0^3x_2 + (3b+4e)x_0^2x_1x_2 + (-b-4e)x_0x_1^2x_2 +$$
$$ex_1^3x_2 + bx_0^2x_2^2 + (-b-4e)x_0x_1x_2^2 + 2ex_1^2x_2^2 + ex_1x_2^3 = 0. \quad (27)$$

e algorithm, we find the following Weierstrass equations for elliptic n $S_{5,5}$ induced by splitting genus 1 pencils.

Elliptic fibrations
$\frac{3}{8}\beta^8 - \frac{5}{12}\beta^6(-\beta^2+2\beta\epsilon+\epsilon^2) - \frac{1}{8}\beta^4(-\beta^2+2\beta\epsilon+\epsilon^2)^2 +$ $\frac{1}{12}\beta^2(-\beta^2+2\beta\epsilon+\epsilon^2)^3 - (1/48)(-\beta^2+2\beta\epsilon+\epsilon^2)^4$ $(1/108)\beta^{12} - (5/9)\beta^{11}\epsilon - (7/18)\beta^{10}\epsilon^2 + (28/27)\beta^9\epsilon^3 + (7/12)\beta^8\epsilon^4$ $(11/12)\beta^7\epsilon^5 + -(35/72)\beta^6\epsilon^6 + (1/3)\beta^5\epsilon^7 + (5/24)\beta^4\epsilon^8 - (5/108)\beta^3\epsilon^9$ $(1/18)\beta^2\epsilon^{10} - (1/72)\epsilon^{11}\beta - (1/864)\epsilon^{12}$ $1/16)\beta^{18}(19\beta^4 - 14\beta^3\epsilon - 3\beta^2\epsilon^2 + 4\beta\epsilon^3 + \epsilon^4)(\epsilon+\beta)^2$
$A = -(24\epsilon^4+\beta^4)\beta^4/48,\ B = -(216\epsilon^8 + 36\beta^4\epsilon^4 + \beta^8)\beta^4/864$ $\Delta = \epsilon^{12}(27\epsilon^4+\beta^4)\beta^8/16$
$-\beta^8 + 16\epsilon^4\beta^4 - 16\epsilon^8)/48,\ B = -(\beta^4 - 8\epsilon^4)(-8\epsilon^8 - 16\epsilon^4\beta^4 + \beta^8)/864$ $^{16}\beta^4(2\epsilon-\beta)(2\epsilon+\beta)(4\epsilon^2+\beta^2)/16$

(28)

iptic Fibrations Induced by Generalized Conic Bundles

sider the divisors:

$$= 2\Omega_0^{(1)} + 4Q_0 + 6\Omega_0^{(2)} + 5\Omega_{1,0}^{(2)} + 4\Omega_1^{(2)} + 3\Omega_{2,1}^{(2)} + 2\Omega_2^{(2)} + \Omega_{3,2}^{(2)} + 3\Omega_{4,0}^{(2)},$$
$$= 2\Omega_4^{(2)} + 4\Omega_{4,0}^{(2)} + 6\Omega_0^{(2)} + 5\Omega_{1,0}^{(2)} + 4\Omega_1^{(2)} + 3\Omega_{2,1}^{(2)} + 2\Omega_2^{(2)} + \Omega_{3,2}^{(2)} + 3Q_0,$$
$$= Q_0 + \Omega_{4,0}^{(1)} + Q_4 + \Omega_{4,3}^2 + 2\left(\Omega_0^{(1)} + \Omega_{1,0}^{(1)} + \Omega_1^{(1)} + \Omega_{2,1}^{(1)} + \right.$$
$$\left. + \ \Omega_2^{(1)} + \Omega_{3,2}^{(1)} + \Omega_3^{(1)} + Q_3 + \Omega_1^{(2)} + \Omega_{2,1}^{(2)} + \Omega_2^{(2)} + \Omega_{3,2}^{(2)} + \Omega_3^{(2)}\right),$$
$$= \Omega_{1,0}^{(2)} + 2\Omega_0^{(2)} + 3Q_0 + 4\Omega_0^{(1)} + 3\Omega_{1,0}^{(1)} + 2\Omega_1^{(1)} + \Omega_{2,1}^{(1)} + 2\Omega_{4,0}^{(1)},$$
$$= \Omega_{1,0}^{(1)} + Q_0 + \Omega_{4,0}^{(2)} + \Omega_{4,3}^{(2)} + 2\left(\Omega_0^{(1)} + \Omega_{4,0}^{(1)} + \Omega_4^{(1)} + \Omega_{4,3}^{(1)} + \Omega_3^{(1)}\right.$$
$$\left. + \ \Omega_{3,2}^{(1)} + \Omega_2^{(1)} + Q_2 + \Omega_4^{(2)}\right).$$

6.1.2 Splitting Genus 1 Fibration: An Algorithm

The aim of this section is to generalize the previous construction to other splitting genus 1 pencils.

Setup Let $\pi : V \to \mathbb{P}^1$ be a K3 surface which is a double cover of a (not necessarily minimal) rational elliptic surface $\tilde{R}$ branched over two fibers. If $\mathscr{H} : \tilde{R} \to \mathbb{P}^1_{(b:e)}$ is a splitting genus 1 pencil, then the induced elliptic fibration on V comes via pullback from a double cover $\mathbb{P}^1_{(\beta:\epsilon)} \to \mathbb{P}^1_{(b:e)}$. Hence given an equation for the fibration $\mathscr{H}$, it suffices to find the branch points of the map $\mathbb{P}^1_{(\beta:\epsilon)} \to \mathbb{P}^1_{(b:e)}$ in order to find the Weierstrass equation for the elliptic fibration on V. We now explain how to do this in general, when the branch curves and equations for $\mathscr{H}$ are given in $\mathbb{P}^2$.

Assume that the equation of V as double cover of $\mathbb{P}^2$ is given by

$$w^2 = f_3(x_0 : x_1 : x_2)g_3(x_0 : x_1 : x_2).$$

Let $h((x_0 : x_1 : x_2), (b : e))$ be the equation of the pushforward of a splitting genus 1 pencil from $\tilde{R}$ to $\mathbb{P}^2$. The equation h is homogeneous of some degree in the coordinate $(x_0 : x_1 : x_2)$ on $\mathbb{P}^2$ and linear in the coordinate $(b : e)$ of the base $\mathbb{P}^1$ of the pencil. For every $(b : e)$, the curve with equation $h((x_0 : x_1 : x_2), (b : e))$ is of arithmetic genus 1.

Write $\cup_{k \in K} C_k$ for the irreducible components of the branch curves $f_3g_3 = 0$. For $D \subset \mathbb{P}^2$, write $\mathrm{mult}_{C_k}(D)$ for the multiplicity of the component C_k in D. Then we have the following.

Lemma 1 *A plane curve $D_{(b:e)} \subset \mathbb{P}^2$ that is member of the pencil $\mathscr{H}$ is a branch curve for the double cover*

$$\begin{array}{ccc} V & \longrightarrow & \tilde{R} \\ \downarrow & & \downarrow \mathscr{H} \\ \mathbb{P}^1_{(\beta:\varepsilon)} & \xrightarrow{2:1} & \mathbb{P}^1_{(b:e)} \end{array}$$

induced by the splitting genus 1 pencil if and only if there exists $k \in K$ such that $\mathrm{mult}_{C_k}(D_{(b:e)}) \neq 0$.

Proof If D meets the branch curves transversely, it does so only in points of even multiplicity and so splits in the double cover as two disjoint elliptic curves. It suffices, therefore, to show that if D contains at least one component C_k, the support of the preimage of D under the 2 : 1 map $\pi : V \to \mathbb{P}^2$ is connected. This follows from the fact that the preimage of C_k is a double curve, the support of which maps isomorphically onto C_k, and so every component of π^*D must meet this curve.

We will also make use of the following elementary fact. Let $D_{(b_0:e_0)} \subset \mathbb{P}^2$ be a plane curve in the pencil $\mathscr{H}$ with equation $h((x_0 : x_1 : x_2), (b_0 : e_0))$. Denote by $r((x_1 : x_2)(b : e))$ the resultant of $f_3(x_0 : x_1 : x_2)g_3(x_0 : x_1 : x_2)$ and $h((x_0 : x_1 : x_2), (b : e))$ with respect to x_0.

Lemma 2 *The resultant* $r((x_1 : x_2)(b : e))$ *vanishes to order*

$$= \sum_k \text{mult}_{C_k}(D_{(b_0:e_0)}) \cdot \begin{cases} \deg C_k & : (1:0:0) \notin C_k \\ (\deg C_k - 1) & : (1:0:0) \in C_k \end{cases}$$

at $(b : e) = (b_0 : e_0)$.

Proof This follows from the geometric description of the zeros of the resultant in terms of projecting the scheme-theoretic intersection of $h = 0$ and $f_3 g_3 = 0$ from the point $(1 : 0 : 0)$.

Writing $\text{ord}_{(b_0:e_0)}$ for the order of vanishing at $(b_0 : e_0)$, we may combine these two Lemmas to show the following:

Corollary 1 *If* $\text{ord}_{(b_0:e_0)}\left(r((x_1 : x_2)(b : e))\right) > 0$, *then the curve* $D_{(b_0:e_0)}$ *is a branch curve for the double cover induced by the splitting genus 1 pencil.*

If $(1 : 0 : 0) \notin \cup_k C_k$ *and* $D_{(b_0:e_0)}$ *is a branch curve for the double cover induced by the splitting genus 1 pencil, then* $\text{ord}_{(b_0:e_0)}\left(r((x_1 : x_2)(b : e))\right) > 0$.

We can therefore determine the relevant branch points from the resultant. This leads to the following algorithm.

Algorithm

1. Compute the resultant $r((x_1 : x_2)(b : e))$ of the two polynomials

$$f_3(x_0 : x_1 : x_2) g_3(x_0 : x_1 : x_2)$$

and

$$h((x_0 : x_1 : x_2), (b : e))$$

in one variable, say x_0.
2. Observe that $r((x_1 : x_2)(b : e)) = c_1(b : e) c_2(b : e) s((x_1 : x_2), (b : e))^2$, where $c_i(b : e)$ are homogeneous polynomials each with a unique root, denoted by $(b_i : e_i)$. (If $(1 : 0 : 0)$ is in the branch curves, it may be necessary to change coordinates first.)
3. Write the Weierstrass form of $h((x_0 : x_1 : x_2), (b : e))$, by applying the standard transformations. This is the equation of a rational elliptic surface, and the base of the fibration is $\mathbb{P}^1_{(b:e)}$.
4. Consider the base change $\mathbb{P}^1_{(\beta:\epsilon)} \to \mathbb{P}^1_{(b:e)}$ given by $(b = \beta^2(b_1/e_1) + \epsilon^2$, $e = \beta^2 + (e_2/b_2)\epsilon^2)$ (cf. (11)). Substituting this base change in the previous Weierstrass equation, one finds the Weierstrass equation of the elliptic fibration on the K3 surface V whose base is $\mathbb{P}^1_{(\beta:\epsilon)}$.

6.2 *The Elliptic Fibrations on* $S_{5,5}$

In this section we want to describe all the elliptic fibrations on $S_{5,5}$ giving equations for each of them.

In [2] a model of $S_{5,5}$ as a double cover of $\mathbb{P}$
induced by (generalized) conic bundles are a
context. Here we explicitly describe in our co
generalized conic bundles and the ones induce
also a Weierstrass equation for each of them.

6.2.1 Elliptic Fibrations Induced by Splitti

The fibration 11 of Table (22) is induced by the

$$M_1 := Q_0 + \Omega_0^{(1)} + \Omega_{1,0}^{(1)} + \Omega_1^{(1)} + Q_1 + \Omega_2^{(2)} + \Omega_3$$

which is the class of a fiber of type I_{12}. The cu
Q_3, and $\Omega_1^{(2)}$ are orthogonal to the components
type IV^*. The classes Q_2, Q_4, $\Omega_{1,2}^{(1)}$, $\Omega_{4,0}^{(1)}$, Ω_1
fibration induced by $|M_1|$. We observe that the
the components of the fiber of type I_{12} (the cur
(2, 2), (3, 2), (4, 2), (0, 2)) and four among the
curves $\Omega_j^{(j)}$ for $(i, j) = (3, 1), (2, 1), (4, 1), (1,$

The push-down of the class M_1 is

$$3h - E_1 - F_1 - G_1 - 2H_1 - E_2 - 2F_2 - G_2 - H_2 - E$$

which corresponds to a pencil of cubics passing
through Q_2 with tangent line m_2, and through Q_3
of this pencil is

$$b(x_0^2 x_1 - x_0 x_1^2 + x_1^2 x_2 + x_1 x_2^2 - 2x_0^2 x_2) + e$$

The Weierstrass form of the (rational) elliptic
(26) is

$$y^2 = x^3 - \frac{1(28b^2 + 4eb + e^2)(2b + e}{48}$$
$$- \frac{(376b^4 + 176eb^3 + 60e^2b^2 + 8}{864}$$

whose discriminant is $b^6(31b^2 + 4eb + e^2)(2b + e$
a fiber of type I_6 on $b = 0$ and one of type IV on (

The fibration 10 of Table (22) is induced by the

$$M_2 := Q_0 + \Omega_0^{(1)} + \Omega_{1,0}^{(1)} + \Omega_1^{(1)} + \Omega_{2,1}^{(1)} + \Omega_2^{(1)} + \Omega_{3,2}^{(1)} +$$
$$+ \Omega_{4,3}^{(2)} + \Omega_4^{(2)} + \Omega_{4,0}^{(2)} + \Omega_0^{(2)},$$

which is the class of a fiber of type I_{16}.

The pus

4h

which corr
and l_3 (in
pencil is

$bx_0^4 - 2$

Using
fibrations

i	
6	$A =$ $B =$ $\Delta =$
11	
10	$A =$ $\Delta =$

6.2.2 E

Let us co

N_1
N_2
N_3

N_4
N_8

The linear system $|N_i|$ induces the elliptic fibration number i of Table (22); indeed the divisor N_i corresponds to an elliptic fibration with a fiber of type II^*, II^*, I_{12}^*, III^*, I_8^* if $i = 1, 2, 3, 4, 8$, respectively, so N_3 (resp. N_8) corresponds to the unique fibration in Table (22) with a fiber of type I_{12}^* (resp. I_8^*), i.e., the fibration 3 (resp. 8). A priori N_1 could correspond either to the fibration 1 or to the fibration 2 (which are the fibrations which admit at least a fiber of type II^*). To distinguish among these cases, we consider the other reducible fibers: the curves orthogonal to N_1 are $\Omega_{h,j}^{(1)}$, for $h, j \in \{1, 2, 3, 4\}$, $h < j$, $\Omega_k^{(1)}$ for $k = 1, 2, 3, 4$, Q_2 and Q_4, and they span the lattice $\widetilde{E_8}$, so N_1 corresponds to the fibration 1 in Table (22); the curves orthogonal to N_2 are $\Omega_{h,j}^{(1)}$, for $h, j \in \{0, 1, 2, 3, 4\}$, $h < j$, $\Omega_k^{(1)}$ for $k = 1, 2, 3, 4$ and Q_2, and they span the lattice D_{10}, so N_2 corresponds to the fibration 2 in Table (22).

Let us now consider the fibration N_4. The fiber associated with N_4 is a fiber of type III^*. The curves $\Omega_2^{(1)}$ and $\Omega_1^{(1)}$ are sections of the fibration. The curves $\Omega_{2,1}^{(2)}$, Ω_2^2, $\Omega_{3,2}^{(2)}$, $\Omega_3^{(2)}$, $\Omega_{4,3}^{(2)}$, $\Omega_4^{(2)}$, Q_2, Q_4 are orthogonal to N_4 and are the components of another fiber of type III^*. This implies that $|N_4|$ is the fibration 4 in Table (22) and that the unique other reducible fiber is of type I_0^*. The curves Q_3, $\Omega_3^{(1)}$, $\Omega_{4,3}^{(1)}$, and $\Omega_{3,2}^{(1)}$ are orthogonal to N_4 and span a lattice isometric to D_4. So they are components of the I_0^*-fiber in the fibration $|N_4|$. A I_0^*-fiber has five components; four of them are Q_3, $\Omega_3^{(1)}$, $\Omega_{4,3}^{(1)}$, and $\Omega_{3,2}^{(1)}$, and the fifth component is another curve, say V_1. So that we have a special fiber of the fibration $|N_4|$, which is $2\Omega_3^{(1)}+\Omega_{4,3}^{(1)}+\Omega_{3,2}^{(1)}+Q_3+V_1$. Hence we can express the class of the curve V_1 as $N_4-(2\Omega_3^{(1)}+\Omega_{4,3}^{(1)}+\Omega_{3,2}^{(1)}+Q_3)$. From this expression one can compute all the intersection numbers of V_1 with all the curves $\Omega_{i,j}^k$, Ω_m^n, and Q_k. One can also observe that since V_1 is a component of a fiber of the fibration $|N_4|$, it is orthogonal to all the components of the two other reducible fibers of the same fibration, i.e., to the components of the III^*-fibers.

Let us now consider the class:

$$
\begin{aligned}
N_7 := {} & \Omega_{1,0}^{(1)} + \Omega_{4,0}^{(1)} + V_1 + \Omega_{4,3}^{(1)} + 2(\Omega_0^{(1)} + Q_0 + \Omega_0^{(2)} + \Omega_{1,0}^{(2)} + \Omega_1^{(2)} + \Omega_{2,1}^{(2)} + \\
& + \Omega_2^{(2)} + \Omega_{3,2}^{(2)} + \Omega_3^{(2)} + \Omega_{4,3}^{(2)} + \Omega_4^{(2)} + Q_2 + \Omega_2^{(1)} + \Omega_{3,2}^{(1)} + \Omega_3^{(1)}).
\end{aligned}
$$

It is the class of the elliptic fibration 7 in Table (22) since it is the class of a fiber of an elliptic fibration on $S_{5,5}$ with a fiber of type I_{14}^*.

All the classes N_i, for $i = 1, 2, 3, 4, 7, 8$ considered above correspond to generalized conic bundles which can be written in the following way in terms of the classes on $\widetilde{R}$:

i	$(\pi_*(N_i))/2$
1	$4h - (E_2 + F_2 + G_2 + H_2) - (2E_3 + 3F_3 + 3G_3 + 2H_3) - 2(E_4 + 2F_4 + G_4 + H_4) - E_{T_2}$
2	$4h - (2E_2 + 2F_2 + 2G_2 + 3H_2) - 2(E_4 + 2F_4 + G_4 + H_4) - (E_3 + 2F_3 + G_3 + H_3) - E_5$
3	$4h - (2E_1 + 2F_1 + 2G_1 + 3H_1) - (E_4 + F_4 + G_4 + 2H_4) - 2(E_3 + 2F_3 + G_3 + H_3) - E_5$
4	$2h - (E_1 + F_1 + 2G_1 + H_1) - (E_3 + 2F_3 + G_3 + H_3)$
7	$7h - 3(E_1 + F_1 + 2G_1 + H_1) - (E_2 + 2F_2 + G_2 + H_2)+$ $-2(E_4 + 2F_4 + G_4 + H_4) - (4E_3 + 6F_3 + 4G_3 + 5H_3)$
8	$4h - 2(E_+ - F_1 + G_1 + H_1) - (2E_2 + 3F_2 + 2G_2 + 2H_2)+$ $-(E_4 + 2F_4 + G_4 + H_4) - (2E_3 + 2F_3 + 2G_3 + 3H_3)$

This allows to compute explicitly the equations of pencils of singular rational curves in $\mathbb{P}^2$ corresponding to our elliptic fibrations on $S_{5,5}$.

i	Pencils in $\mathbb{P}^2$
1	$a(x_0^4 - x_0^3x_1 + 3x_0^2x_1x_2 - 4x_0^3x_2 + 6x_0^2x_2^2 - 3x_0x_1x_2^2 - 4x_0x_2^3 + x_1x_2^3 + x_2^4) + gx_0x_1^2x_2$
2	$n(x_0^4 - 2x_0^3x_1 + x_0^2x_1^2 + 7x_0^2x_1x_2 - 3x_1^2x_0x_2 - 4x_2x_0^3 + 6x_2^2x_0^2 - 8x_0x_1x_2^2)+$ $+n(2x_1^2x_2^2 - 4x_2^3x_0 + 3x_1x_2^3 + x_2^4) + m(8x_1^2x_2x_0 - 8x_1^3x_2)$
3	$f(-x_0^4 + x_0^2x_1x_2 + x_0^3x_2 - x_1^2x_2^2) + n(x_0^4 + x_2^2x_1^2 - x_0^3x_2 - x_0x_1x_2^2)$
4	$\tau x_0^2 + \sigma(x_0x_2 - x_1x_2)$
7	$\tau x_0x_1x_2(x_2 - x_0)(x_0 - x_1)(x_0^2 - x_0x_2 + x_1x_2) + \sigma(-x_1x_0^6 + 2x_0x_1^3x_2^3 - 12x_0^4x_1x_2^2+$ $+7x_0^3x_1^2x_2^2 - 8x_0^2x_1^2x_2^3 + 6x_0^5x_1x_2 - 3x_0^4x_1^2x_2 + x_0^7 - x_2^4x_1^3 - x_0^2x_1^3x_2^2 - 3x_2^4x_1x_0^2+$ $+10x_0^3x_1x_2^3 + 6x_2^2x_0^5 - 4x_2x_0^6 + 3x_2^4x_1^2x_0 + x_2^4x_0^3 - 4x_2^3x_0^4)$
8	$s(x_0^4 - 2x_0^3x_1 + x_0^2x_1^2 + 3x_0^2x_1x_2 - x_0x_1^2x_2 - 2x_0^3x_2 + x_0^2x_2^2 - x_0x_1x_2^2)+$ $+t(-x_0^2x_1x_2 + x_0x_1^2x_2 + 2x_0x_1x_2^2 - x_1^2x_2^2)$

Now it remains to find the Weierstrass equations of the elliptic fibrations on $S_{5,5}$ corresponding to the linear systems $\big(\pi_*(N_j)/2\big)$ on $\widetilde{R}$. In case 4, the curves which are fibers of the conic bundle have degree 2, so we can directly apply the first algorithm in Sect. 5.2. In cases 1, 2, 3, and 8, the curves are quartics which satisfy condition (†), and so we may apply the second algorithm in Sect. 5.2. The rational parameterizations and induced Weierstrass equations are given by

i	Rational parameterization	Elliptic fibration
1	$x_0 = -agp^3 - 2agp^2 - agp$ $x_1 = -g^2p^4 - 2agp^2 - a^2$ $x_2 = g^2p^4 - agp^3 + g^2p^3 - agp^2$	$A = -3a^4g^4,\ B = a^7g^5 + a^5g^7$ $\Delta = -432g^{10}a^{10}(a-g)^2(a+g)^2$
2	$x_0 = -64m^2p^4 + 8nmp^3 - 8nmp$ $x_1 = -64m^2p^4 + 16nmp^3$ $-n^2p^2 - 16nmp^2 + 2n^2p - n^2$ $x_2 = -64m^2p^4 + 8nmp^3 + 64m^2p^3 - 8nmp^2$	$A = 108m^2n^4(-n^2 + 384m^2)$ $B = -432m^3n^5(n^4 - 576n^2m^2 + 55296m^4)$ $\Delta = 570630428688384n^{10}m^{12}(n^2 - 432m^2)$
3	$x_0 = p(-np + fp + n)(-np^2 + fp^2 + fp - n + f)$ $x_1 = (p+1)(-n+f)p^2(-np + fp + n)$ $x_2 = (-np^2 + fp^2 + fp - n + f)^2$	$A = (3f^6 - 18f^5n + 36f^4n^2 - 24f^3n^3 - 3f^2n^4$ $+6fn^5 - n^6)\cdot 27(f-n)^2$ $B = (9f^6 - 54f^5n + 117f^4n^2 - 108f^3n^3 + 39f^2n^4$ $-6fn^5 + n^6)\cdot 27n(f-n)^3(3f^2 - 6fn + 2n^2)$ $\Delta = -8503056(n-2f)(3n-2f)f^2(2n-f)^2(n-f)^{18}$
4		$A = \tau^3(1+\tau)^2,\ B = 0,$ $\Delta = 4\tau^9(1+\tau)^6$
8	$x_0 = (p-1)(sp+t)(tp^2 - sp - t)$ $x_1 = sp^2(tp^2 - sp - t)$ $x_2 = (p-1)(-sp + tp - t)(sp + t)$	$A = -27t^2s^2(s^4 + 4s^2t^2 + t^4)$ $B = -27t^3s^3(s^2 + 2t^2)(2s^4 + 8s^2t^2 - t^4)$ $\Delta = 8503056s^8t^{14}(s^2 + 4t^2)$

(29)

The fibration induced by $|N_7|$ is the unique elliptic fibration on a K3 surface with a fiber of type I_{14}^* (which is a maximal fiber, since if a K3 surface admits an elliptic fibration with this reducible fiber, then this is the unique reducible fiber). So it suffices to know the equation of this elliptic fibration, which is classically known [13, Theorem 1.2]. Hence for $|N_7|$ we only rewrite here the known equation, a part from the fact that we chose the parameters over $\mathbb{P}^1_{(\tau:\sigma)}$ in such a way that four fibers of type I_1 are over the points $(\tau : \sigma) = ((\pm\sqrt{-26 \pm 14i\sqrt{7}})/4 : 1)$. These values correspond to the septic in $|N_7|$ which are tangent to m_2 in a smooth point. We therefore have equation:

$$\begin{aligned} A &= \tau^2(-12\sigma^6 - (3/8)\sigma^4\tau^2 - (15/2048)\sigma^2\tau^4 - (9/262144)\tau^6), \\ B &= \tau^3\sigma(-16\sigma^8 - (3/4)\tau^2\sigma^6 - (21/1024)\sigma^4\tau^4 - (35/131072)\tau^6\sigma^2 \\ &\quad -(63/33554432)\tau^8), \\ \Delta &= -(729/4503599627370496)\tau^{20}(2048\sigma^4 + 52\tau^2\sigma^2 + \tau^4). \end{aligned} \tag{30}$$

We note that it is possible to obtain such an equation using our techniques. Indeed one may find a rational parameterization of the septic plane curves of the generalized conic bundle by rational cubic curves, and from there the computation follows in the same way as above. We omit this computation here as the equation is already in the literature.

7 The K3 Surface $X_{5,5}$ and Its Elliptic Fibrations

The K3 surface $X_{5,5}$ is very well-known and studied, in particular since its Néron–Severi group allows to describe the moduli spaces of K3 surfaces with an elliptic fibration with a 5-torsion section in terms of L-polarized K3 surfaces (see [3]). Indeed, if a K3 surface admits an elliptic fibration with a 5-torsion section, then the lattice $U \oplus M$ has to be primitively embedded in its Néron–Severi lattice, where M is an overlattice of index 5 of a root lattice. The lattice M is known to be an overlattice of index 5 of A_4^4, and so the lattice $U \oplus M$ is isometric to the Néron–Severi lattice of $X_{5,5}$. Hence the moduli space of the K3 surfaces which are $NS(X_{5,5})$-polarized coincides with the moduli space of the K3 surfaces which admit an elliptic fibration with a 5-torsion section.

7.1 The List of All the Elliptic Fibrations

The transcendental lattice of $X_{5,5}$ is known to be $T_X \simeq U \oplus U(5)$ (see, e.g., [5]). This allows to apply the Nishiyama method in order to classify (at least lattice theoretically) the elliptic fibrations on $X_{5,5}$. This method is studied and applied in several papers, and we do not intend to describe it in details. Here we just observe that we can apply the method using $A_4 \oplus A_4$ as the lattice T, so that T is a negative-definite lattice with the same discriminant form as the one of the transcendental lattice T_X and whose rank is $\mathrm{rank}(T) = \mathrm{rank}(T_X) + 4$. Then one has to find the primitive embeddings of $A_4 \oplus A_4$ in the Niemeier lattice up to isometries, and this can be done by embedding T into the root lattice of the Niemeier lattices. The list of the Niemeier lattices and the possible embeddings of root lattices in Niemeier lattices up to the Weyl group can be found in [9] (see also [7]). In particular one has that A_4 embeds primitively in a unique way (up to the action of the Weyl group) in A_n for $n > 4$, in D_m for $m > 4$, and in E_h for $h = 6, 7, 8$ (see [9, Lemmas 4.2 and 4.3]). On the other hand, $A_4 \oplus A_4$ has a primitive embedding in A_n for $n \geq 9$ and in D_m for $m \geq 10$ and has no primitive embeddings in E_h for $h = 6, 7, 8$ (see [9, Lemma 4.5]). All these primitive embeddings are unique with the exception of $A_4 \oplus A_4 \hookrightarrow D_{10}$, for which there are two possible primitive embeddings (see [9, Page 325, just before Step 3]). The orthogonal complement of the embedded copy of $A_4 \oplus A_4$ in the root lattice of each Niemeier lattices is a lattice L which can be computed by Nishiyama [9, Corollary 4.4] and which encodes information about both the reducible fibers and the rank of the Mordell–Weil group of the elliptic fibrations. In particular the root lattice of L is the lattice spanned by the irreducible components of the reducible fibers orthogonal to the zero section. More precise information on the sections can be obtained by a deeper analysis of these embeddings, but this is outside the scope of this paper. So in the following list, we give the root lattice of the Niemeier lattice that we are considering, the embeddings of $A_4 \oplus A_4$ in this root lattice, the root lattices of the orthogonal complement, and in the last two columns the properties of the associated elliptic fibration. This gives the complete list of the types of elliptic fibrations on $X_{5,5}$:

n^o	Niemeier	embedding(s)	roots orthogonal	singular fibers	rk(MW)
1	E_8^3	$A_4 \subset E_8$ $A_4 \subset E_8$	$A_4 \oplus A_4 \oplus E_8$	$2I_5 + II^* + 2I_1$	0
2	$E_8 \oplus D_{16}$	$A_4 \subset E_8$ $A_4 \subset D_{16}$	$A_4 \oplus D_{11}$	$I_5 + I_7^* + 6I_1$	1
3	$E_8 \oplus D_{16}$	$A_4 \oplus A_4 \subset D_{16}$	$E_8 \oplus D_6$	$II^* + I_2^* + 6I_1$	2
4	$E_7^2 \oplus D_{10}$	$A_4 \subset E_7$ $A_4 \subset E_7$	$A_2^2 \oplus D_{10}$	$2I_3 + I_6^* + 6I_1$	2
5	$E_7^2 \oplus D_{10}$	$A_4 \subset E_7$ $A_4 \subset D_{10}$	$E_7 \oplus A_2 \oplus D_5$	$III^* + I_3 + I_1^* + 5I_1$	2
6	$E_7^2 \oplus D_{10}$	$A_4 \oplus A_4 \subset D_{10}$	E_7^2	$2III^* + 6I_1$	2
7	$E_7^2 \oplus D_{10}$	$A_4 \oplus A_4 \subset D_{10}$	E_7^2	$2III^* + 6I_1$	2
8	$E_7 \oplus A_{17}$	$A_4 \subset E_7$ $A_4 \subset A_{17}$	$A_2 \oplus A_{12}$	$I_3 + I_{13} + 8I_1$	2
9	$E_7 \oplus A_{17}$	$A_4 \oplus A_4 \subset A_{17}$	$E_7 \oplus A_7$	$III^* + I_8 + 7I_1$	2
10	D_{24}	$A_4 \oplus A_4 \subset D_{24}$	D_{14}	$I_{10}^* + 8I_1$	2
11	$D_{12} \oplus D_{12}$	$A_4 \subset D_{12}$ $A_4 \subset D_{12}$	$D_7 \oplus D_7$	$2I_3^* + 6I_1$	2
12	$D_{12} \oplus D_{12}$	$A_4 \oplus A_4 \subset D_{12}$	$D_{12} \oplus A_1^2$	$I_8^* + 2I_2 + 6I_1$	2
13	D_8^3	$A_4 \subset D_8$ $A_4 \subset D_8$	$D_8 \oplus A_3 \oplus A_3$	$I_4^* + 2I_4 + 6I_1$	2
14	$D_9 \oplus A_{15}$	$A_4 \subset D_9$ $A_4 \subset A_{15}$	$D_4 \oplus A_{10}$	$I_0^* + I_{11} + 7I_1$	2
15	$D_9 \oplus A_{15}$	$A_4 \oplus A_4 \subset A_{15}$	$D_9 \oplus A_5$	$I_5^* + I_6 + 7I_1$	2
16	E_6^4	$A_4 \subset E_6$ $A_4 \subset E_6$	$A_1^2 \oplus E_6^2$	$2I_2 + 2IV^* + 4I_1$	2
17	$E_6 \oplus D_7 \oplus A_{11}$	$A_4 \subset E_6$ $A_4 \subset D_7$	$A_1^3 \oplus A_{11}$	$3I_2 + I_{12} + 6I_1$	2
18	$E_6 \oplus D_7 \oplus A_{11}$	$A_4 \subset E_6$ $A_4 \subset A_{11}$	$A_1 \oplus D_7 \oplus A_6$	$I_2 + I_3^* + I_7 + 6I_1$	2
19	$E_6 \oplus D_7 \oplus A_{11}$	$A_4 \subset D_7$ $A_4 \subset A_{11}$	$E_6 \oplus A_1^2 \oplus A_6$	$IV^* + 2I_2 + I_7 + 5I_1$	2
20	$E_6 \oplus D_7 \oplus A_{11}$	$A_4 \oplus A_4 \subset A_{11}$	$E_6 \oplus D_7 \oplus A_1$	$IV^* + I_3^* + I_2 + 5I_1$	2
21	D_6^4	$A_4 \subset D_6$ $A_4 \subset D_6$	D_6^2	$2I_2^* + 8I_1$	4
22	$D_6 \oplus A_9^2$	$A_4 \subset D_6$ $A_4 \subset A_9$	$A_4 \oplus A_9$	$I_5 + I_{10} + 9I_1$	3
23	$D_6 \oplus A_9^2$	$A_4 \subset A_9$ $A_4 \subset A_9$	$D_6 \oplus A_4 \oplus A_4$	$I_2^* + 2I_5 + 6I_1$	2
24	$D_6 \oplus A_9^2$	$A_4 \oplus A_4 \subset A_9$	$D_6 \oplus A_9$	$I_2^* + I_{10} + 6I_1$	1
25	$D_5^2 \oplus A_7^2$	$A_4 \subset D_5$ $A_4 \subset D_5$	A_7^2	$2I_8 + 8I_1$	2
26	$D_5^2 \oplus A_7^2$	$A_4 \subset D_5$ $A_4 \subset A_7$	$A_7 \oplus D_5 \oplus A_2$	$I_8 + I_1^* + I_3 + 6I_1$	2
27	$D_5^2 \oplus A_7^2$	$A_4 \subset A_7$ $A_4 \subset A_7$	$D_5^2 \oplus A_2^2$	$2I_1^* + 2I_3 + 4I_1$	2
28	A_8^3	$A_4 \subset A_8$ $A_4 \subset A_8$	$A_3^2 \oplus A_8$	$2I_4 + I_9 + 7I_1$	2
29	A_{24}	$A_4 \oplus A_4 \subset A_{24}$	A_{14}	$I_{15} + 9I_1$	2
30	A_{12}^2	$A_4 \subset A_{12}$ $A_4 \subset A_{12}$	A_7^2	$2I_8 + 8I_1$	2
31	A_{12}^2	$A_4 \oplus A_4 \subset A_{12}$	$A_2 \oplus A_{12}$	$I_3 + I_{13} + 8I_1$	2
32	$D_4 \oplus A_5^4$	$A_4 \subset A_5$ $A_4 \subset A_5$	$D_4 \oplus A_5^2$	$I_0^* + 2I_6 + 6I_1$	2
33	A_6^4	$A_4 \subset A_6$ $A_4 \subset A_6$	$A_1^2 \oplus A_6^2$	$2I_2 + 2I_7 + 6I_1$	2
34	A_4^6	$A_4 \subset A_4$ $A_4 \subset A_4$	A_4^4	$4I_5 + 4I_1$	0

(31)

Observe that lines 6 and 7 correspond to the two different embeddings of $A_4 \oplus A_4$ in D_{10}. The K3 surface $X_{5,5}$ is obtained as double cover of a rational surface $R_{5,5}$ branched on two smooth fibers, so there are no elliptic fibrations induced by generalized conic bundles or by splitting genus 1 pencils (see [4]). So an elliptic fibration on $X_{5,5}$ is either induced by a conic bundle on $R_{5,5}$ or it is of type 3.

Putting together these considerations with the results of Sects. 4 and 5, we proved the following proposition.

Proposition 3 *The elliptic fibrations on $X_{5,5}$ are of 34 types, listed in Table* (7.1)*. The fibration in line 34 of Table* (7.1) *is induced by $\mathscr{E}_R$, and its equation is given in Sect. 4.2.2; the fibrations of lines 22, 30, and 32 are induced by conic bundles on $R_{5,5}$ and their equations are given in Sect. 5.3.2. The other fibrations on $X_{5,5}$ are of type 3.*

7.2 Fibration of Type 3: An Example, the Fibration 26

The aim of this section is to construct explicitly an example of an elliptic fibration of type 3 and to discuss the geometry of the noncomplete linear system on $\mathbb{P}^2$ which induces this fibration.

The divisor

$$D_1 := \Omega_0^{(2,2)} + \Omega_0^{(1,1)} + 2Q_0 + 2\Omega_0^{(2,1)} + \Omega_1^{(2,1)} + \Omega_4^{(2,1)}$$

corresponds to the class of the fiber of a fibration which has one reducible fiber of type I_1^*; thus $|D_1|$ is one of the fibrations 5,26,27 in Table (7.1). The curves $\Omega_2^{(2,1)}$, $\Omega_3^{(2,1)}$, Q_2, Q_3, $\Omega_1^{(2,2)}$, $\Omega_4^{(2,2)}$, $\Omega_1^{(1,1)}$, $\Omega_4^{(1,1)}$ are sections of the fibration $|D_1|$. Assuming that $\Omega_2^{(2,1)}$ is the zero section, there is a fiber whose nontrivial components are $\Omega_2^{(2,2)}$, $\Omega_3^{(2,2)}$, Q_4, $\Omega_4^{(1,2)}$, $\Omega_2^{(1,2)}$, $\Omega_3^{(1,2)}$, and $\Omega_1^{(1,2)}$. So there is a fiber of type I_8 and $|D_1|$ is the fibration 26 in (7.1), and there is a fiber of type I_3 whose nontrivial components are $\Omega_2^{(1,1)}$ and $\Omega_3^{(1,1)}$.

Denoted by $\pi : X_{5,5} \to R_{5,5}$, we have $\pi(Q_i) = P_i$ and $\pi(\Omega_i^{(j,1)}) = \pi(\Omega_i^{(j,2)}) = \Theta_i^{(j)}$, for $j = 1, 2$. In particular $\iota(\Omega_i^{(j,1)}) = \Omega_i^{(j,2)}$, thus, $\iota(D_1) \neq D_1$ and indeed $D_2 := \iota(D_1)$ is

$$D_2 := \Omega_0^{(2,1)} + \Omega_0^{(1,2)} + 2Q_0 + 2\Omega_0^{(2,2)} + \Omega_1^{(2,2)} + \Omega_4^{(2,2)}.$$

We observe that $\Omega_0^{(2,1)} D_1 = 0$, $\Omega_0^{(1,2)} D_1 = 2$, $Q_0 D_1 = 0$, $\Omega_0^{(2,2)} D_1 = 0$, $\Omega_1^{(2,2)} D_1 = 1$, $\Omega_4^{(2,2)} D_1 = 1$, and thus $D_2 D_1 = 0 + 2 + 0 + 0 + 1 + 1 = 4$.

So $D_1 D_2 = 4$ and $(D_1 + D_2)^2 = 8$. In particular a smooth member of the linear system $|D_1 + D_2|$ is a curve of genus 5.

We are interested in the class of the curve $\pi(D_1+D_2)$. By the projection formula, $\pi_*(D_1 + D_2) = 2\pi(D_1 + D_2)$, so we are looking for $\frac{1}{2}\pi_*(D_1 + D_2)$. Since

$$\pi(D_1) = \pi(D_2), \frac{1}{2}\pi_*(D_1 + D_2) = \pi_*(D_1) = \pi_*(D_2).$$

We recall that the map π restricted to $\Omega_i^{(j,1)}$ is a $1:1$ map to $\Theta_i^{(j)}$, and similarly the map π restricted to $\Omega_i^{(j,2)}$ is a $1:1$ map to $\Theta_i^{(j)}$. On the other hand, the map π restricted to the sections Q_i is a $2:1$ map to the section P_i. So

$$\pi_*(D_1)=\Theta_0^{(2)}+\Theta_0^{(1)}+4P_0+2\Theta_0^{(2)}+\Theta_1^{(2)}+\Theta_4^{(2)}=3\Theta_0^{(2)}+\Theta_1^{(2)}+\Theta_4^{(2)}+\Theta_0^{(1)}+4P_0.$$

Hence by (4)

$$\pi_*(D_1)=3E_1+\ell_1+\ell_2+m_1+4F_1=3h-E_2-F_2-E_{Q_5}-E_4-2F_4-E_3-F_3,$$

which is the class of the strict transforms of cubics in $\mathbb{P}^2$ passing through Q_2, Q_3, Q_4 with tangent ℓ_2, and Q_5.

The equation of the cubics satisfying these properties is

$$\begin{aligned} &ax_0^3+bx_0^2x_1+(-a-b)x_0x_1^2+dx_0^2x_2+\\ &ex_0x_1x_2+fx_1^2x_2+(-3a-2d)x_0x_2^2+(a-e-f)x_1x_2^2+(d+2a)x_2^3=0. \end{aligned} \tag{32}$$

This equation depends on five parameters (four projective parameters) and specializes to equations of cubics which split on the double cover and induces elliptic fibrations on $X_{5,5}$.

The generic cubic c_3 as in Eq. (32) is a cubic in $\mathbb{P}^2$. We recall that $X_{5,5}$ is the double cover of $\mathbb{P}^2$ branched along the reducible sextic c_6 whose equation is $f_3g_3=0$ for two cubics f_3 and g_3. So c_3 and c_6 meet in 18 points in $\mathbb{P}^2$ counted with multiplicity. Recall that c_6 is singular at Q_2, Q_3, Q_4, and Q_5, and in Q_4, c_6 is the union of two cubics, both with tangent direction ℓ_2. Hence c_3 intersects c_6 in Q_2, Q_3, Q_5 with multiplicity 2 and in Q_4 with multiplicity 4. Outside these points, c_3 and c_6 intersect in $18-2-2-2-4=8$ points. The inverse image of c_3 on $X_{5,5}$ is a double cover of c_3 branched in eight points. Since generically c_3 is a smooth cubic in $\mathbb{P}^2$, it has genus 1, and then its inverse image on $X_{5,5}$ has genus g such that $2g-2=2(0)+8$. So $g=5$.

We already observed that c_6 is the union of two smooth cubics in $\mathbb{P}^2$, which are the image of the branch curves of the double cover $X_{5,5}\to R_{5,5}$. Denoting by $b_1: f_3=0$ and $b_2:=g_3=0$ these two cubics, c_3 intersects b_1 in nine points, five of which are Q_2, Q_3, Q_4 with tangent ℓ_2, and Q_5. So c_3 and b_1 generically intersect in four other points. In the case c_3 splits in the double cover (which is the case in which c_3 is the image both of D_1 and D_2), these four points are either one point with multiplicity 4 or two points with multiplicity 2. The strict transform of b_1 (resp. b_2) on $R_{5,5}$ is a smooth fiber of the fibration on $R_{5,5}$, and its pullback to $X_{5,5}$ is a smooth fiber, denoted by B_1 (resp. B_2),[1] of the fibration induced on $X_{5,5}$ by the one on $R_{5,5}$. Since a section of this fibration is Q_0, $B_1Q_0=B_2Q_0=1$, $B_1\Omega_i^{(k,j)}=0$ and thus $D_1B_1=D_1B_2=D_2B_1=D_2B_2=2$. In particular D_1 and D_2 intersect in four points, two on B_1 and two on B_2. Considering the image of these curves in $\mathbb{P}^2$, one realizes that if c_3 is the image of D_1, then it intersects b_1 (resp. b_2) in two points each with multiplicity 2.

[1]Not to be confused with the conic bundles B_1 and B_2 in Sect. 3.

So, theoretically, in order to find the specializations of a curve c_3 which is the image of D_1, one has to require that the intersection $c_3 \cap b_1$ consists of the points Q_2, Q_3, Q_4 with tangent ℓ_2, and Q_5, and of two other points each with multiplicity 2.

As in the previous context one can compute the resultant between the equation of the cubics c_3 and the branch locus of the double cover $X_{5,5} \to \mathbb{P}^2$, given in (14). Since not all the curves in the linear system $|c_3|$ split in the double cover, the resultant of the equation of c_3 and the equation of the branch locus of $X \to \mathbb{P}^2$ is not the square of a polynomial. Nevertheless one recognizes some factors with even multiplicity (which correspond to the conditions that c_3 passes with a certain multiplicity through a certain base point of the pencil of cubics from which $R_{5,5}$ arises), and one can also observe that for certain choices of the values (a, d, e, f), the resultant becomes a square. For example, one observes that for $f = -a - d - e$ the resultant with respect to x_1 is

$$x_2^2(x_0 - x_2)^6(x_0x_2(em + am - a + dm) + x_0^2(am + bm) - x_2^2(d - 2a))^2$$
$$(x_0x_2(el + al - a + dl) + x_0^2(al + bl) - x_2^2(d - 2a))^2$$

so in this case we know that the intersection between the generic member of $|c_3|$ and the branch curve is always with even degree, which is the necessary condition to have a splitting in any point.

Acknowledgements This work was initiated during the Women in Numbers Europe 2 workshop. We thank the organizers and the Lorentz Center in Leiden for providing such a stimulating research environment. We also thank Bernd Sturmfels for drawing our attention to the connections with tropical geometry and the referee for many helpful suggestions and comments that have improved this article.

Francesca Balestrieri was partially supported by the EPSRC Scholarship EP/L505031/1. Alice Garbagnati was partially supported by FIRB 2012 "Moduli Spaces and their Applications". Cecília Salgado was partially supported by Cnpq grant 446873/2014 – 4 and by Faperj Jovem Cientista do Nosso Estado grant E 10/2016/226621. Isabel Vogt was partially supported by National Science Foundation Graduate Research Fellowship Program DMS-1122374, as well as DMS-1601946.

References

1. A. Beauville, Les familles stables de courbes elliptiques sur $\mathbb{P}^1$ admettant 4 fibres singulières. CR Acad. Sci. Paris **294**, 657–660 (1982)
2. P. Comparin, A. Garbagnati, Van Geemen–Sarti involutions and elliptic fibrations on K3 surfaces double cover of $\mathbb{P}^2$. J. Math. Soc. Jpn. **66**, 479–522 (2014)
3. A. Garbagnati, Elliptic K3 surfaces with abelian and dihedral groups of symplectic automorphisms. Commun. Algebra **41**, 583–616 (2013)
4. A. Garbagnati, C. Salgado, Linear systems on rational elliptic surfaces and elliptic fibrations on K3 surfaces. Pure and Applied Algebra. arXiv:1703.02783. https://doi.org/10.1016/j.jpaa.2018.03.010
5. A. Garbagnati, A. Sarti, Symplectic automorphisms of prime order on K3 surfaces. J. Algebra **318**, 323–350 (2007)

6. R. Miranda, The basic theory of elliptic surfaces, Dip. di matematica-Univ. Pisa. Available on line at http://www.math.colostate.edu/~miranda/BTES-Miranda.pdf
7. H.-V. Niemeier, Definite quadratische Formen der Dimension 24 und Diskriminante 1. J. Number Theory **5**, 142–178 (1973)
8. V.V. Nikulin, Factor groups of groups of automorphisms of hyperbolic forms with respect to subgroups generated by 2-reflections. J. Sov. Math. **22**, 1401–1475 (1983)
9. K. Nishiyama, The Jacobian fibrations on some K3 surfaces and their Mordell-Weil groups. Japan. J. Math. (N.S.) **22**, 293–347 (1996)
10. Q. Ren, K. Shaw, B. Sturmfels, Tropicalization of del Pezzo surfaces. Adv. Math. **300**, 156–189 (2016)
11. M. Schuett, T. Shioda, Elliptic surfaces, algebraic geometry in East Asia - Seoul 2008. Adv. Stud. Pure Math. **60**, 51–160 (2010)
12. I. Shimada, On elliptic K3 surfaces. Mich. Math. J. **47**, 423–446 (2000). arXiv version with the complete Table arXiv:math/0505140
13. T. Shioda, The elliptic K3 surfaces with a maximal singular fibre CR Acad. Sci. Paris, Ser. I **337**, 461–466 (2003)
14. E.B. Vinberg, The two most algebraic K3 surfaces. Math. Ann. **265**, 1–21 (1983)
15. D.Q. Zhang, Quotients of K3 surfaces modulo involutions. Japan J. Math. (N.S.) **24**, 335–366 (1998)

On Birch and Swinnerton-Dyer's Cubic Surfaces

Mckenzie West

Abstract In a 1975 paper of Birch and Swinnerton-Dyer, a number of explicit norm form cubic surfaces are shown to fail the Hasse principle. They make a correspondence between this failure and the Brauer–Manin obstruction, recently discovered by Manin. We extend their work to show that a larger set of cubic surfaces has a Brauer–Manin obstruction to the Hasse principle, thus verifying the Colliot-Thélène–Sansuc conjecture for infinitely many cubic surfaces.

Keywords Cubic · Surface · Brauer–Manin · Obstruction

2010 Mathematics Subject Classification. Primary 11G35; Secondary 14J20

1 Introduction

Suppose X is a smooth projective variety over a number field k. Denote by $X(k)$ and $X(\mathbf{A}_k)$ the rational and adèlic points of X, respectively. The natural inclusion of k into $\mathbf{A}_k$ gives $X(k) \neq \emptyset \Rightarrow X(\mathbf{A}_k) \neq \emptyset$. The variety X satisfies the *Hasse principle* if the converse is true, $X(\mathbf{A}_k) \neq \emptyset \Rightarrow X(k) \neq \emptyset$. There are many examples of varieties which do not satisfy the Hasse principle; Cassels and Guy, in [3], provided one of the original counterexamples in the case of cubic surfaces,

$$5x^3 + 12y^3 + 9z^3 + 10w^3 = 0\,. \tag{1}$$

As a cohomological generalization of quadratic reciprocity, Manin constructs, in [12], the *Brauer set* $X(\mathbf{A}_k)^{\mathrm{Br}}$ which lies between $X(k)$ and $X(\mathbf{A}_k)$. We say that X has a *Brauer–Manin obstruction to the Hasse principle* if $X(\mathbf{A}_k)^{\mathrm{Br}} = \emptyset$ while $X(\mathbf{A}_k) \neq \emptyset$. Around this time, Birch and Swinnerton-Dyer, in [1], considered

M. West (✉)
Kalamazoo College, Kalamazoo, MI, USA
e-mail: Mckenzie.West@kzoo.edu

I. I. Bouw et al. (eds.), *Women in Numbers Europe II*, Association for Women in Mathematics Series 11, https://doi.org/10.1007/978-3-319-74998-3_10

counterexamples to the Hasse principle for rational surfaces via very direct arguments. They comment that Manin's method should apply and provide a brief sketch to this effect. We will examine the cubic surfaces constructed by Birch and Swinnerton-Dyer:

> Let K_0/k be a non-abelian cubic extension and L/k its Galois closure. Suppose K/k is the unique quadratic extension which lies in L. We will assume that $(1, \phi, \psi)$ are any linearly independent generators for K_0/k and K/k is generated by θ. Then consider the Diophantine equation given by
>
> $$m \operatorname{Norm}_{L/K}(ax + by + \phi z + \psi w) = (cx + dy) \operatorname{Norm}_{K/k}(x + \theta y), \quad (2)$$
>
> where the m, a, b, c, d are suitably chosen k-rational integers.

Throughout, we will refer to surfaces given by an equation of the form (2) as *BSD cubics*. Birch and Swinnerton-Dyer show that as long as a, b, c, d have "certain" divisibility properties, the projective cubic surface in $\mathbb{P}^3_k$ defined by Eq. (2) do not satisfy the Hasse principle. Their examples of failure are very specific, choosing values for almost all of the constants. This is done by considering a rational solution $[x : y : z : w]$ and examining the possible factorizations of the ideal $(x + \theta y)$ in $\mathcal{O}_K$. They find two possible reasons the Hasse principle may fail and give an example computation of the Brauer–Manin obstruction for each. This paper re-examines the BSD surfaces, extending their work using more explicit formulas for the lines on the surfaces.

Theorem 1 *Let k be a number field and X is as in* (2). *Let L'/k be the minimal extension over which the 27 lines, or exceptional curves, on $\overline{X}$ are defined. If $9 \mid [L' : k]$ then $\operatorname{Br} X/\operatorname{Br} k \simeq \mathbb{Z}/3\mathbb{Z}$ otherwise X satisfies the Hasse Principle.*

Furthermore, one may compute the existence of Brauer–Manin obstructions within this family of surfaces, as in the following result.

Theorem 2 *Let k be a number field, L/K be unramified, and the ϕ_i and ψ_i be integral units in K_0 with the minimal polynomial of ψ_i/ϕ_i being separable modulo $\mathfrak{p} \mid 3\mathcal{O}_k$. Let $p\mathcal{O}_k$ be a prime for which $p\mathcal{O}_L = \mathcal{P}_1\mathcal{P}_2$ such that $\mathfrak{p} \| \theta\overline{\theta}$ and the $\mathcal{P}_i$ are primes in $\mathcal{O}_L$. Assume no prime in $\mathcal{O}_K$ divides both θ and $\overline{\theta}$. Then the variety defined by:*

$$\prod_{i=0}^{2} (x + \phi_i z + \psi_i w) = py(x + \theta y)(x + \overline{\theta} y)$$

has a Brauer–Manin obstruction to the Hasse principle.

1.1 The Hasse Principle for Cubic Surfaces

After the paper of Birch and Swinnerton-Dyer, Colliot-Thélène and Sansuc conjectured that the Brauer–Manin obstruction is the only obstruction to the Hasse principle for arbitrary smooth projective geometrically rational surfaces

[5, Questions j_1, k_1, page 233]. Some motivation for this conjecture came from the study of conic bundles. In 1987, Colliot-Thélène et al. [6] systematically studied diagonal cubic surfaces over $\mathbb{Q}$ having integral coefficients up to 100, verifying the conjecture for each one of these surfaces. They were the first to prove that the Cassels and Guy cubic (1) had a Brauer–Manin obstruction. The conjecture of Colliot-Thélène and Sansuc has been extended by Colliot-Thélène to all rationally connected varieties [4]; evidence for such a generalization has been seen recently in works of Harpaz and Wittenberg (see, e.g., [10]).

Work by Elsenhans and Jahnel has shown general construction for lines on cubic surfaces as well as the resulting elements in $\mathrm{Br}\, k(X)$ [8]. More recently, they prove that cubic surfaces violating the Hasse principle are Zariski dense in the moduli space of all cubic surfaces [9].

1.2 Outline

The notation for the paper will be fixed in Sect. 2. We will explicitly describe the Brauer group for the BSD cubic surfaces in Sect. 3 and subsequently prove Theorem 1. This computation will exploit the exceptional geometry of cubic surfaces and the previous results of Corn [7], and Swinnerton-Dyer [16, 17].

In Sect. 4, we prove the existence of an adèlic point for a family of surfaces followed by general computations of the Brauer set. Proposition 1 and Theorem 2 show that local solvability can fail only at primes dividing the coefficient d. This section concludes with a proof of Theorem 2.

Lastly, in Sect. 5, we first look back at an example given in [1] and verify that its obstruction is given by the results of Sect. 4. A second example with a Brauer–Manin obstruction given by two nonzero invariant summands is then presented.

2 Setup and Background

Let k be a number field with absolute Galois group G_k, and let L/k be any Galois extension with $\mathrm{Gal}(L/k) \simeq S_3$. Fix K/k as the unique quadratic extension of k in L. Let $\mathcal{O}_F$ be the ring of integers for the field F.

Lemma 1 *Every BSD cubic is isomorphic to one of the form*

$$\prod_{i=0}^{2}(x + \phi_i z + \psi_i w) = dy(x + \theta y)(x + \overline{\theta} y) , \tag{3}$$

where $d \in \mathcal{O}_k$ and $\{\phi_0, \phi_1, \phi_2\}$, $\{\psi_0, \psi_1, \psi_2\} \subseteq \mathcal{O}_L$, and $\{\theta, \overline{\theta}\} \subseteq \mathcal{O}_K$ are respective Galois conjugates over k with $(1, \phi_i, \psi_i)$ being a k-basis for a degree 3 extension of k.

Proof There is an isomorphism of varieties given by

$$[x\colon y\colon z\colon w] \mapsto \left[ax+by\colon \frac{cx+dy}{(d-c\theta)(d-c\overline{\theta})}\colon z\colon w\right],$$

from the surface in $\mathbb{P}^3_k$ defined by (2), to the surface in $\mathbb{P}^3_k$ defined by

$$m(ad-bc)^2\prod_{i=0}^{2}(x+\phi_i z+\psi_i w) = ((d-c\theta)(d-c\overline{\theta}))^2 y(x+\theta' y)(x+\overline{\theta}' y),$$

where $\theta' = (-b+a\theta)(d-c\overline{\theta})$. A subsequent isomorphism given by scaling variables results in (3).

Let X be the closed subvariety of $\mathbb{P}^3_k$ defined by (3) and let $\overline{X}$ be the base change of X to the algebraic closure of k. Denote the Picard group of $\overline{X}$ by $\operatorname{Pic}\overline{X}$. For a fixed $\mathscr{A} \in \operatorname{Br} X := \mathrm{H}^2_{\text{ét}}(X, \mathbb{G}_m)$, there is a commutative diagram

$$\begin{array}{ccccccccc}
 & & X(k) & \hookrightarrow & X(\mathbf{A}_k) & & & & \\
 & & \downarrow \mathrm{ev}_{\mathscr{A}} & & \downarrow \mathrm{ev}_{\mathscr{A}} & \searrow^{\Phi_{\mathscr{A}}} & & & \\
0 & \longrightarrow & \operatorname{Br} k & \longrightarrow & \bigoplus_v \operatorname{Br} k_v & \xrightarrow[\mathrm{inv}]{} & \mathbb{Q}/\mathbb{Z} & \longrightarrow & 0
\end{array}$$

where $\mathrm{ev}_{\mathscr{A}}$ is the specialization of $\mathscr{A}$ and $\mathrm{inv} = \sum_v \mathrm{inv}_v$ is the *invariant map*. The *Brauer set* is defined as $X(\mathbf{A}_k)^{\mathrm{Br}} := \bigcap_{\mathscr{A}\in \operatorname{Br} X} \Phi_{\mathscr{A}}^{-1}(0)$.

There is an inclusion of $\operatorname{Br} X$ into $\operatorname{Br} k(X)$, so elements of $\operatorname{Br} X$ can be realized as central simple algebras over the field $k(X)$. Moreover, Azumaya algebras over a field such as $k(X)$ are exactly the central simple algebras over $k(X)$. Let F_1/F_2 be fields with F_1 a cyclic extension of degree n, and let $a \in F_2^*$, then the *cyclic algebra* $(F_1/F_2, a)$ is defined to be the quotient $F_1[T]_\sigma/(T^n - a)$. Here $F_1[T]_\sigma$ is the twisted polynomial ring, i.e., $Tb = \sigma(b)T$ for all $b \in F_1$. Further such cyclic algebras are central simple algebras over F_2, so $(F_1/F_2, a)$ can be used to represent an equivalence class in $\operatorname{Br} F_2$.

3 Proof of Main Theorem

Lemma 2 *Let X and L/k be as in Sect. 2. The field of definition, of the 27 lines on $\overline{X}$, is the splitting field for the system of equations over L,*

$$\begin{cases} 1+A\phi_i+C\psi_i = 0\,, \\ \theta(1+A\phi_j+C\psi_j) = (B\phi_j+D\psi_j)\,, \\ \overline{\theta}(1+A\phi_k+C\psi_k) = (B\phi_k+D\psi_k)\,, \\ (B\phi_0+D\psi_0)(B\phi_1+D\psi_1)(B\phi_2+D\psi_2) = d\theta\overline{\theta}\,, \end{cases} \tag{4}$$

where i, j, and k are fixed and satisfy $\{i, j, k\} = \{0, 1, 2\}$. Moreover

$$\mathrm{Gal}(L'/k) \cong \mathrm{Gal}(L'/L) \times \mathrm{Gal}(L/k) \cong H \times S_3 ,$$

where H is some subgroup of S_3.

Proof There are 9 lines, $L_{i,j}$, defined by $0 = x + \phi_i z + \psi_i w$ and

$$0 = \begin{cases} y & \text{if } j = 0 , \\ x + \theta y & \text{if } j = 1 , \\ x + \overline{\theta} y & \text{if } j = 2 , \end{cases}$$

clearly these are defined over L. The remaining 18 lines, $L_{(i,j,k),n}$, are of the form $z = Ax + By$ and $w = Cx + Dy$ where A, B, C, and D satisfy the system of equations (4). Here the value of n in $L_{(i,j,k),n}$ corresponds to the three possible solutions of the system (4) for a fixed triple (i, j, k).

The fact that the field of definition for all of these lines can be determined by a single set of values for i, j, and k is a result of the facts that action by elements of $\mathrm{Gal}(\overline{k}/k)$ on the system in (4) permutes the values of i, j, and k as A_3 and that action by Galois on the lines preserves intersection pairings.

From Galois theory, we have a short exact sequence:

$$0 \longrightarrow \mathrm{Gal}(L'/L) \longrightarrow \mathrm{Gal}(L'/k) \longrightarrow \mathrm{Gal}(L/k) \longrightarrow 0 .$$

To prove that $\mathrm{Gal}(L'/k) \cong \mathrm{Gal}(L'/L) \times \mathrm{Gal}(L/k)$, it suffices to show that this sequence splits. Let $D_{(i,j,k),n}$ denote solutions to the system (4). Then from above, we know that $L' = L(D_{(0,1,2),0}, D_{(0,1,2),1}, D_{(0,1,2),2})$. Given any σ in $\mathrm{Gal}(L/k)$, we can define $s(\sigma) \in \mathrm{Gal}(L'/k)$ using these generators. Specifically act via σ on the equations of (4) and denote this by $\sigma(i, j, k)$. Then define $s(\sigma)(D_{(0,1,2),n}) = D_{\sigma(0,1,2),n}$, extend this map to L' in the natural way to obtain $s(\sigma) \in \mathrm{Gal}(L'/k)$.

Using the Groebner basis methods within Magma [2], we find that the field L' is the splitting field of the following polynomial over L:

$$\begin{aligned}
&(\theta - \overline{\theta})(\phi_1\psi_2 - \phi_2\psi_1)^2(\phi_2\psi_0 - \phi_0\psi_2)^2(\phi_0\psi_1 - \phi_1\psi_0)^2 T^3 \\
&- (\phi_1\psi_2 - \phi_2\psi_1)(\phi_2\psi_0 - \phi_0\psi_2)(\phi_0\psi_1 - \phi_1\psi_0)(\phi_0\psi_1 - \phi_0\psi_2 - \phi_1\psi_0 + \phi_1\psi_2 \\
&+ \phi_2\psi_0 - \phi_2\psi_1)(\phi_0\phi_1\psi_2\theta\overline{\theta} - 1/2\phi_0\phi_1\psi_2\overline{\theta}^2 + 1/2\phi_0\phi_2\psi_1\theta^2 - \phi_0\phi_2\psi_1\theta\overline{\theta} \\
&- 1/2\phi_1\phi_2\psi_0\theta^2 + 1/2\phi_1\phi_2\psi_0\overline{\theta}^2)T^2 \\
&+ \overline{\theta}\theta(\phi_0\psi_1 - \phi_0\psi_2 - \phi_1\psi_0 + \phi_1\psi_2 + \phi_2\psi_0 - \phi_2\psi_1)^2(\phi_0^2\phi_1^2\psi_2^2\overline{\theta} + 2\phi_0^2\phi_1\phi_2\psi_1\psi_2\theta \\
&- 2\phi_0^2\phi_1\phi_2\psi_1\psi_2\overline{\theta} - \phi_0^2\phi_2^2\psi_1^2\theta - 2\phi_0\phi_1^2\phi_2\psi_0\psi_2\theta + 2\phi_0\phi_1\phi_2^2\psi_0\psi_1\overline{\theta} + \phi_1^2\phi_2^2\psi_0^2\theta
\end{aligned}$$

$$- \phi_1^2\phi_2^2\psi_0^2\overline{\theta})T$$

$$+ \theta^2\overline{\theta}^2\phi_0\phi_1\phi_2(\phi_0\psi_1 - \phi_0\psi_2 - \phi_1\psi_0 + \phi_1\psi_2 + \phi_2\psi_0 - \phi_2\psi_1)^3$$

$$+ d(\phi_0\phi_1\psi_2\overline{\theta} - \phi_0\phi_2\psi_1\theta + \phi_1\phi_2\psi_0\theta - \phi_1\phi_2\psi_0\overline{\theta})^3 .$$

The code and output for this computation is available in the Appendix.

Since X is rational, $\ker(\operatorname{Br} X \to \operatorname{Br} \overline{X}) = \operatorname{Br} X$, [13, Thm. 42.8], and there is an isomorphism given by the Hochschild–Serre spectral sequence,

$$\operatorname{Br} X / \operatorname{Br} k \xrightarrow{\sim} \mathrm{H}^1(G_k, \operatorname{Pic} \overline{X}) . \tag{5}$$

Moreover, $\Phi_{\mathscr{A}}$ factors through this quotient. Therefore it is sufficient to calculate this finite group rather than determining the entirety of $\operatorname{Br} X$.

Swinnerton-Dyer provides a classification of possible $\mathrm{H}^1(G_k, \operatorname{Pic} \overline{X})$ via the following structures existing within the set of exceptional curves on $\overline{X}$ (see also [7]).

Definition 1 A *double-six* on X is a set $\{L_1, \ldots, L_6\} \cup \{M_1, \ldots, M_6\}$ of 12 exceptional curves on X such that the L_i are pairwise skew, the M_i are pairwise skew, and L_i intersects M_j exactly when $i \neq j$.

A *nine* on X is a set consisting of three skew curves together with six curves each of which intersect exactly two of those three. A *triple-nine* is a partitioning of the 27 exceptional curves on X into three nines.

We summarize the results of Swinnerton-Dyer, and later Corn, in the following Lemma.

Lemma 3 *Let X be a cubic surface defined over the number field k.*

1. *([7, Lem. 1.3.19],[16, Lem. 1]) Elements of* $\operatorname{Br} X / \operatorname{Br} k$ *of order 2 correspond to G_k-stable double-sixes such that neither six is itself G_k-stable, no pair $\{L_i, M_i\}$ is G_k-stable and no set of three such pairs is G_k-stable.*
2. *([16, Lem. 6]) There is a one-to-one correspondence between order 3 elements of* $\operatorname{Br} X / \operatorname{Br} k$ *and triple-nines on X such that each nine is itself G_k-stable but no set of three skew lines within any nine is G_k-stable.*
3. *([16, Lem. 5]) In order for* $\operatorname{Br} X / \operatorname{Br} k$ *to have more than two elements of order 3, it is necessary and sufficient that the field of definition for the exceptional curves be of order 3 over k.*
4. *([16]) The quotient* $\operatorname{Br} X / \operatorname{Br} k$ *is isomorphic to one of* $\{1\}$, $\mathbb{Z}/2\mathbb{Z}$, $(\mathbb{Z}/2\mathbb{Z})^2$, $\mathbb{Z}/3\mathbb{Z}$, *or* $(\mathbb{Z}/3\mathbb{Z})^2$. □

As a result of the definition and intersection properties of lines on cubic surfaces, we can write a nine as a 3×3 matrix

$$\begin{pmatrix} \ell_{1,1} & \ell_{1,2} & \ell_{1,3} \\ \ell_{2,1} & \ell_{2,2} & \ell_{2,3} \\ \ell_{3,1} & \ell_{3,2} & \ell_{3,3} \end{pmatrix} ,$$

such that the intersection pairing satisfies

$$(\ell_{i,j}, \ell_{m,n}) = \begin{cases} 1 & \text{if } i \neq m \text{ and } j \neq n \, , \\ -1 & \text{if } i = m \text{ and } j = n \, , \\ 0 & \text{otherwise} \, . \end{cases}$$

Proof (Proof of Theorem 1) Let L'/k be the field of definition for the 27 lines. A triple-nine for which the individual nines are fixed by G_k is

$$\begin{pmatrix} L_{0,0} & L_{1,1} & L_{2,2} \\ L_{1,2} & L_{2,0} & L_{0,1} \\ L_{2,1} & L_{0,2} & L_{1,0} \end{pmatrix}, \begin{pmatrix} L_{(0,1,2),0} & L_{(0,1,2),1} & L_{(0,1,2),2} \\ L_{(1,2,0),0} & L_{(1,2,0),1} & L_{(1,2,0),2} \\ L_{(2,0,1),0} & L_{(2,0,1),1} & L_{(2,0,1),2} \end{pmatrix}, \begin{pmatrix} L_{(0,2,1),0} & L_{(0,2,1),1} & L_{(0,2,1),2} \\ L_{(1,0,2),0} & L_{(1,0,2),1} & L_{(1,0,2),2} \\ L_{(2,1,0),0} & L_{(2,1,0),1} & L_{(2,1,0),2} \end{pmatrix} . \tag{6}$$

The Galois group G_k permutes the first nine, fixing no skew triple. The rows of the second two nines will be permuted via the permutation action on the roots (ϕ_0, ϕ_1, ϕ_2). The action of G_k on the columns of the second nines will determine whether or not any skew triple is fixed.

If $9 \mid [L' : k]$, then $3 \mid [L' : L]$. In particular, all three columns of the second two nines have a nontrivial action via $\mathrm{Gal}(\overline{k}/L)$, and thus no skew triple can be fixed. Thus by Lemma 3, we have $\mathrm{H}^1(G_k, \operatorname{Pic} \overline{X}) \simeq \mathbb{Z}/3\mathbb{Z}$.

Otherwise $9 \nmid [L' : k]$ and $[L' : L] \mid 2$. Let K'/k be the subfield of L' such that $[L' : K'] = 3$. Note that K' is unique in L', contains K, is Galois over k, and $[K' : k] \mid 4$. This is because $\mathrm{Gal}(L'/k)$ is one of S_3, the dihedral group of order 12, or the dicyclic group of order 12, each of which have a unique, normal subgroup of order 3. Whenever a cubic surface has a quadratic point, then it has a point over the original extension. Therefore, if the base change of X to K', $X_{K'}$, satisfies the Hasse Principle, then so does X.

Assume first that the columns of the second two nines are fixed by $\mathrm{Gal}(L'/K')$. We can assume that the lines were numbered so that every line in the first column of the second nine has intersection number 0 with every line in the first column of the third line. In general, every element of a given column of the second nine will intersect the same three lines in exactly one column of the third nine, and they will not intersect any of the other six lines in the nine. Since the columns of the second two nines are fixed by $\mathrm{Gal}(L'/K')$, we can blow down X along these six skew lines to obtain a degree 9 del Pezzo surface $X'_{K'}$ defined over K'. It is well-known that degree 9 del Pezzo surfaces satisfy the Hasse principle. So by the Lang–Nishimura lemma, $X_{K'}$ must also satisfy the Hasse principle (cf. [14]).

Secondly assume that the columns of the second two nines are not fixed by $\mathrm{Gal}(L'/K')$. Specifically, let α be a generator of $\mathrm{Gal}(L'/K')$. Without loss of generality, we assume $\alpha(L_{(0,1,2),0}) = L_{(1,2,0),1}$ and $\alpha^2(L_{(0,1,2),0}) = L_{(2,0,1),2}$. Then the combined intersection of $L_{(0,1,2),0}$, $L_{(1,2,0),1}$, and $L_{(2,0,1),2}$ is a single point fixed by $\mathrm{Gal}(L'/K')$. That is, $X(K')$ is nonempty, and $X_{K'}$ satisfies the Hasse principle.

The map in (5) is generally difficult to invert. We achieve this via a result of Swinnerton-Dyer in the following corollary.

Corollary 1 *If* $\mathrm{H}^1(G_k, \operatorname{Pic}\overline{X}) \simeq \mathbb{Z}/3\mathbb{Z}$, *then it is generated by an algebra* $\mathscr{A}$ *such that* $\mathscr{A} \otimes_k K \simeq \left(L(X)/K(X), \frac{x+\theta y}{y}\right)$.

Proof Let $D = L_{0,0} + L_{1,1} + L_{1,0} - V(y)$ where $V(y)$ corresponds to the divisor on X given by intersecting it with the hyperplane $y = 0$. Clearly D is not principal as the intersection pairing between D and $L_{1,1}$ is nonzero. Then

$$\begin{aligned}\operatorname{Norm}_{L/K}(D) &= (L_{0,0} + L_{1,1} + L_{1,0}) + (L_{1,0} + L_{2,1} + L_{2,0}) + \\ &\quad (L_{2,0} + L_{0,1} + L_{0,0}) - 3V(y)\,, \\ &= L_{1,1} + L_{2,1} + L_{0,1} - V(y)\,, \\ &= V(x+\theta y) - V(y)\,, \\ &= \operatorname{div}\left(\frac{x+\theta y}{y}\right)\,.\end{aligned}$$

Using [17, Lem. 2], we can conclude that the class of the cyclic algebra

$$\left(L(X)/K(X), \frac{x+\theta y}{y}\right)$$

in the Brauer group of X_K is nontrivial and that it is the extension of an element of $\operatorname{Br} X$. (See also [7, Prop. 2.2.5].)

4 Invariant Map Computations

Since the $\mathscr{A} \in \operatorname{Br} X/\operatorname{Br} k$ are explicit, one may compute the map $\phi_{\mathscr{A}}$ more easily. However, before doing so, we would like to verify the existence of an adèlic point.

Lemma 4 *We continue using the notation set in Sect. 2. Assume the following conditions hold:*

1. L/K *is unramified,*
2. $\phi_0\phi_1\phi_2 = \psi_0\psi_1\psi_2 = \pm 1$,
3. *for all primes* $\mathfrak{p}$ *of* $\mathscr{O}_k$ *lying over 3,* ψ_i *and* ϕ_i *are nonzero modulo* $\mathfrak{p}$ *and the minimal polynomial of* $\psi_i\phi_i^{-1}$ *is separable, and*
4. *for any prime* $\mathfrak{p}$ *of* $\mathscr{O}_k$ *dividing* $d\theta\overline{\theta}$, *there exists a valuation* w *of* $\mathscr{O}_L$ *lying over* $\mathfrak{p}$ *such that* $\mathrm{w}(\theta) \geq \mathrm{w}(d)$ *and* $\mathrm{w}(\overline{\theta}) = 0$.

Then $X(\mathbf{A}_k) \neq \emptyset$.

Proof For all $\mathfrak{p} \nmid 3d\theta\overline{\theta}\mathcal{O}_k$, the modulo $\mathfrak{p}$ reduction of the scheme $X \cap V(x)$ is a geometrically irreducible genus 1 curve and will subsequently have a $k_\mathfrak{p}$ point by the Hasse bound.

Suppose $\mathfrak{p} \mid 3\mathcal{O}_k$ and $\mathfrak{p} \nmid d\theta\overline{\theta}\mathcal{O}_k$. Then $X \cap V(x) \rightarrow \mathbb{P}^1$ defined by $[0\colon y\colon z\colon w] \mapsto [z\colon w]$ is one-to-one and surjective on $k_\mathfrak{p}$ points. Assumption 3 provides that at least one of these points is smooth.

Recall, as in the proof of Theorem 1, a cubic surface will have a point in $k_\mathfrak{p}$ if and only if it has one in any quadratic extension of $k_\mathfrak{p}$. Thus for the primes $\mathfrak{p}$ of k dividing $d\theta\overline{\theta}$, to show $X(k_\mathfrak{p}) \neq \emptyset$, it will suffice to find a $K_\mathscr{P}$ point for a prime $\mathscr{P}$ of K dividing $\mathfrak{p}$.

If $\mathscr{P} \mid d\mathcal{O}_K$ and $\mathscr{P}$ splits over L, then $X_\mathscr{P}$ is the union of 3 lines all defined over $K_\mathscr{P}/\mathscr{P}$ and has many $K_\mathscr{P}$ points.

Assume $\mathscr{P} \mid d\mathcal{O}_K$, $\mathscr{P}$ remains prime in L, and that it satisfies $\mathrm{w}(\theta) \geq \mathrm{w}(d)$ and $\mathrm{w}(\overline{\theta}) = 0$ where $\mathrm{w} = \mathrm{v}_\mathscr{P}$. Consider the cubic surface $\mathscr{X}$ in $\mathbb{P}^3_k$ defined by the equation,

$$d\prod_{i=0}^{2}(x + \phi_i z + \psi_i w) - y\,(x + (\theta/d)y)\,(dx + \overline{\theta}y) = 0\,,$$

which is isomorphic to X. Note that this equation has $\mathcal{O}_\mathscr{P}$ coefficients since $\mathrm{w}(\theta) \geq \mathrm{w}(d)$. Modulo $\mathscr{P}$, the defining equation for $\mathscr{X}$ becomes

$$\mathscr{X}_\mathscr{P}\colon \overline{\theta}_\mathscr{P} y^2\,(x + (\theta/d)_\mathscr{P} y)\ ,$$

where $\overline{\theta}_\mathscr{P}$ and $(\theta/d)_\mathscr{P}$ are the restriction of the respective constants to the quotient $\mathcal{O}_\mathscr{P}/\mathscr{P}$. The surface $\mathscr{X}_\mathscr{P}$ has a smooth point $[(\theta/d)_\mathscr{P}\colon -1\colon 1\colon 1]$ which will lift to a $K_\mathscr{P}$ point, $[x_0\colon y_0\colon z_0\colon w_0] \in \mathscr{X}(K_\mathscr{P})$. Via the isomorphism, we have $[dx_0\colon y_0\colon dz_0\colon dw_0] \in X(K_\mathscr{P})$.

Lastly, we consider the case that $\mathscr{P} \mid \theta\overline{\theta}\mathcal{O}_K$ and $\mathscr{P} \nmid d\mathcal{O}_K$. By assumption 4, we can assume that $\mathscr{P} \mid \theta$ and $\mathscr{P} \nmid \overline{\theta}$. Then the defining polynomial for X modulo $\mathscr{P}$ is

$$\prod_{i=0}^{2}(x + \phi_i z + \psi_i w) - dxy^2\overline{\theta}\ ,$$

which has the smooth point $[0, 1, 0, 0]$.

Remark 1 In the case of $\mathrm{w}(d) > \mathrm{w}(\theta)$, a similar argument can be made with the additional assumption of the surjectivity of the cube map in $\mathcal{O}_\mathscr{P}/\mathscr{P}$.

Of course Lemma 4 is not comprehensive; there are surfaces in the class which have adèlic points but do not satisfy the conditions listed above. The intention of this lemma is to provide proof that there are indeed infinitely many surfaces of this form which have an adèlic point.

There is a classical formula for inv_v provided L_v/K_v is unramified given by the local Artin map. That is, for all places v unramified in L/K,

$$\mathrm{inv}_v\left((L_v/K_v, f(P_v))_\sigma\right) = \frac{ij}{k} \pmod 1 ,$$

where $i = \mathrm{v}_v(f(P))$, $\sigma^j = \mathrm{Frob}_{L_v/K_v}$, and $k = [L_v\colon K_v]$ (cf. [15, XIV.2]).

Proposition 1 *Assume the notation of Sect. 2. Suppose v is a finite place of K which is unramified in L/K such that one of the following is true:*

1. $\mathrm{v}_v(d) = 0 \pmod 3$, *and that θ or $\overline{\theta}$ has valuation 0,*
2. *or v splits completely in L/K.*

Let $\mathscr{A} = \left(L(X)/K(X), \frac{x+\theta y}{y}\right) \in \mathrm{Br}(X_K)$. *Then* $\mathrm{inv}_v(\mathscr{A}(P_v)) = 0$ *for all P_v in $X(K_v)$. Moreover,* $\mathrm{inv}_\infty(\mathscr{A}(P_\infty)) = 0$.

Proof The structure of this proof follows that of [11, III.5.18]. In the infinite case, we must have $\mathrm{inv}_\infty(\mathscr{A}(P_\infty)) = 0$, as $[L\colon K] = 3$ and $\mathrm{inv}_\infty(\mathscr{A}(P_\infty)) = 0$ or $1/2$.

Suppose that v splits completely in L. Then $L_v = K_v$ and $(L_v/K_v, f(P_v))$ is trivial. Thus we must have $\mathrm{inv}_v(\mathscr{A}(P_v)) = 0$.

If v remains prime in L, then $[L_v\colon K_v]{=}3$. Take $P_v{=}[x_0\colon y_0\colon z_0\colon w_0]{\in}X(K_v)$. Via scaling, assume that $x_0,\ y_0,\ z_0$, and w_0 are integral and at least one has valuation 0. Since $\prod_{i=0}^{2}(x_0 + \phi_i z_0 + \psi_i w_0)$ is a norm from L to K, $y_0 = 0$ would imply $x_0 = z_0 = w_0 = 0$, which is not possible. Thus $y_0 \neq 0$. In particular, $f(P_v) = \frac{x_0+\theta y_0}{y_0}$ is defined for all $P_v \in X(K_v)$.

For simplicity, set $\mathrm{v} = \mathrm{v}_v$ and $N = \prod_{i=0}^{2}(x_0 + \phi_i z_0 + \psi_i w_0)$. If $\mathrm{v}(N) = 0$, then $\mathrm{v}(y_0) = \mathrm{v}(x_0 + \theta y_0) = 0$. Hence $\mathrm{inv}(\mathscr{A}(P_v)) = 0$. On the other hand, suppose $\mathrm{v}(N) > 0$. Here the restriction of N modulo v is a norm on the residue class field of L_v to that of K_v. Thus N having positive valuation implies that the restriction $\overline{N} = 0$. Hence $x_0, z_0, w_0 \equiv 0 \pmod v$ so $\mathrm{v}(y_0) = 0$. In fact, $3 \mid \mathrm{v}(N)$. Thus, $\mathrm{v}(d) + \mathrm{v}(x_0 + \theta y_0) + \mathrm{v}(x_0 + \overline{\theta} y_0) \equiv 0 \pmod 3$. However, $\mathrm{v}(x_0 + \theta y_0) = 0$ or $\mathrm{v}(x_0 + \overline{\theta} y_0) = 0$, since v does not divide both θ and $\overline{\theta}$. In particular, $\mathrm{v}(x_0 + \theta y_0) \equiv 0 \pmod 3$ and $\mathrm{inv}_v(\mathscr{A}(P_v)) = 0$.

Given the result of Proposition 1, in all cases where L/K is unramified, we simply need to consider the places of k over which d has valuation that is nonzero modulo 3 when computing the invariant map.

Theorem 3 *Continue using the notation established in Sect. 2. Assume L/K is unramified. Let $\mathfrak{p}$ be a prime of $\mathscr{O}_k$ dividing d such that $\mathfrak{p}$ factors as the product of two primes in $\mathscr{O}_L$ and one of those primes has corresponding valuation* w *satisfying* $\mathrm{w}(\theta) \not\equiv 0 \pmod 3$ *and* $\mathrm{w}(\overline{\theta}) = 0$. *Let* $\mathscr{A} = \left(L(X)/K(X), \frac{x+\theta y}{y}\right) \in \mathrm{Br}(X_K)$. *Then* $\sum_{\mathscr{P}|\mathfrak{p}} \mathrm{inv}_{\mathscr{P}} \mathscr{A}_K(P_{\mathscr{P}}) \neq 0$ *for all* $(P_{\mathscr{P}}) \in \prod_{\mathscr{P}|\mathfrak{p}} X(K_{\mathscr{P}})$ *if and only if* $\mathrm{w}(d) \not\equiv 0 \pmod 3$.

Proof Let $P_{\mathrm{w}} = [x_0 : y_0 : w_0, z_0] \in X(K_{\mathrm{w}})$. Naturally

$$\mathrm{w}(dy_0(x+\theta y_0)(x+\overline{\theta}y_0)) > 0,$$

so we must be in the case that $\mathrm{w}(x_0)$, $\mathrm{w}(w_0)$, $\mathrm{w}(z_0) > 0$ and $\mathrm{w}(y_0) = 0$. Since the left-hand side of (3) is a cubic norm over K, its valuation must be divisible by 3, so the same is true for the right-hand side. And so

$$\mathrm{w}(x_0+\theta y_0) = \mathrm{w}(dy_0(x+\theta y_0)(x+\overline{\theta}y_0)) - \mathrm{w}(d) \equiv -\,\mathrm{w}(d) \pmod 3\,.$$

In particular $\mathrm{inv}_{\mathrm{w}}(\mathscr{A}(P_{\mathrm{w}})) = 1/3$ or $2/3$ if and only if $\mathrm{w}(d) \not\equiv 0 \pmod 3$.

On the other hand, if $\overline{\mathrm{w}}$ is the conjugate valuation of w over L and $P_{\overline{\mathrm{w}}} = [x_1 : y_1 : w_1, z_1] \in X(K_{\overline{\mathrm{w}}})$, then $\overline{\mathrm{w}}(x_1+\theta y_1) = 0$, so $\mathrm{inv}_{\overline{\mathrm{w}}}(\mathscr{A}(P_{\overline{\mathrm{w}}})) = 0$. Thus, for all $(P_{\mathscr{P}}) \in \prod_{\mathscr{P}|\mathfrak{p}} X(K_{\mathscr{P}})$,

$$\sum \mathrm{inv}_{\mathscr{P}}(\mathscr{A}(P_{\mathscr{P}})) = \mathrm{inv}_{\mathrm{w}}(\mathscr{A}(P_{\mathrm{w}})) = 0$$

if and only if $\mathrm{w}(d) \equiv 0 \pmod 3$.

Corollary 2 *Continue using the notation established in Sect. 2. Assume L/K is unramified. Fix θ so that no primes of $\mathscr{O}_K$ divide both θ and $\overline{\theta}$. Let $\mathfrak{p}$ be a prime of $\mathscr{O}_k$ such that $\mathrm{v}_{\mathfrak{p}}(d) = n$ for some $n \not\equiv 0 \pmod 3$ and $\mathfrak{p} \mid \theta\overline{\theta}$. Assume all other primes dividing $d\mathscr{O}_K$ split in L/K. Let $\mathscr{A} = \left(L(X)/K(X), \frac{x+\theta y}{y}\right) \in \mathrm{Br}(X_K)$. If $X(\mathbf{A}_K) \neq \emptyset$ and $\mathfrak{p} = \mathscr{P}_1\mathscr{P}_2$ in $\mathscr{O}_K$, each of which is inert in $\mathscr{O}_L$, then for all $(P_v) \in X(\mathbf{A}_K)$, we have $\sum_v \mathrm{inv}_v(\mathscr{A}_K(P_v)) \neq 0$.*

Proof By Proposition 1

$$\sum_v \mathrm{inv}_v(\mathscr{A}(P_v)) = \mathrm{inv}_{\mathscr{P}_1}(\mathscr{A}(P_{\mathscr{P}_1})) + \mathrm{inv}_{\mathscr{P}_2}(\mathscr{A}(P_{\mathscr{P}_2}))\,.$$

Then as in the proof of Theorem 3, one of these summands is 0, and the other is nonzero, depending on which one of $\mathscr{P}_i$ divides θ.

Proof (Proof of Theorem 2) Lemma 4 guarantees that $X(\mathbf{A}_k) \neq \emptyset$. Via Proposition 1 and Corollary 2, we conclude that $X(\mathbf{A}_K)^{\mathrm{Br}} = \emptyset$. As $X(\mathbf{A}_k)^{\mathrm{Br}} \subseteq X(\mathbf{A}_K)^{\mathrm{Br}}$, it is then clear that X has a Brauer–Manin obstruction to the Hasse principle.

5 Examples

Examples that fit the situation of Corollary 2 are easy to come by. Given any L/K unramified, we can find many θ satisfying the conditions of Corollary 2. Then it is a quick check via Hensel's Lemma and the Weil Conjectures to show that there is an adèlic point. In fact the original example of BSD fits this case.

Example 1 Suppose $\theta' = \frac{1}{2}(1 + \sqrt{-23})$ and ϕ_i so that $\phi_i^3 = \phi_i + 1$ and $\psi_i = \phi_i^2$. Define X_{BSD} by

$$2\prod_{i=0}^{2}(x + \phi_i z + \phi_i^2 w) = (x - y)(x + \theta' y)(x + \overline{\theta}' y) .$$

Via the isomorphisms in the proof of Lemma 1, we have the isomorphic X given by

$$\prod_{i=0}^{2}(x + \phi_i z + \psi_i^2 w) = 32y(x + \theta y)(x + \overline{\theta} y) ,$$

where $\theta = -\theta' - 6$.

We find that X has adèlic points but no rational points. Moreover, X has a Brauer–Manin obstruction to rational points as described in Corollary 2.

There are few published examples where the invariant map has two or more nonzero summands. Given the theorems above, examples of this can be found quickly.

Example 2 Suppose the ϕ_i satisfy $\phi_i^3 + \phi_i + 1 = 0$ and $\theta,\ \overline{\theta}$ are the roots of $T^2 - 4T + 35$.

Then

$$X\colon \prod_{i=0}^{2}(x + \phi_i z + \phi_i^2 w) \ = \ 5 \cdot 7y(x + \theta y)(x + \overline{\theta} y) ,$$

has a Brauer–Manin obstruction to the Hasse principle with the invariant map being

$$1/3 \ + \ 1/3 \quad \text{or} \quad 2/3 \ + \ 2/3 ,$$

depending on the choice of algebra $\mathscr{A}$. The case of $1/3 + 2/3$ does not appear as the $\sigma \in \mathrm{Gal}(L/K)$ that reduces to Frob_5 also reduces to Frob_7.

Appendix

Magma Code for Defining Equation Computation

```
> R<phi_0,phi_1,phi_2,psi_0,psi_1,psi_2,theta,
thetabar,d> := PolynomialRing(Rationals(),9);
> Q := FieldOfFractions(R);
> A4<A,B,C,D> := AffineSpace(Q,4);
> polys := [1+A*phi_0+C*psi_0,
```

```
> theta*(1+phi_1*A+psi_1*C)-(B*phi_1+D*psi_1),
> thetabar*(1+phi_2*A+psi_2*C)-(B*phi_2+D*psi_2),
> (B*phi_0+D*psi_0)*(B*phi_1+D*psi_1)*
(B*phi_2+D*psi_2)-d*theta*thetabar];
>
> S := Scheme(A4,polys);
> f := ClearDenominators(GroebnerBasis(S))[4];
> cos := Coefficients(f);
>
> Factorization(R!cos[1]);
[
    <theta - thetabar, 1>,
    <phi_1*psi_2 - phi_2*psi_1, 2>,
    <phi_0*psi_2 - phi_2*psi_0, 2>,
    <phi_0*psi_1 - phi_1*psi_0, 2>
]
> Factorization(R!cos[2]);
[
    <phi_1*psi_2 - phi_2*psi_1, 1>,
    <phi_0*psi_2 - phi_2*psi_0, 1>,
    <phi_0*psi_1 - phi_1*psi_0, 1>,
    <phi_0*psi_1 - phi_0*psi_2 - phi_1*psi_0 +
    phi_1*psi_2 + phi_2*psi_0 - phi_2*psi_1, 1>,
    <phi_0*phi_1*psi_2*theta*thetabar -
    1/2*phi_0*phi_1*psi_2*thetabar^2 +
    1/2*phi_0*phi_2*psi_1*theta^2 -
    phi_0*phi_2*psi_1*theta*thetabar -
    1/2*phi_1*phi_2*psi_0*theta^2 +
    1/2*phi_1*phi_2*psi_0*thetabar^2, 1>
]
> Factorization(R!cos[3]);
[
    <thetabar, 1>,
    <theta, 1>,
    <phi_0*psi_1 - phi_0*psi_2 - phi_1*psi_0 +
    phi_1*psi_2 + phi_2*psi_0 - phi_2*psi_1, 2>,
    <phi_0^2*phi_1^2*psi_2^2*thetabar +
    2*phi_0^2*phi_1*phi_2*psi_1*psi_2*theta -
    2*phi_0^2*phi_1*phi_2*psi_1*psi_2*thetabar -
    phi_0^2*phi_2^2*psi_1^2*theta -
    2*phi_0*phi_1^2*phi_2*psi_0*psi_2*theta +
    2*phi_0*phi_1*phi_2^2*psi_0*psi_1*thetabar +
    phi_1^2*phi_2^2*psi_0^2*theta -
    phi_1^2*phi_2^2*psi_0^2*thetabar, 1>
]
```

```
> Factorization(&+ [t : t in Terms(R!cos[4]) |
not IsDivisibleBy(t,d)]);
[
    <thetabar, 2>,
    <theta, 2>,
    <phi_2, 1>,
    <phi_1, 1>,
    <phi_0, 1>,
    <phi_0*psi_1 - phi_0*psi_2 - phi_1*psi_0 +
    phi_1*psi_2 + phi_2*psi_0 - phi_2*psi_1, 3>
]
> Factorization(&+ [t : t in Terms(R!cos[4]) |
IsDivisibleBy(t,d)]);
[
    <d, 1>,
    <phi_0*phi_1*psi_2*thetabar -
    phi_0*phi_2*psi_1*theta +
    phi_1*phi_2*psi_0*theta -
    phi_1*phi_2*psi_0*thetabar, 3>
]
```

References

1. B.J. Birch, P. Swinnerton-Dyer, The Hasse problem for rational surfaces. J. Reine Angew. Math. **274**(275), 164–174 (1975)
2. W. Bosma, J. Cannon, C. Playoust, The Magma algebra system. I. The user language. J. Symb. Comput. **24**(3-4), 235–265 (1997); Computational algebra and number theory (London, 1993)
3. J.W.S. Cassels, M.J.T. Guy, On the Hasse principle for cubic surfaces. Mathematika **13**(02), 111–120 (1966)
4. J.-L. Colliot-Thélène, Points rationnels sur les fibrations, in *Higher Dimensional Varieties and Rational Points*, ed. by K.J. Böröczky, J. Kollár, T. Szamuely (Springer, Heidelberg, 2003), pp. 171–221
5. J.-L. Colliot-Thélène, J.-J. Sansuc, La descente sur les variétés rationnelles, in *Journées de Géométrie Algébrique d'Angers, Juillet 1979*, ed. by A. Beauville (Sijthof and Noordhof, Leiden, 1980), pp. 223–237
6. J.-L. Colliot-Thélène, D. Kanevsky, J.-J. Sansuc, Arithmétique des surfaces cubiques diagonales, in *Diophantine Approximation and Transcendence Theory* (Springer, Heidelberg, 1987), pp. 1–108
7. P. Corn, Del Pezzo surfaces and the Brauer–Manin obstruction. PhD thesis, University of California, Berkely (2005)
8. A.S. Elsenhans, J. Jahnel, On the order three Brauer classes for cubic surfaces. Open Math. **10**(3), 903–926 (2012)
9. A.S. Elsenhans, J. Jahnel, Cubic surfaces violating the Hasse principle are Zariski dense in the moduli scheme. Adv. Math. **280**, 360–378 (2015)

10. Y. Harpaz, O. Wittenberg, On the fibration method for zero-cycles and rational points. Ann. Math. **183**(1), 229–295 (2016)
11. J. Jahnel, *Brauer Groups, Tamagawa Measures, and Rational Points on Algebraic Varieties*, vol. 198 (American Mathematical Society, Providence, 2014)
12. Y.I. Manin, Le groupe de Brauer-Grothendieck en géométrie diophantienne, in *Actes du Congres International des Mathématiciens*, vol. 1 (World Scientific, Singapore, 1971), pp. 401–411
13. Y.I. Manin, *Cubic Forms: Algebra, Geometry, Arithmetic*, vol. 4 (North-Holland, Amsterdam, 1974)
14. H. Nishimura, Some remarks on rational points. Mem. Coll. Sci. Univ. Kyoto, Ser. A Math. **29**(2), 189–192 (1955)
15. J.P. Serre, *Local Fields*, vol. 67 (Springer, New York, 1979)
16. H.P.F. Swinnerton-Dyer, The Brauer group of cubic surfaces. Math. Proc. Camb. Philos. Soc. **113**(03), 449–460 (1993)
17. H.P.F. Swinnerton-Dyer, Brauer–Manin obstructions on some del Pezzo surfaces. Math. Proc. Camb. Philos. Soc. **125**(02), 193–198 (1999)

MIX
Papier aus verantwortungsvollen Quellen
Paper from responsible sources
FSC® C105338

If you have any concerns about our products,
you can contact us on
ProductSafety@springernature.com

In case Publisher is established outside the EU,
the EU authorized representative is:
Springer Nature Customer Service Center GmbH
Europaplatz 3, 69115 Heidelberg, Germany

Printed by Libri Plureos GmbH
in Hamburg, Germany